DATE DUE

Non-Viral Vectors for Gene Therapy

Second Edition
Part II

Advances in Genetics, Volume 54

Serial Editors

Jeffery C. Hall
Waltham, Massachusetts

Jay C. Dunlap
Hanover, New Hampshire

Theodore Friedmann
La Jolla, California

Non-Viral Vectors for Gene Therapy

Second Edition
Part II

Edited by

Leaf Huang
Center for Pharmacogenetics
University of Pittsburg School of Pharmacy
Pittsburg, Pennsylvania

Mien-Chie Hung
Department of Molecular and Cellular Oncology
The University of Texas
M.D. Anderson Cancer Center
Houston, Texas

Ernst Wagner
Ludwig-Maximilians-Universität München
Munich, Germany

WITHDRAWN
FAIRFIELD UNIVERSITY
LIBRARY

ELSEVIER
ACADEMIC
PRESS

AMSTERDAM • BOSTON • HEIDELBERG • LONDON
NEW YORK • OXFORD • PARIS • SAN DIEGO
SAN FRANCISCO • SINGAPORE • SYDNEY • TOKYO

Elsevier Academic Press
525 B Street, Suite 1900, San Diego, California 92101-4495, USA
84 Theobald's Road, London WC1X 8RR, UK

This book is printed on acid-free paper.

Permissions may be sought directly from Elsevier's Science & Technology Rights
Department in Oxford, UK: phone: (+44) 1865 843830, fax: (+44) 1865 853333,
E-mail: permissions@elsevier.co.uk. You may also complete your request on-line
via the Elsevier homepage (http://elsevier.com), by selecting
"Customer Support" and then "Obtaining Permissions."

For all information on all Elsevier Academic Press publications
visit our Web site at www.books.elsevier.com

ISBN-13: 978-0-1201-7654-0
ISBN-10: 0-12-017654-8

PRINTED IN THE UNITED STATES OF AMERICA
05 06 07 08 09 9 8 7 6 5 4 3 2 1

To Our Families

Contents

1 NAKED DNA, OLIGONUCLEOTIDE AND PHYSICAL METHODS

2   GENE REGULATION

Contributors

Numbers in parentheses indicate the pages on which the authors' contributions begin.

Mohammed S. Al-Dosari (65) Department of Pharmaceutical Sciences, University of Pittsburgh School of Pharmacy, Pittsburgh, Pennsylvania 15261

Eric W. F. W Alton (291) Department of Gene Therapy, Imperial College London, National Heart and Lung Institute, London SW3 6LR, United Kingdom

Franck André (83) Laboratory of Vectorology and Gene Transfer, UMR 8121 CNRS Institut Gustave-Roussy, F-94805 Villejuif Cédex, France

Vladimir Budker (3) Departments of Pediatrics and Medical Genetics, Waisman Center, University of Wisconsin–Madison, Madison, Wisconsin 53705

Michele P. Calos (179) Department of Genetics, Stanford University School of Medicine, Stanford, California 94305

Weihsu Claire Chen (315) Center for Pharmacogenetics, School of Pharmacy, University of Pittsburgh, Pittsburgh, Pennsylvania 15261

Zhengrong Cui (257) Department of Pharmaceutical Sciences, College of Pharmacy, Oregon State University, Corvallis, Oregon 97331

Jane C. Davies (291) Department of Gene Therapy, Imperial College London, National Heart and Lung Institute, London SW3 6LR, United Kingdom

Chi-Ping Day (235) Department of Molecular and Cellular Oncology, The University of Texas M.D. Anderson Cancer Center, Houston, Texas 77030

Stephen C. Ekker (189) Department of Genetics, Cell Biology and Development, Arnold and Mabel Beckman Center for Transposon Research, University of Minnesota, Minneapolis, Minnesota 55455 and Discovery Genomics, Inc., Minneapolis, Minnesota 55413

Julie Gehl (83) Department of Oncology 54B1, Herlev University Hospital, University of Copenhagen, DK-2730 Herlev, Denmark

Daniel S. Ginsburg (179) Department of Genetics, Stanford University School of Medicine, Stanford, California 94305

Perry B. Hackett (189) Department of Genetics, Cell Biology and Development, Arnold and Mabel Beckman Center for Transposon

Research, University of Minnesota, Minneapolis, Minnesota 55455 and Discovery Genomics, Inc., Minneapolis, Minnesota 55413

Fengtian He (21) Center for Pharmacogenetics and Department of Pharmaceutical Sciences, School of Pharmacy, University of Pittsburgh, Pittsburgh, Pennsylvania 15261

Leaf Huang (315) Center for Pharmacogenetics, School of Pharmacy, University of Pittsburgh, Pittsburgh, Pennsylvania 15261

Mien-Chie Hung (235) Department of Molecular and Cellular Oncology, The University of Texas M.D. Anderson Cancer Center, Houston, Texas 77030

Joseph E. Knapp (65) Department of Pharmaceutical Sciences, University of Pittsburgh School of Pharmacy, Pittsburgh, Pennsylvania 15261

David A. Largaespada (189) Department of Genetics, Cell Biology and Development, Arnold and Mabel Beckman Center for Transposon Research, University of Minnesota, Minneapolis, Minnesota 55455 and Discovery Genomics, Inc., Minneapolis, Minnesota 55413

Jiang Li (21) Center for Pharmacogenetics and Department of Pharmaceutical Sciences, School of Pharmacy, University of Pittsburgh, Pittsburgh, Pennsylvania 15261

Song Li (21) Center for Pharmacogenetics and Department of Pharmaceutical Sciences, School of Pharmacy, University of Pittsburgh, Pittsburgh, Pennsylvania 15261

Dexi Liu (65) Department of Pharmaceutical Sciences, University of Pittsburgh School of Pharmacy, Pittsburgh, Pennsylvania 15261

Feng Liu (43) Center for Pharmacogenetics, School of Pharmacy, University of Pittsburgh, Pittsburgh, Pennsylvania 15261

Hui-Wen Lo (235) Department of Molecular and Cellular Oncology, The University of Texas M.D. Anderson Cancer Center, Houston, Texas 77030

Douglas W. Losordo (339) Divisions of Cardiology, Vascular Medicine, and Cardiovascular Research, Caritas St. Elizabeth's Medical Center, Boston, Massachusetts 02135

Patrick Y. Lu (117) Intradigm Corporation, Rockville, Maryland 20852

Zheng Ma (21) Center for Pharmacogenetics and Department of Pharmaceutical Sciences, School of Pharmacy, University of Pittsburgh, Pittsburgh, Pennsylvania 15261

R. Scott McIvor (189) Department of Genetics, Cell Biology and Development, Arnold and Mabel Beckman Center for Transposon Research University of Minnesota Minneapolis, Minnesota 55455 and Discovery Genomics, Inc., Minneapolis, Minnesota 55413

Carol H. Miao (143) Department of Pediatrics, University of Washington and Children's Hospital and Regional Medical Center, Seattle, Washington 98195

Lluis M. Mir (83) Laboratory of Vectorology and Gene Transfer, UMR 8121 CNRS Institut Gustave-Roussy, F-94805 Villejuif Cédex, France

Pernille H. Moller (83) Department of Oncology 54B1, Herlev University Hospital, University of Copenhagen, DK-2730 Herlev, Denmark

Bruce Pitt (21) Department of Environmental and Occupational Health, Graduate School of Public Health, School of Pharmacy, University of Pittsburgh, Pittsburgh, Pennsylvania 15261

Pinak B. Shah (339) Divisions of Cardiology, Vascular Medicine, and Cardiovascular Research, Caritas St. Elizabeth's Medical Center, Boston, Massachusetts 02135

Pradeep Tyagi (43) Center for Pharmacogenetics, School of Pharmacy, University of Pittsburgh, Pittsburgh, Pennsylvania 15261

Annette Wilson (21) Department of Environmental and Occupational Health, Graduate School of Public Health, School of Pharmacy, University of Pittsburgh, Pittsburgh, Pennsylvania 15261

Jon A. Wolff (3) Departments of Pediatrics and Medical Genetics, Waisman Center, University of Wisconsin–Madison, Madison, Wisconsin 53705

Martin C. Woodle (117) Intradigm Corporation, Rockville, Maryland 20852

Frank Xie (117) Intradigm Corporation, Rockville, Maryland 20852

Preface

Since the pioneering discovery by Felgner *et al.* (1987) that cationic lipid can efficiently transfect cells, there was a surge of research activity in this area. The field received another boost when Nabel *et al.* (1993) successfully completed a small phase I clinical trial in gene therapy of melanoma using cationic liposome as a vector. The data strongly suggest that non-viral vectors may be efficacious and safe in humans. Since then, many different cationic lipids and polymers have been developed as vectors, some of them also entered into clinical trials and others became commercial transfection agents.

An equally important event occurred in 1990 (Wolff *et al.*). The work indicated that naked DNA can transfect muscle cells when injected intramuscularly. This is the beginning of using physical methods to introduce DNA into cells. Since then, all major physical techniques, including pressure, electricity, sound, light, heat, and particle bombardment, have been attempted. Some of these methods are quite efficient. For example, the hydrodynamic injection methods developed independently by Zhang *et al.*, 1998 and Liu *et al.*, 1998, is the best method of transfecting liver cells among all viral and non-viral vectors.

There was also much progress made in the molecular biological design of the transgene expression system. For example, site-specific integration of the transgene is now possible for prolonged gene expression without the threat of insertional mutagenesis.

Since the publication of the first edition of *Non-viral Vectors for Gene Therapy* in 1999, the field has experienced significant progress in both chemical and physical vectors. More importantly, many mechanistic studies have appeared to address how the vector works and why the vector produces toxicity. It is safe to state that 18 years after the Felgner's publication, the field of non-viral vector for gene therapy is approaching maturity.

Due to much expansion of the field, it is not possible to include all chapters of this edition in a single book. The two volumes are roughly divided into chemical and physical methods emphasizing mechanistic aspects of the vector. The new edition is then ended with a high note of delivering siRNA for therapeutic purpose. RNA interference is definitely a new dimension in non-viral gene therapy which will attract much attention in the years to come.

We wish to thank Pat Gonzalez of Elsevier and Nicole Sebula of the University of Pittsburgh for excellent editorial assistances throughout various phases of production for these two volumes.

Leaf Huang
Mien-Chie Hung
Ernst Wagner

Section 1

NAKED DNA, OLIGONUCLEOTIDE AND PHYSICAL METHODS

1

The Mechanism of Naked DNA Uptake and Expression

Jon A. Wolff and Vladimir Budker

Departments of Pediatrics and Medical Genetics
Waisman Center, University of Wisconsin–Madison
Madison, Wisconsin 53705

ABSTRACT

The administration of naked nucleic acids into animals is increasingly being used as a research tool to elucidate mechanisms of gene expression and the role of genes and their cognate proteins in the pathogenesis of disease in animal models (Herweijer and Wolff, 2003; Hodges and Scheule, 2003). It is also being used in several human clinical trials for genetic vaccines, Duchenne muscular dystrophy, peripheral limb ischemia, and cardiac ischemia (Davis *et al.*, 1996; Romero *et al.*, 2002; Tsurumi *et al.*, 1997). Naked DNA is an attractive non-viral vector because of its inherent simplicity and because it can easily be produced in bacteria and manipulated using standard recombinant DNA

Advances in Genetics, Vol. 54
Copyright 2005, Elsevier Inc. All rights reserved.

0065-2660/05 $35.00
DOI: 10.1016/S0065-2660(05)54001-X

techniques. It shows very little dissemination and transfection at distant sites following delivery and can be readministered multiple times into mammals (including primates) without inducing an antibody response against itself (i.e., no anti-DNA antibodies generated) (Jiao et al., 1992). Also, contrary to common belief, long-term foreign gene expression from naked plasmid DNA (pDNA) is possible even without chromosome integration if the target cell is postmitotic (as in muscle) or slowly mitotic (as in hepatocytes) and if an immune reaction against the foreign protein is not generated (Herweijer et al., 2001; Miao et al., 2000; Wolff et al., 1992; Zhang et al., 2004). With the advent of intravascular and electroporation techniques, its major restriction—poor expression levels—is no longer limiting and levels of foreign gene expression in vivo are approaching what can be achieved with viral vectors.

Direct in vivo gene transfer with naked DNA was first demonstrated when efficient transfection of myofibers was observed following injection of mRNA or pDNA into skeletal muscle (Wolff et al., 1990). It was an unanticipated finding in that the use of naked nucleic acids was the control for experiments designed to assess the ability of cationic lipids to mediate expression in vivo. Subsequent studies also found foreign gene expression after direct injection in other tissues such as heart, thyroid, skin, and liver (Acsadi et al., 1991; Hengge et al., 1996; Kitsis and Leinwand, 1992; Li et al., 1997; Sikes and O'Malley 1994; Yang and Huang, 1996). However, the efficiency of gene transfer into skeletal muscle and these other tissues by direct injection is relatively low and variable, especially in larger animals such as nonhuman primates (Jiao et al., 1992).

After our laboratory had developed novel transfection complexes of pDNA and amphipathic compounds and proteins, we sought to deliver them to hepatocytes in vivo via an intravascular route into the portal vein. Our control for these experiments was naked pDNA and we were once again surprised that this control group had the highest expression levels (Budker et al., 1996; Zhang et al., 1997). High levels of expression were achieved by the rapid injection of naked pDNA in relatively large volumes via the portal vein, the hepatic vein, and the bile duct in mice and rats. The procedure also proved effective in larger animals such as dogs and nonhuman primates (Eastman et al., 2002; Zhang et al., 1997). The next major advance was the demonstration that high levels of expression could also be achieved in hepatocytes in mice by the rapid injection of naked DNA in large volumes simply into the tail vein (Liu et al., 1999; Zhang et al., 1999). This hydrodynamic tail vein (HTV) procedure is proving to be a very useful research tool not only for gene expression studies, but also more recently for the delivery of small interfering RNA (siRNA) (Lewis et al., 2002; McCaffrey et al., 2002).

The intravascular delivery of naked pDNA to muscle cells is also attractive particularly since many muscle groups would have to be targeted for intrinsic muscle disorders such as Duchenne muscular dystrophy. High levels of gene expression were first achieved by the rapid injection of naked DNA in large volumes via an artery route with both blood inflow and outflow blocked surgically (Budker *et al.*, 1998; Zhang *et al.*, 2001). Intravenous routes have also been shown to be effective (Hagstrom et al., 2004; Liang *et al.*, 2004; Liu *et al.*, 2001). For limb muscles, the ability to use a peripheral limb vein for injection and a proximal, external tourniquet to block blood flow renders the procedure to be clinically viable.

This review concerns itself with the mechanism by which naked DNA is taken up by cells *in vivo*. A greater understanding of the mechanisms involved in the uptake and expression of naked DNA, and thus connections between postulated mechanisms and expression levels, is emphasized. Inquiries into the mechanism not only aid these practical efforts, but are also interesting on their own account with relevance to viral transduction and cellular processes. The delivery to hepatocytes is first discussed given the greater information available for this process, and then uptake by myofibers is discussed. © 2005, Elsevier Inc.

I. OVERVIEW OF HEPATOCYTE DELIVERY

As with studies involving viral transduction, the delivery of naked nucleic acids can be broken down into several transport steps, including (1) transport of nucleic acids from the site of injection to the liver, (2) exit of nucleic acids out of the liver vessel lumen into contact with hepatocytes, (3) uptake by hepatocytes, (4) entry into the cytoplasm, and (5) entry into the nucleus.

The key destination for the journey of typical plasmid DNA expression vectors is the nucleus where the plasmids utilize RNA polymerase II promoters to transcribe messenger RNA for expression of the desired foreign gene. An important observation that has to be incorporated into any mechanism hypothesis is that foreign gene expression from a variety of promoters has only been observed in hepatocytes. This suggests that the plasmid DNA only enters the nucleus of hepatocytes and not any of the nonparenchymal cells, such as endothelial cells or Kupffer cells that line the sinusoidal space. For siRNA, only cytoplasmic entry is required to access the RISC complex and RNA interference effects have been best documented in hepatocytes.

Except for actual liver delivery, studies involving the direct injection of nucleic acids into liver vessels (portal vein, hepatic vein, and bile duct) are considered together with studies using tail vein injections because it is assumed that the basic mechanism is the same for the two different routes of administration.

II. TRANSPORT OF NUCLEIC ACIDS FROM INJECTION SITE TO LIVER

For tail vein injections, this point requires some discussion but the process of injecting nucleic acid directly into liver vessels obviously avoids this delivery hurdle. Direct observation of the liver shows that the liver swells during the tail vein injection procedure (Zhang *et al.*, 1999, 2004) (Fig. 1.1). Efficient gene expression and delivery of the plasmid DNA to the liver both require that the DNA be injected rapidly and in a large volume. Similarly, siRNA or morpho-linooligonucleotide suppression of gene expression also requires that the nucleic acid be injected into the tail vein using the HTV procedure. For example, fluorescently labeled pDNA or siRNA was only associated with the liver when injected using the HTV procedure but not after normal tail vein (NTV) injections using low volumes and injection speeds.

Southern blot and polymerase chain reaction (PCR) analysis provided quantitative information concerning the effect of the HTV procedure on liver

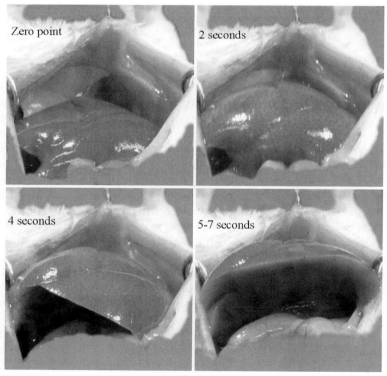

Figure 1.1. Liver at various times after HTV injection. The liver immediately swells and becomes pale, indicating that it fills with the injection fluid. (See Color Insert.)

delivery (Kobayahi *et al.*, 2001). Immediately after HTV injection of radioactive DNA, 50% of the radioactive signal was associated with the liver and 20% remained in the plasma, whereas after NTV injection, ~30% of the radioactive DNA was associated with the liver and ~55% was in the plasma. As explained later, the high baseline liver uptake of 30% after NTV is probably due to Kupffer and endothelial cell uptake.

The most likely mechanism by which the HTV procedure enables efficient liver delivery is that rapid injection of the large volume causes a transient right-sided CHF that leads to back pressure of the nucleic solution retrograde from the IVC to the hepatic vein and back up into the liver. Consistent with this postulated mechanism is that there is a transient decrease in heart rate and an increase in IVC pressure immediately after injection (Zhang *et al.*, 2004).

III. EXTRAVASTION OF INJECTED NUCLEIC ACID

Given that plasmid DNA has a gyration ratio of 0.1 μm and that sinusoid fenestrae have an average diameter also of ~0.1 μm, simple diffusion of the pDNA would not enable sufficient pDNA extravasation for expression. If fluorescently labeled pDNA is injected into the portal vein without substantially increasing the intravascular pressure, then little pDNA is observed out of the sinusoids and is associated with parenchymal cells. However, if injected under high-pressure conditions, then much more labeled pDNA becomes associated with hepatocytes (Budker *et al.*, 2000). Similarly, much more fluorescently labeled pDNA becomes associated with hepatocytes if the pDNA is injected under HTV conditions. For greater quantitative assessment, twofold more radioactively labeled DNA becomes associated with hepatocytes if injected under HTV conditions (Lecocq *et al.*, 2003).

For injections into liver vessels, blocking the outflow of the injected fluid enables the intravascular pressure to rise and extravasation of pDNA. The ability of the HTV procedure to extravasate pDNA may be related to the tortuousness of the hepatic veins coming into the sinusoids (Zhang *et al.*, 2004). This anatomical feature may provide increased resistance to outflow, thereby enabling an increased intravascular pressure. Thus, increased intravascular pressure within the liver, an effect common to both direct liver blood vessel and HTV injections, appears to be a prerequisite for pDNA delivery to hepatocytes.

The resulting increased intravascular pressure has been postulated to enhance the extravasation of pDNA by enlarging the sinusoidal fenestrae. Consistent with this hypothesis are scanning electron micrograph (EM) images that show enlarged fenestrae after HTV injection (Zhang *et al.*, 2004). In addition, we have found large amounts of DNA in the space of Disse after HTV injection (Fig. 1.2A). Convective flow of pDNA within the injection fluid out of the intravascular space would also be required to account for the increased hepatocyte delivery. Viral vectors such as lentiviral vectors also have had more expression when delivered in increased volume in mice (Condiotti *et al.*, 2004).

After portal vein, hepatic vein, or HTV injection, not all of the hepatocytes do not take up the injected nucleic acid. Depending on the exact injection conditions and the type of nucleic acid (pDNA or siRNA), somewhere between 5 and 70% of the hepatocytes can be seen to take up the labeled nucleic acid (Rossmanith *et al.*, 2002). This is consistent with the general experience that less than 50% of hepatocytes can be made to express a foreign gene or be inhibited by siRNA. The restriction of nucleic acid delivery to a subset of hepatocytes may be related to limited pDNA extravasation in certain areas of the liver. For example, pDNA expression is greater around the central vein after HTV and hepatic vein injections. Interestingly, we have not found expression greater around the portal vein after portal vein injections.

IV. CYTOPLASMIC ENTRY

After HTV injection, pDNA could enter the cytoplasm via endocytic vesicles or transient membrane pores; evidence in support and against each of these hypotheses is discussed.

While the initial phenomenon was first described using pDNA, a variety of molecules can be taken up after HTV injection, including PCR fragments (Hofman *et al.*, 2001). Concerning nucleic acids and its analogues, gene expression has been blocked using either siRNA or morpholino antisense molecules (Lewis *et al.*, 2002; McCaffrey *et al.*, 2002, 2003). Fluorescently labeled siRNA is observed in up to 60% of hepatocytes, mostly in their nuclei. Proteins such as *Escherichia coli* ß-galactosidase, IgG, and bovine serum albumin can also enter hepatocytes after HTV (Kobayahi *et al.*, 2001; Zhang *et al.*, 2004). In addition, small molecules such as Evans blue can also be taken up (Zhang *et al.*, 2004). The ability of a variety of molecules to be taken up by hepatocytes after HTV injection suggests that the uptake process is not receptor mediated. However, the HTV procedure did not increase liver uptake of radioactive PEG (4000 molecular weight), suggesting that some molecules may not enter or are not retained by hepatocytes (Kobayahi *et al.*, 2001).

The possibility of a receptor-mediated uptake process was raised on the basis of competition experiments. Luciferase expression was inhibited when a luciferase pDNA vector was coinjected with a variety of polyanions into the portal vein (under high pressure conditions) or by HTV injection (Budker *et al.*, 2000). A putative, 75- to 80-kDa receptor for oligonucleotide uptake has been

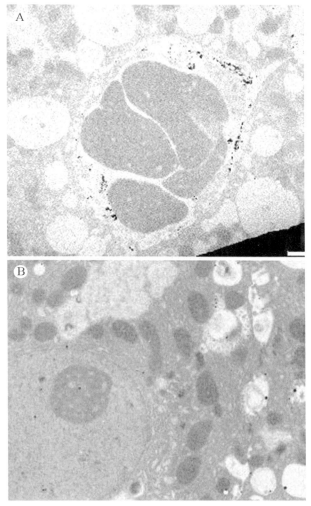

Figure 1.2. Electron micrographs showing pDNA distribution in liver after HTV injection. DNP-modified pDNA was injected in ICR mouse by HTV and 5 min after injection liver was isolated and fixed with formaldehyde/glutaraldehyde. EM sections were stained with anti-DNP antibodies and gold-labeled secondary antibodies. DNA can be seen as small black spots. (A) DNA in space of Disse and large hepatic vesicles. (B) DNA in hepatic nuclei and vesicles.

noted to be present on hepatocytes (Vlassov *et al.*, 1994). Another study found that a variety of polyanions, dextran sulfate, heparin, poly I. and poly C failed to inhibit the uptake of radioactive DNA by the liver nor luciferase expression (Kobayahi *et al.*, 2001). However, this study did not coinject these excess polyanions with the DNA under high pressure but preinjected them under NTV conditions. Nonetheless, we have repeated our competition experiments using HTV injections and found that polyanions had a less inhibitory effect than previously noted, but excess nonexpressing DNA still consistently inhibited luciferase expression. It has also been found that reporter gene expression is a saturating process, which is consistent with a receptor-mediated process (Rossmanith *et al.*, 2002). While we took care to exclude the possibility that excess DNA and polyanions decreased pDNA uptake or expression by inhibiting transport steps other than cellular uptake (e.g., nuclear entry), such postuptake effects may still be operative.

Hepatocyte membrane pores after HTV injection have been observed using scanning EM (Zhang *et al.*, 2004). The quick onset of elevated ALT and AST blood levels 2 h after injection is also consistent with the appearance of membrane pores (Rossmanith *et al.*, 2002). It has also been proposed that increased expression is generated via pores that are induced either by massaging the liver or by repetitive injections (Hickman *et al.*, 1995; Liu and Huang, 2002; Liu *et al.*, 2004).

EM studies also show that the HTV procedure causes an increased number of vesicles to appear in hepatocytes (V. Budker *et al.*, manuscript in preparation). In fact, gold-labeled pDNA can be seen in these vesicles (Fig. 1.2). Subcellular fractionation studies (done by centrifugation) indicated that radioactively labeled DNA after HTV injection becomes associated with vesicles associated with hepatocytes (Lecocq *et al.*, 2003). Curiously, the DNA is sensitive to DNase degradation (Lecocq *et al.*, 2003). It has been proposed that adenovirus entry into cells occurs via macropinosomes that are induced by the adenovirus interaction with specific cellular receptors and that are intrinsically leaky (Meier *et al.*, 2002). Perhaps pDNA entry after HTV injection occurs by a similar process in which the HTV procedure induces vesicles that take up pDNA and from which the pDNA leaks out.

One interesting observation is that if a mouse is first injected with only fluid using the HTV procedure and is then immediately injected with pDNA using the NTV, substantial luciferase expression can still be observed (Andrianaivo *et al.*, 2004; Zhang *et al.*, 2004). In fact, the NTV injection of the expression pDNA vector can be delayed for several minutes with expression still possible, albeit at lower levels than if injected just using the HTV procedure. For example, if the expressing pDNA is injected 12 or 45 s after HTV injection, expression falls to 30- or 10,000-fold, respectively, lower than the value achieved when the pDNA is injected by the HTV procedure

(Andrianaivo *et al.*, 2004). Another study found that a 30-s delay causes a ~20-fold drop in expression (Zhang *et al.*, 2004). Furthermore, Evans blue uptake can still be taken up by hepatocytes if injected by the NTV procedure for several minutes after HTV injection (Zhang *et al.*, 2004). These results indicate that hepatocytes after HTV injection remain in a "competent" state for several minutes. This "competent" state may be due to the persistence of membrane pores or increased vesicular uptake.

Delayed cellular entry has also been observed using fluorescently labeled pDNA (Budker *et al.*, 2000). At less than 5 min after HTV injection, fluorescently labeled pDNA was located at the plasma membrane of hepatocytes and was associated with sinusoidal cells. At 1 h after HTV, the labeled pDNA was inside of more than 10% of the hepatocytes (Eastman *et al.*, 2002). This delay in hepatocyte uptake could be due to the slow transport of DNA through membrane pores or uptake into vesicles. With other fluorescently labeled molecules, such as proteins and siRNA, delayed uptake was less apparent (V. Budker *et al.*, manuscript in preparation). Another study using a different DNA fluorescent-labeling method did not observe increased hepatocyte uptake after HTV injection (Kobayahi *et al.*, 2001).

V. ENTRY OF NUCLEIC ACID INTO THE NUCLEUS

All eukaryotic cells are divided into two main functionally distinct, membrane-bound compartments: the cytoplasm and nucleus. The two compartments are separated by the nuclear envelope—two concentric membrane layers punctured by pores. The pores, called nuclear pore complexes (NPCs), are formed by gigantic supramolecular assemblies of multiple copies of some 30–50 different proteins (Fahrenkrog and Aebi, 2003). NPCs allow the selective, active transport of macromolecules in both directions, provided they carry specific signals, or "addresses," that are recognized by receptor molecules, which in turn mediate their translocation through the central channel of the pore. Macromolecules larger than 50–60 kDa cannot cross this barrier efficiently without displaying such signals. This is why oligonucleotides can enter the nucleus readily after microinjection into the cytoplasm of cultured cells or after HTV injection. Once in the nucleus they are retained by interaction with positively charged chromatin components and thereby accumulate in the nucleus.

In contrast, the nuclear entry for much larger sized pDNA is more challenging. Cytoplasmic microinjections injections in asynchronized, cultured cells required 50- to 250-fold higher concentrations of pDNA than nuclear injections did to achieve similar levels of expression (Ludtke *et al.*, 2002). In addition, approximately 0.6 and 2.9 molecules injected into the nucleus enabled 10 and 50% of the cells to express detectable levels of GFP, respectively. This indicates that only a few pDNA molecules are necessary to achieve expression in a cell, thus greatly increasing the difficulty of studying the cellular transport steps necessary for foreign gene expression.

Our study also found that pDNA was expressed at reasonably high levels following its microinjection into the cytoplasm of HeLa, BHK 21, and 3T3 cells that had not divided (Ludtke et al., 2002). For example, after 10 and 1000 ng/μl of pDNA were injected cytoplasmically, 28 and 50% of the cells that had not divided expressed GFP, respectively, as compared to 50 and 90% for cells that had divided. This result suggested that pDNA could enter the nonmitotic nuclei of mononucleated cells, albeit at a lower efficiency than mitotic nuclei. Other studies using different protocols have observed that cell division enhanced cytoplasmic-injected pDNA expression up to 15-fold (Escriou et al., 2001; Jiang et al., 1998). The ability of pDNA to enter intact nuclei of nondividing cells is consistent with our previous experience using multinucleated myotubes and digitonin-permeabilized cells in culture (Hagstrom et al., 1997; Wolff et al., 1992). Also consistent with these observations is a report on the nuclear accumulation of long, linear λ DNA molecules in reconstituted Xenopus nuclei (Salman et al., 2001). The final conclusion of that paper is that the DNA enters the nucleus through the NPC, without using any energy source, but with very slow kinetics. The first step is especially slow until the DNA end becomes associated with a NPC and starts to thread through it. Then there is a slow inward movement that is thought to be driven by linear diffusion and a weak "pulling force" from inside, which may be generated by retention of the end inside (Salman et al., 2001). The measured femto-newton force range is two orders of magnitude smaller than the pico-newton forces associated with typical motor protein activities, thus it seems very unlikely that a molecular motor mechanism was involved in the translocation step.

These studies in cultured cells are relevant to several observations concerning pDNA delivery to hepatocytes in vivo. For one, is hepatocyte mitosis required for pDNA expression in vivo? After portal vein injections, nuclei with BrDU incorporation increased to approximately 1% of the hepatocytes 2 days after injection but remained near basal levels of 0.1% 6 h and 1 day after injection (Herweijer et al., 2001). Given that the percentage of transfected hepatocytes is substantially greater than the percentage of dividing cells, it is unlikely that hepatocyte mitosis is required for pDNA to enter the nucleus from the cytoplasm.

The transport of pDNA through the cytoplasm may also be limiting. The disparity between the ability for naked DNA to enter isolated nuclei and the inefficiency of its nuclear entry from the cytoplasm of intact cells can be explained by structural and biochemical differences between the reconstituted nuclei in a cell-free environment and the complex membrane and cytoskeletal network present in the intact cytoplasm. For example, addition of a cytoplasm extract decreases the entry of naked DNA into the nuclei of digitonin-treated cells (Hagstrom *et al.*, 1997). This phenomenon of "cytoplasmic sequestration" was observed even after heavily labeling the pDNA with hundreds of NLS peptides (Sebestyen *et al.*, 1998). The nature of the molecular interactions leading to this cytoplasmic retention has not been fully characterized yet. Furthermore, pDNA that is held up in the cytoplasm can be degraded by cytoplasmic DNases (Lechardeur and Lukacs, 2002). After HTV injection, many of the hepatocytes are swollen. Perhaps this attenuates the injected sequestration or degradation of the pDNA in the cytoplasm (Zhang *et al.*, 2004). Consistent with this hypothesis is the observation that expression from cytoplasmically microinjected pDNA increases when it is injected within a larger volume (Zhang *et al.*, 2004).

An alternative possibility is that the pDNA enters the nucleus by a more direct route. We have observed labeled pDNA in the nucleus 5 min after HTV injection (V. Budker *et al.*, manuscript in preparation) (Fig. 1.2B). The time course for gene expression may have some relevance to this question of pDNA nuclear transport after HTV injection. The onset of expression was more rapid after the nuclear microinjection of pDNA than after the cytoplasmic injection (Ludtke *et al.*, 2002). The observation that expression from a CMV-LacZ construct appeared within 2 h after HTV injection and peaked at 8 h (Rossmanith *et al.*, 2002) also raises the possibility of a more direct route to the nucleus. One possibility is that HTV causes a convective flow into the cyto-plasm and breaches the nuclear barrier transiently.

VI. UPTAKE OF NAKED DNA BY MUSCLE CELLS AFTER DIRECT INTRAMUSCULAR INJECTION

The following section highlights several features of naked pDNA expression in muscle that are of relevance to the mechanism of entry. After the direct injection or intravascular delivery of naked pDNA, foreign gene expression is mainly limited to myofibers (Hagstrom *et al.*, 2004; Wolff *et al.*, 1990). It is noteworthy that pDNA expression only occurs in the parenchymal cells of either muscle or liver, despite the fact that nonparenchymal cells (such as endothelial cells) come into contact with the pDNA and the high pressure and increased flow. In addition, all types of striated muscle, including both type I and type II skeletal myofibers and cardiac muscle cells, can express naked pDNA, indicating that the uptake process is common among all types of striated muscles (Acsadi *et al.*, 1991; Jiao *et al.*, 1992).

Muscles such as the rectus femoris or tibialis anterior that are circum-scribed by a well-defined epimysium may enable the highest levels, as they provide the best distribution and retention of the injected pDNA. Similarly, high levels of expression have been observed in diaphragm muscle, a well-marcated thin muscle (Davis and Jasmin, 1993; Liang et al., 2004; Liu et al., 2001). Expression can also be aided by enhanced distribution of the pDNA, which has been accomplished by preinjection of muscles with large volumes of hypertonic solutions and polymers or by an improved injection technique, such as positioning the needle along the longitudinal axis of the muscle (Davis et al., 1993; Mumper et al., 1996). Taken together, these results suggest that expression is enhanced if the muscle can be swelled during direct, intramuscular injection. However, implantation with forceps of pellets of dried pDNA also yields high expression (Jiao et al., 1992; Wolff et al., 1991). Also, myofibers distant from the implantation site are often transfected, suggesting that muscle swelling is not absolutely necessary for expression.

The state of muscle appears to affect the levels of expression. High levels of foreign gene expression were obtained in 2-week-old mouse and rat muscles (Danko et al., 1997). For example, approximately 50% of the myofibers were intensely blue following the intramuscular injection of a β-galactosidase expression vector in 2-week-old Balb/C mice. Also, muscle regeneration induced by myotoxic agents enabled higher levels of expression (Danko et al., 1994; Vitadello et al., 1994; Wells and Wells, 2002). Muscle regeneration induced by ischemia also enhanced pDNA expression (Takeshita et al., 1996; Tsurumi et al., 1996). Myotoxic agents such as cardiotoxin and amide local anesthetics such as bupivacaine (Marcaine) are advantageous because they selectively destroy myofibers without harming myoblasts or the vascular endothelial cells, thus enabling complete recovery of the muscle. Optimal luciferase expression was obtained when the pDNA was injected 3 to 7 days after bupivacaine or cardiotoxin injection or initiation of ischemia, a time when a substantial number of muscle cells have begun to recover from the effects of the myotoxic agent. In these studies involving young muscle or regenerating muscle, increased foreign gene activity may be due to either enhanced pDNA uptake or expression such as from transcriptional activation.

A general trend is that expression decreases as the size of the animal increases. Expression is slightly less in rats than in mice but is substantially less in animals larger than rodents such as rabbits, cats, and monkeys (Jiao et al., 1992). This was observed in both younger and adult larger animals, including nonhuman primates. The connective tissue in primate muscle may prevent the distribution of the pDNA or its contact with sarcolemma. Histochemical and fluorescent stains indicated that primate muscle had substantially more connective tissue within the perimysium. The thicker perimysium in primate muscle may restrict the distribution of the pDNA or may serve as a conduit and an increased potential space for the pDNA to be dispersed without coming in contact with the myofibers.

While expression following liver delivery is rapid, the onset of expression in muscle appears slower. In muscle, expression from RNA vectors peaks within the first day after injection, whereas expression from pDNA vectors peaks at 14 days or even longer (Davis et al., 1993; Manthorpe et al., 1993; Wolff et al., 1990, 1992). This could be the result of delayed pDNA uptake or transcriptional expression. Southern blot analysis indicates that the majority of the injected pDNA is degraded rapidly within hours but that a small percentage of the injected pDNA persists in an open, circular form (Manthorpe et al., 1993; Wolff et al., 1991). Increased expression is obtained with increasing amounts of pDNA, but expression plateaus at different amounts of injected pDNA depending on the muscle type, species, and pDNA vector.

The state of the injected DNA also affects expression. Injection of linearized pDNA yields much less expression, presumably due to its enhanced degradation (Buttrick et al., 1992; Wolff et al., 1992). Larger sized pDNA express less efficiently on a molar basis when injected intramuscularly (or intravascularly) into muscle but less so in liver when injected using HTV injections (Zhang et al., 2004). These results suggest that differences exist in the upake of naked DNA by myofibers as compared to hepatoctyes.

VII. DELIVERY OF NAKED DNA TO MUSCLE VIA INTRAVASCULAR ROUTES

Given that the expression of naked DNA was inefficient in larger animals and humans, efforts were sought to increase expression. The intravascular delivery of naked pDNA to muscle cells was attractive because it avoids the limited distribution of pDNA though the interstitial space following intramuscular injection. Muscle has a high density of capillaries (Browning et al., 1996) that are in close contact with the myofibers (Lee and Schmid-Schonbein, 1995). Delivery of pDNA to muscle via capillaries puts the pDNA into direct contact with every myofiber and substantially decreases the interstitial space the pDNA has to traverse in order to access a myofiber. However, the endothelium in muscle capillaries is of the continuous, nonfenestrated type and has low solute permeability, especially to large macromolecules. This is in contrast to the fenestrated endothelium of liver.

Thus, it was surprising that high levels of expression can be achieved following the intravascular injection of naked pDNA to limb and diaphragm muscles via either artery or venous routes of administrations (Hagstrom et al., 2004; Liang et al., 2004). There is critical dependence on the volume and speed of injection, suggesting that increased hydrostatic pressure, rapid flow, or both are required for efficient expression. Use of fluorescently labeled DNA provided direct evidence that these injection conditions enabled extravasation of the injected DNA (Budker *et al.*, 1998). Muscle swelling appears to correlate with expression levels.

In order to understand the mechanism by which pDNA can be extravasated, it is instructive to consider the mechanisms by which smaller macromolecules such as proteins can traverse the endothelial barrier. One postulated mechanism is transcytosis, which can utilize plasmalemmal vesicles or transient transendothelial channels formed by the fusion of vesicles. These anatomical features may be responsible for the physiologic transcytosis observations modeled by a large number of small pores with radii of about 4 nm. Physiology experiments also suggest that muscle endothelium has a very low number of large pores with radii of 20–30 nm. Although the radius of gyration of 6 kb pDNA is ~100 nm (Fishman and Patterson, 1996), supercoiled DNA in plectonomic form has superhelix dimensions of approximately 10 nm (Rybenkov *et al.*, 1997). This implies that pDNA is capable of crossing microvascular walls by stringing through the large pores. Presumably, the rate of pDNA extravasation is increased by enhancing fluid convection through these large pores by raising the transmural pressure difference in selective regions.

VIII. CONCLUSIONS

As is apparent from this review, studies to elucidate the mechanism of naked pDNA entry into cells have been quite stimulating. It remains remarkable that naked DNA can in fact be expressed efficiently in a variety of cells *in vivo*. Given that viruses have evolved over millions of years of evolution to accomplish DNA transfer, it is surprising that just naked DNA can in fact traverse the several steps in the pathway to expression. Most likely naked DNA delivery did not evolve specifically for polynucleic acid uptake. Instead the pDNA may exploit a transport system that evolved for other purposes. For example, polynucleic acid uptake may provide an evolutionary advantage on the basis of its exquisite ability for immune activation. The conditions required for optimal naked DNA uptake, such as tissue swelling, may mimic the tissue conditions associated with infection. DNA uptake may also play a role in clearing the large amount of released polynucleic acid that occurs physiologically and in disease states.

The ability for naked DNA to enter cells may also have implications for the evolution of viruses. It casts the ability of viruses in a different light; their accomplishment—the ability to transduce cells—may not have been so remarkable.

As further studies continue to elucidate the mechanisms of nucleic acid transport, one can look forward to greater insight into this very interesting phenomenon. These efforts are also likely to lead to improved methods that enable greater expression efficiencies and have greater clinical utility. The eventual impact will be on more powerful gene therapies to treat disease and to alleviate suffering.

References

Acsadi, G., Jiao, S., *et al.* (1991). Direct gene transfer and expression into rat heart *in vivo. New Biol.* **3**(1), 71–81.

Andrianaivo, F., Lecocq, M., *et al.* (2004). Hydrodynamics-based transfection of the liver: Entrance into hepatocytes of DNA that causes expression takes place very early after injection. *J. Gene Med.* **6**, 877–883.

Browning, J., Hogg, N., *et al.* (1996). Capillary density in skeletal muscle of Wistar rats as a function of muscle weight and body weight. *Microvas. Res.* **52**(3), 281–287.

Budker, V., Budker, T., *et al.* (2000). Hypothesis: Naked plasmid DNA is taken up by cells *in vivo* by a receptor-mediated process. *Gene Ther.* **2**, 76–88.

Budker, V., Zhang, G., *et al.* (1996). Naked DNA delivered intraportally expresses efficiently in hepatocytes. *Gene Ther.* **3**(7), 593–598.

Budker, V., Zhang, G., *et al.* (1998). The efficient expression of intravascularly delivered DNA in rat muscle. *Gene Ther.* **5**(2), 272–276.

Buttrick, P. M., Kass, A., *et al.* (1992). Behavior of genes directly injected into the rat heart *in vivo. Circ. Res.* **70**(1), 193–198.

Condiotti, R., Curran, M., *et al.* (2004). Prolonged liver-specific transgene expression by a non-primate lentiviral vector. *Biochem. Biophys. Res. Commun.* **320**, 998–1006.

Danko, I., J., Fritz, D., *et al.* (1994). Pharmacological enhancement of *in vivo* foreign gene expression in muscle. *Gene Ther.* **1**, 114–121.

Danko, I., Williams, P., *et al.* (1997). High expression of naked plasmid DNA in muscles of young rodents. *Hum. Mol. Genet.* **6**(9), 1435–1443.

Davis, H. L., Demeneix, B. A., *et al.* (1993). Plasmid DNA is superior to viral vectors for direct gene transfer into adult mouse skeletal muscle. *Hum. Gene Ther.* **4**(6), 733–740.

Davis, H. L., and Jasmin, B. J. (1993). Direct gene transfer into mouse diaphragm. *FEBS Lett.* **333** (1–2), 146–150.

Davis, H. L., McCluskie, M. J., *et al.* (1996). DNA vaccine for hepatitis B: Evidence for immunogenicity in chimpanzees and comparison with other vaccines. *Proc. Natl. Acad. Sci. USA* **93**(14), 7213–7218.

Eastman, S. J., Baskin, K. M., *et al.* (2002). Development of catheter-based procedures for transducing the isolated rabbit liver with plasmid DNA. *Hum. Gene Ther.* **13**(17), 2065–2077.

Escriou, V., Carriere, M., *et al.* (2001). Critical assessment of the nuclear import of plasmid during cationic lipid-mediated gene transfer. *J. Gene Med.* **3**, 179–187.

Fahrenkrog, B., and Aebi, U. (2003). The nuclear pore complex: Nucleocytoplasmic transport and beyond. *Nature Rev. Mol. Cell Biol.* **4**(10), 757–766.

Fishman, D. M., and Patterson, G. D. (1996). Light scattering studies of supercoiled and nicked DNA. *Biopolymers* **38**, 535–552.

Hagstrom, J., Ludtke, J., *et al.* (1997). Nuclear import of DNA in digitonin-permeabilized cells. *J. Cell Sci.* **110**(Part 18), 2323–2331.

Hagstrom, J. E., *et al.* (2004). A facile non-viral method for delivering genes and siRNA to skeletal muscle of mammalian limbs. *Mol. Ther. J. Am. Soc. Gene Ther.* **10**, 386–398.

Hengge, U. R., Walker, P. S., *et al.* (1996). Expression of naked dna in human, pig, and mouse skin. *J. Clin. Invest.* **97**(12), 2911–2916.

Herweijer, H., and Wolff, J. A. (2003). Progress and prospects: Naked DNA gene transfer and therapy. *Gene Ther.* **10**(6), 453–458.

Herweijer, H., Zhang, G., *et al.* (2001). Time course of gene expression after plasmid DNA gene transfer to the liver. *J. Gene Med.* **3**(3), 280–291.

Hickman, M. A., Malone, R. W., *et al.* (1995). Hepatic gene expression after direct dna injection. *Adv. Drug Deliv. Rev.* **17**(3), 265–271.

Hodges, B. L., and Scheule, R. K. (2003). Hydrodynamic delivery of DNA. *Expert Opin. Biol. Ther.* **3**(6), 911–918.

Hofman, C. R., Dileo, J. P., *et al.* (2001). Efficient *in vivo* gene transfer by PCR amplified fragment with reduced inflammatory activity. *Gene Ther.* **8**, 71–74.

Jiang, C., O'Conner, S., *et al.* (1998). Efficiency of cationic lipid-mediated transfection of polarized and differentiated airway epithelial cells *in vitro* and *in vivo*. *Hum. Gene Ther.* **9**, 1531–1542.

Jiao, S., Williams, P., *et al.* (1992). Direct gene transfer into nonhuman primate myofibers *in vivo*. *Hum. Gene Ther.* **3**(1), 21–33.

Kitsis, R. N., and Leinwand, L. A. (1992). Discordance between gene regulation *in vitro* and *in vivo*. *Gene Express* **2**(4), 313–318.

Kobayahi, N., Kuramoto, T., *et al.* (2001). Hepatic uptake and gene expression mechanisms following intravenous administration of plasmid DNA by conventional and hydrodynamics-based procedures. *J. Pharm. Exp. Ther.* **297**, 853–860.

Lechardeur, D., and Lukacs, G. L. (2002). Intracellular barriers to non-viral gene transfer. *Curr.Gene Ther.* **2**(2), 183–194.

Lecocq, M., Andrianaivo, F., *et al.* (2003). Uptake by mouse liver and intracellular fate of plasmid DNA after a rapid tail vein injection of a small or a large volume. *J. Gene Med.* **5**, 142–156.

Lee, J., and Schmid-Schonbein, G. W. (1995). Biomechanics of skeletal muscle capillaries: Hemodynamic resistance, endothelial distensibility, and pseudopod formation. *Ann. Biomed. Eng.* **23**(3), 226–246.

Lewis, D. L., Hagstrom, J. E., *et al.* (2002). Efficient delivery of siRNA for inhibition of gene expression in postnatal mice. *Nature Genet.* **32**(1), 107–108.

Li, K., Welikson, R. E., *et al.* (1997). Direct gene transfer into the mouse heart. *J. Mol.Cell. Cardiol.* **29**(5), 1499–1504.

Liang, K. W., Nishikawa, M., *et al.* (2004). Restoration of dystrophin expression in mdx mice by intravascular injection of naked DNA containing full-length dystrophin cDNA. *Gene Ther.* **11**(11), 901–908.

Liu, F., and Huang, L. (2002). Noninvasive gene delivery to the liver by mechanical massage. *Hepatology* **35**(6), 1314–1319.

Liu, F., Lei, J., *et al.* (2004). Mechanism of liver gene transfer by mechanical massage. *Mol. Ther. J. Am. Soc. Gene Ther.* **9**(3), 452–457.

Liu, F., Nishikawa, M., *et al.* (2001). Transfer of full-length Dmd to the diaphragm muscle of Dmd (mdx/mdx) mice through systemic administration of plasmid DNA. *Mol.Ther. J. Am. Soc. Gene Ther.* **4**(1), 45–51.

Liu, F., Song, Y., *et al.* (1999). Hydrodynamics-based transfection in animals by systemic administration of plasmid DNA. *Gene Ther* **6**(7), 1258–1266.

Ludtke, J. J., Sebestyen, M. G., *et al.* (2002). The effect of cell division on the cellular dynamics of microinjected DNA and dextran.[erratum appears in *Mol. Ther.* 6(1),134 (2002)]. *Mol. Ther. J. Am. Soc. Gene Ther.* **5**(5 Pt 1), 579–588.

Manthorpe, M., Cornefert-Jensen, F., *et al.* (1993). Gene therapy by intramuscular injection of plasmid DNA: Studies on firefly luciferase gene expression in mice. *Hum.Gene Ther.* **4**(4), 419–431.

McCaffrey, A. P., Meuse, L., *et al.* (2002). RNA interference in adult mice. *Nature* **418**(6893), 38–39.

McCaffrey, A. P., Meuse, L., *et al.* (2003). A potent and specific morpholino antisense inhibitor of hepatitis C translation in mice. *Hepatology* **38**(2), 503–508.

Meier, O., Boucke, K., *et al.* (2002). Adenovirus triggers macropinocytosis and endosomal leakage together with its clathrin-mediated uptake. *J. Cell Biol.* **158**(6), 1119–1131.

Miao, C. H., Ohashi, K., *et al.* (2000). Inclusion of the hepatic locus control region, an intron, and untranslated region increases and stabilizes hepatic factor IX gene expression *in vivo* but not *in vitro*. *Mol. Ther. J. Am. Soc. Gene Ther.* **1**(6), 522–532.

Mumper, R. J., Duguid, J. G., *et al.* (1996). Polyvinyl derivatives as novel interactive polymers for controlled gene delivery to muscle. *Pharm. Res.* **13**(5), 701–709.

Romero, N. B., Benveniste, O., *et al.* (2002). Current protocol of a research phase I clinical trial of full-length dystrophin plasmid DNA in Duchenne/Becker muscular dystrophies. II. Clinical protocol. *Neuromusc. Disord.* **12**(Suppl. 1), S45–S48.

Rossmanith, W., Chabicovsky, M., *et al.* (2002). Cellular gene dose and kinetics of gene expression in mouse livers transfected by high-volume tail-vein injected of naked DNA. *DNA Cell Biol.* **21**, 847–853.

Rybenkov, V. V., Vologodskii, A. V., *et al.* (1997). The effect of ionic conditions on the conformations of supercoiled DNA.1. Sedimentation analysis. *J. Mol. Biol.* **267**, 299–311.

Salman, H., Zbaida, D., *et al.* (2001). Kinetics and mechanism of DNA uptake into the cell nucleus. *Proc. Natl. Acad. Sci. USA* **98**, 7247–7252.

Sebestyen, M. G., Ludtke, J. J., *et al.* (1998). DNA vector chemistry: The covalent attachment of signal peptides to plasmid DNA. *Nature Biotechnol.* **16**(1), 80–85.

Sikes, M. L., O'Malley, B. W., Jr., *et al.* (1994). *In vivo* gene transfer into rabbit thyroid follicular cells by direct DNA injection. *Hum. Gene Ther.* **5**, 837–844.

Takeshita, S., Tsurumi, Y., *et al.* (1996). Gene transfer of naked DNA encoding for three isoforms of vascular endothelial growth factor stimulates collateral development *in vivo*. *Lab. Invest.* **75**(4), 487–501.

Tsurumi, Y., Kearney, M., *et al.* (1997). Treatment of acute limb ischemia by intramuscular injection of vascular endothelial growth factor gene. *Circulation* **96**(9 Suppl. S), 382–388.

Tsurumi, Y., Takeshita, S., *et al.* (1996). Direct intramuscular gene transfer of naked DNA encoding vascular endothelial growth factor augments collateral development and tissue perfusion. *Circulation* **94**(12), 3281–3290.

Vitadello, M., Schiaffino, M. V., *et al.* (1994). Gene transfer in regenerating muscle. *Hum. Gene Ther.* **5**(1), 11–18.

Vlassov, V., Deeva, E., *et al.* (1994). Transport of oligonucleotides across natural and model membranes. *Biochim. Biophys. Acta* **1197**, 95–108.

Wells, D. J., and Wells, K. E. (2002). Gene transfer studies in animals: What do they really tell us about the prospects for gene therapy in DMD. *Neuromusc. Disord.* **12**, S11–S22.

Wolff, J. A., Dowty, M. E., *et al.* (1992a). Expression of naked plasmids by cultured myotubes and entry of plasmids into T tubules and caveolae of mammalian skeletal muscle. *J. Cell Sci.* **103** (Pt. 4), 1249–1259.

Wolff, J. A., Ludtke, J. J., *et al.* (1992b). Long-term persistence of plasmid DNA and foreign gene expression in mouse muscle. *Hum. Mol. Genet.* **1**(6), 363–369.

Wolff, J. A., Malone, R. W., *et al.* (1990). Direct gene transfer into mouse muscle *in vivo*. *Science* **247** (4949 Pt 1), 1465–1468.

Wolff, J. A., Williams, P., *et al.* (1991). Conditions affecting direct gene transfer into rodent muscle *in vivo*. *Biotechniques* **11**(4), 474–485.

Yang, J. P., and Huang, L. (1996). Direct gene transfer to mouse melanoma by intratumor injection of free DNA. *Gene Ther.* **3**(6), 542–548.

Zhang, G., Budker, V., *et al.* (1999). High levels of foreign gene expression in hepatocytes after tail vein injections of naked plasmid DNA. *Hum. Gene Ther.* **10**(10), 1735–1737.

Zhang, G., Budker, V., *et al.* (2001). Efficient expression of naked DNA delivered intraarterially to limb muscles of nonhuman primates. *Hum. Gene Ther.* **12,** 427–438.

Zhang, G., Fetterly, B., *et al.* (2004a). Intraarterial delivery of naked plasmid DNA expressing full-length mouse dystrophin in the mdx mouse model for Duchenne muscular dystrophy. *Hum. Gene Ther.*

Zhang, G., Gao, X., *et al.* (2004b). Hydroporation as the mechanism of hydrodynamic delivery. *Gene Ther.* **11**(8), 675–682.

Zhang, G., Vargo, D., *et al.* (1997). Expression of naked plasmid DNA injected into the afferent and efferent vessels of rodent and dog livers. *Hum. Gene Ther.* **8**(15), 1763–1772.

2

Targeted Delivery of Therapeutic Oligonucleotides to Pulmonary Circulation

Annette Wilson,* Fengtian He,[†] Jiang Li,[†] Zheng Ma,[†] Bruce Pitt,* and Song Li[†]

*Department of Environmental and Occupational Health
Graduate School of Public Health, School of Pharmacy
University of Pittsburgh, Pittsburgh, Pennsylvania 15261
[†]Center for Pharmacogenetics and Department of Pharmaceutical Sciences
School of Pharmacy, University of Pittsburgh, Pittsburgh, Pennsylvania 15261

Advances in Genetics, Vol. 54
Copyright 2005, Elsevier Inc. All rights reserved.

0065-2660/05 $35.00
DOI: 10.1016/S0065-2660(05)54002-1

ABSTRACT

Functional oligodeoxynucleotides (ODN) such as antisense ODN (AS-ODN) show promise as new therapeutics for the treatment of a number of pulmonary diseases. They also hold potential to serve as a research tool for the study of gene function related to lung physiology. The success of their application is largely dependent on the development of an efficient delivery vehicle. This chapter summarizes work toward the development of lipidic vectors for targeted ODN delivery to pulmonary circulation. Recent advancements in the development of novel ODN are also discussed briefly.   © 2005, Elsevier Inc.

I. INTRODUCTION

Targeted delivery of macromolecules such as proteins or genes to the pulmonary endothelium is an important strategy for therapeutic approaches in disorders such as acute lung injury and pulmonary hypertension. A variety of approaches have been used to accomplish these goals, including (a) engraftment of genetically modified smooth muscle cells (Campbell *et al.*, 2001) or endothelial progenitor cells (Nagaya *et al.*, 2003); (b) immunotargeting of protein streptavidin complexes via monoclonal antibodies to platelet endothelial cell adhesion molecule-1 (PECAM-1) (Christofidou-Solomidou *et al.*, 2003), intercellular adhesion molecule-1 (ICAM-1) (Muro *et al.*, 2003), and thrombomodulin (TM) (Christofidou-Solomidou *et al.*, 2002); (c) adenoviral mediated gene delivery using bispecific antibodies to angiotensin-converting enzyme (ACE) (Reynolds *et al.*, 2000) or unmodified standard replication-deficient recombinant adenoviruses redirected after cationic liposome (Ma *et al.*, 2002a); or (d) gene transfer using anionic (Hughes *et al.*, 1989) or cationic (Li *et al.*, 2000; Ma *et al.*, 2002b) liposomes. Recently targeted delivery of small-sized functional ODN has also been explored as a novel approach for the treatment of pulmonary diseases.

Functional ODN such as AS-ODN have been developed as a powerful tool for target validation and therapeutic purposes. After the biological function of a gene is verified, it can be determined if modulation of this target gene can affect a disease process. This can be achieved through an antisense approach. AS-ODN pair and hybridize with their complementary mRNA, preventing its translation into protein or inducing its degradation via activation of RNase H. Considerable progress has been made through the development of novel chemical modifications to ODN to protect them from nuclease degradation and to increase their affinity for their target mRNA. RNA interference has also been developed into a highly efficient method of silencing gene expression by the use of 21- to 23-mer small interfering RNA (siRNA) molecules (Elbashir *et al.*, 2001).

Despite advancements in the development of novel therapeutic ODN, their successful application is largely dependent on the development of a delivery vehicle to efficiently deliver the ODN to the lung with minimal toxicity. Two major synthetic vectors have been developed to achieve this goal, namely polymer- and lipid-based vectors (Li and Ma, 2001).

II. CHEMISTRY OF OLIGODEOXYNUCLEOTIDES

AS-ODN are designed from genomic sequence information and in fact are the reversed complement of the genomic sequence. It is not necessary to know the entire sequence of the gene. AS-ODN are typically made up of 15–20 nucleotides that are complementary to their target RNA. However, secondary and tertiary mRNA structures need to be accounted for in the design of the ODN. These structures may affect the target site of the RNA, making it inaccessible to the ODN. Computer models of RNA structures may be generated but because they may not represent the RNA structure within cells, they are of limited use. Several strategies have been developed to determine the sites of RNA that may be hybridized. The use of random or semirandom ODN libraries and RNase H followed by primer extension has been shown to reveal accessible sites (Kurreck et al., 2002). Another method is to screen a large number of ODN against the target RNA in the presence of RNase H and evaluate the extent of cleavage and gene expression (Ho et al., 1996, 1998). DNA array can also be used to map the target RNA.

When designing the ODN, certain elements in the sequence should be avoided. A string of four contiguous guanosine (G) residues should be avoided as they may form guanine quartets (G-quartets), causing protein binding or non-specific RNA binding, thus decreasing the efficiency of the ODN and potentially leading to side effects (Benimetskaya et al., 1997; Burgess et al., 1995). In addition, CpG motifs should be avoided as they are known to stimulate immune responses in mammalian systems (Krieg, 1999).

Two major mechanisms account for the antisense properties of ODN. First, they may be designed to induce RNase H cleavage of the RNA moiety in the DNA–RNA complex of the target mRNA, leading to its degradation. Second, they may inhibit translation of the mRNA by sterically blocking the action of the ribosome. ODN may undergo modifications to enhance these mechanisms by increasing their stability and potency.

A. Polynucleotide backbone

ODN in their unmodified form, phosphodiester (PO) ODN, are unstable and prone to nuclease degradation in biological fluids. A number of modifications of the polynucleotide backbone of ODN have been shown to improve their

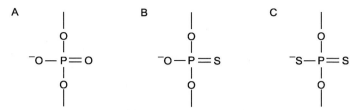

Figure 2.1. Structures of the polynucleotide backbone: (A) phosphodiester, (B) phosphorothioate, and (C) phosphorodithioate.

stability (Fig. 2.1). The phosphothioate (PS) ODN (one nonbridging oxygen substituted with one sulfur) is more stable and less sensitive to nucleases than PO ODN. The phosphorodithioate (PS2) ODN (substitution of two sulfurs) is the most stable polynucleotide and is not digested by nucleases. These backbone analogs not only change the nuclease sensitivity of the ODN, but also affect its function by altering its capacity to bind to proteins and thus affecting ODN uptake (Krieg *et al.*, 1996). PS ODN bind to proteins known to interact with polyanions, such as heparin-binding protein. A study in which primates were treated with PS ODN led to an acute toxicity through the induction of complement cascade (Galbraith *et al.*, 1994; Henry *et al.*, 1997; Wallace *et al.*, 1996). This protein binding also increased their half-life. PS ODN were shown to have a half-life of 9–10 h in human serum in comparison to approximately 1 h for PO ODN (Campbell *et al.*, 1990; Kurreck *et al.*, 2002; Phillips *et al.*, 2000). Stability studies suggest that phosphorothioate substitutions are necessary for maximum resistance to nucleases (Krieg *et al.*, 1996). PS ODN also induce RNase H cleavage of the mRNA when duplexed with the RNA, which enhances its potency. A disadvantage of PS ODN is their slightly reduced affinity for complementary RNA compared to PO ODN. There is, however, enhanced specificity of hybridization for PS ODN in comparison to PO ODN (Crooke, 2000).

B. Modified sugar

AS-ODN containing nucleotides with alkyl modifications at the 2′ position of the ribose (Fig. 2.2) are less toxic than PS ODN and also have a slightly higher affinity for its complementary mRNA (see Table 2.1 for comparison). However, these ODN cannot induce RNase H activity, thus reducing their potency. Because of this, their antisense effect can only be due to the steric blocking of translation. 2′-O-methyl and 2′-O-methoxy-ethyl RNA are the most important members of this class (Zamaratski *et al.*, 2001). A chimeric ODN can be designed to include both PS ODN and 2′-O-methyl RNA to retain RNase H activity while minimizing the PS ODN-associated toxicity (Cole-Strauss *et al.*, 1996; Kmiec, 1999).

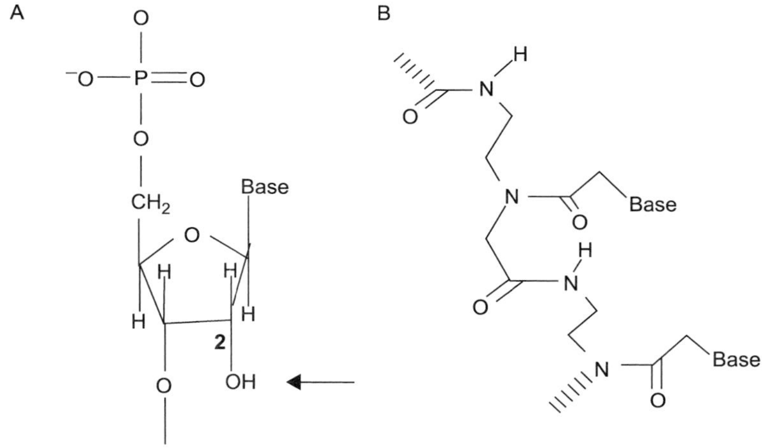

Figure 2.2. Chemical modifications to AS-ODN. (A) Site for chemical modification on ribose of ODN. (B) Structure of peptide nucleic acids (PNA).

Table 2.1. Effects of ODN Composition on Affinity, Potency, Biological Stability, and Toxicity[a]

ODN composition	Affinity	Potency	Biological stability	Toxicity	Ref.
PO ODN	++	+	+	+	Wickstrom (1986)
PS ODN	++	+++	+++	++	Galbraith et al. (1994); Zon (1988);
2-O-alkyl RNA	+++	++	+++	+	Manoharan (1999)
PNA	+++	+++	+++	+	Hamilton et al. (1999); Larsen et al. (1999)
siRNA	++?	++++?	+++	+?	McManus and Sharp (2002); Yu et al. (2002)

[a]+, lowest; ++++, highest.

C. Peptide nucleic acids

Peptide nucleic acids (PNAs) are a new type of oligonucleotide that have attracted increasing attention. As shown in Fig. 2.2, PNAs replace the deoxyribose phosphate backbone of ODN with polyamide linkages (Nielsen et al., 1991). PNAs hybridize efficiently to RNA and are very stable. PNAs bind to their target via two modes. PNAs containing mixed purine/pyrimidines form a duplex with complementary nucleic acids by Watson–Crick hydrogen bonding and are referred to as duplex-forming PNAs (Egholm *et al*., 1993). PNAs containing only pyrimidines bind to complementary nucleic acids by Watson–Crick Hoogsteen hydrogen bonding forming PNA_2/nucleic acid triplexes and are referred to as triplex-forming PNAs (Betts *et al*., 1995). Both of these formations are very stable, exceeding the stability of DNA–RNA and DNA–DNA hybridization. However, they do not activate RNase H cleavage of RNA (Hanvey *et al*., 1992). Additionally, PNAs are neutrally charged, which decreases their cellular uptake. Intracellular delivery has been improved by coupling the PNA with negatively charged oligomers, lipids, or specific peptides known to be internalized by certain cell types (Nielsen, 1999). Pooga *et al*. (1998) developed a 21-mer duplex-forming PNA conjugated to the cell-penetrating peptide transportan. The PNA was found to efficiently downregulate the expression of the human galanin receptor in Bowes membrane both *in vitro* and *in vivo* (Pooga *et al*., 1998). We have shown that PNA hybridized with an inert PO ODN can be targeted efficiently to the pulmonary endothelium via cationic lipids (Yuan *et al*., 2003). Chimeric oligonucleotides composed of both PNA and DNA have been developed (Borgatti *et al*., 2004). These chimeric oligonucleotides may be advantageous over PNAs because they can activate RNase H and can also be formulated in lipid vectors that have been developed for the delivery of DNA ODN. PNA is relatively nontoxic due to its neutral charge and lack of protein binding.

D. Small interfering RNA

Many advances have been made in the area of RNA interference (RNAi). RNAi is initiated by long double-stranded RNAs (dsRNAs), which are processed into 21–23 nucleotide-long RNAs using the RNase III (Dicer) enzyme. This enzyme is believed to act as a dimer that cleaves the strands of dsRNAs, leaving two-nucleotide, $3'$ overhanging ends. These small interfering RNA (siRNAs) are then incorporated into the RNA-induced silencing complex (RISC), a protein:RNA complex that guides a nuclease, which degrades the target mRNA. RNAi technology is thought to be significantly more potent than traditional antisense methods. Short hairpin RNAs (shRNAs), fold-back stem–loop structures that give rise to siRNA intracellularly, are also capable of inducing RNAi (Yu *et al*., 2002). This opened up the possibility of constructing vectors expressing interfering RNA for long-term silencing of gene expression (Tuschl, 2002). Another advantage of siRNA is the ease in its design compared to AS-ODN. The secondary structure of target mRNA seems to be of less concern in the design of siRNA compared to AS-ODN (Reynolds *et al*., 2004). Nevertheless, a number of problems still remain. Randomly selected siRNAs produce knockdown $\geq 50\%$ with a 58–78% success rate, while very effective siRNAs (90–95%) are found by chance 11–18% of the time (Reynolds *et al.*, 2004). In stably transfected cells, gene silencing was observed for up to 2 months (Brummelkamp *et al.*, 2002). Modifications may be made to the nucleotides of synthesized RNA to further prolong the interference capability (Amarzguioui *et al.*, 2003). *In vivo* studies of RNAi in mammals have shown promising results. In one study, siRNA or plasmid coding for shRNA was injected rapidly with a large volume via the tail vein in mice (McCarey *et al.*, 2002). Reporter genes that were either encoded on cotransfected plasmids or in transgenic mouse strains were effectively downregulated in most of the organs (Lewis *et al.*, 2002). Despite its promise as a research tool and as a new type of therapeutics, several studies have shown that siRNA also inhibit gene expression in a sequence-nonspecific manner (off-target effect), which is dose and cell type dependent (Bridge *et al.*, 2003; Scacheri *et al.*, 2004; Sledz *et al.*, 2003). Furthermore, induction of the interferon response has also been observed in mammalian cells (Bridge *et al.*, 2003; Sledz *et al.*, 2003). More studies are warranted to address the potential *in vivo* toxicity of siRNA.

III. PULMONARY PHYSIOLOGY THAT AFFECTS ENDOTHELIUM TARGETING

A. Pulmonary circulation

Normal pulmonary circulation is a low-pressure, high-capacitance bed that receives 100% of the cardiac output. It matches circulation with ventilation to provide substantial oxygen for the blood supply (West *et al.*, 1965). The intimal lining of all blood vessels is made of a continuous single layer of endothelial cells (ECs). In the human lung, ECs occupy a surface area of 130 m^2 (Simionescu, 1991). Approximately 30% of all ECs are found in the pulmonary vasculature. In addition to being a semipermeable layer of nucleated cells separating blood from tissue, and in the lungs from air, vascular endothelium is a highly specialized metabolically active barrier possessing numerous physiological, immunological, and synthetic functions. The pulmonary endothelium functional and structural integrity is essential for both pulmonary and systemic cardiovascular homeostasis. In addition to its large surface area, pulmonary endothelium is also the first vascular bed the vector will interact with following systemic administration. Thus pulmonary ECs represent an ideal target for targeted delivery of therapeutic agents via the vascular route.

B. Mechanisms of uptake of solutes by pulmonary endothelial cells

Solutes and gases mainly diffuse across the pulmonary endothelium. Diffusion of solutes is dependent on their physical properties. Several pathways are available for the transport of solutes and fluid: vesicles (caveolae), interendothelial junctions (tight junctions), and transendothelial channels. Solutes and water can diffuse the endothelium readily by paracellular (between cells) or transcellular (through cells) routes (Renkin, 1985; Rippe and Haraldsson, 1987). Water freely crosses ECs through aqueporin channels (Schnitzer *et al.*, 1996). Diffusion of macromolecules may occur by transcellular transport through small and large endothelial pores (Grotte, 1956; Pappenheimer *et al.*, 1951). Only about 0.02% of the capillary surface is made up of these pores. Diffusion through the pores is dependent on the molecular size of the solute (Predescu and Palade, 1993; Siflinger-Birnboim *et al.*, 1987; Taylor and Granger, 1984). Additionally, many macromolecules are transported by the formation of vesicles via transcytotic transport to the basal surface of the endothelium (Predescu *et al.*, 2004). This may be clathrin- or caveoli-mediated endocytosis (Mukherjee *et al.*, 1997). Despite the abundant mechanisms in vascular ECs that mediate the uptake of solutes, delivery of ODN to the pulmonary endothelium without a vehicle has proven to be inefficient possibly due to a lack of receptors for ODN on ECs and the transient residence times of ODN in pulmonary circulation.

C. Surface markers of pulmonary vascular endothelial cells

The EC surface is composed of sialyl residues that are responsible for its negative charge. The plasmalemmal membrane, vesicles, and paracellular channels contain microdomains of anionic sites consisting of glycosaminoglycans, sialoconjugates, and monosaccharide residues (Simionescu *et al.*, 1982). Thus the cationic lipid vector may effectively interact with the negatively charged EC, which may contribute to the high efficiency in the cationic lipid-mediated delivery of nucleic acids to the pulmonary endothelium via the vascular route.

Endothelial cells from different sites within the vasculature have common properties, but specific properties may differ between ECs found in small and large vessels. For example, micrographs show a larger number of vesicles on the apical surface of ECs of microvessels vs the pulmonary artery (McIntosh *et al.*, 2002). ECs also contain surface antigens and receptors that recognize specific macromolecules. For example, evidence shows that transcytosis of albumin across the endothelium occurs via binding to a receptor that signals formation of a vesicle for transportation to the basolateral membrane (Schnitzer and Oh, 1994). Many of these receptors and antigens are common in different organ systems, but some of these are organ and site specific. This is very important in developing strategies for drug development. For example, small- and medium-sized vesicles are largely involved in the pathogenesis of pulmonary hypertension. Thus therapeutic agents that are targeted to endothelium lining these blood vesicles should be ideal for the treatment of such disease. A description of the potential ligands targeted to different endothelium-specific surface markers for systemic delivery follows.

1. Antibodies

Antibody (Ab)-mediated targeting has the advantages of high specificity and high affinity toward the antigen. Surface antigens specific for endothelium are discussed later. In some disease states, these antigens may be elevated, thus increasing targeting efficiency. Antigens that can be potentially utilized for endothelium targeting include PECAM-1, TM, ACE, and ICAM-1.

a. PECAM-1

PECAM-1 (CD31) is a 130-kDa adhesion molecule constitutively expressed in ECs (Newman, 1997). Cytokines do not appear to affect its expression (Vaporician *et al.*, 1993). Its expression is preferential to the surface of pulmonary ECs and has a high surface density that facilitates leukocyte migration through the endothelium. The internalization efficiency of Ab to PECAM-1 is epitope dependent. Despite similar affinities, monoclonal antibodies (mAbs) to different epitopes of PECAM-1 demonstrated pronounced differences in internalization after binding with the EC surface and, subsequently, in pulmonary uptake (Danilov *et al.*, 2001). Increased internalization of anti-PECAM-1 has been accomplished with the conjugation of streptavidin (Muzykantov *et al.*, 1999). We have shown that anti-PECAM-1 Ab-targeted cationic polymer [(polyethyleneimine (PEI)] selectively mediates delivery of the luciferase reporter gene to both cultured mouse lung endothelial cells (MLEC) and intact mouse lungs following tail vein injection (Li *et al.*, 2000). Anti-PECAM Ab-1-targeted lipidic vectors are also currently being examined for their potential in targeted delivery of functional ODN to lung ECs *in vitro* and *in vivo* (Wilson *et al.*, unpublished data).

b. TM

TM is another attractive antigen to target using an antibody. TM is a 75-kDa glycoprotein (CD141) controlling the enzymatic activities of thrombin and is found abundantly on the luminal surface of the pulmonary capillary wall (Kennel *et al.*, 1988; Maruyama *et al.*, 1990). The tight junction endothelium of the lung expresses more TM than the kidney or liver. Ab to TM (273-34A) has been employed for targeted delivery of various types of agents to mouse pulmonary circulation, including radioisotope, chemotherapeutic agents, and plasmid DNA (Kennel *et al.*, 1988; Maruyama *et al.*, 1990). We have shown that the TM Ab-conjugated lipid vector also efficiently mediates the selective delivery of ODN to cultured lung ECs and pulmonary endothelium of intact mice (Fig. 2.3) (Wilson *et al.*, unpublished data). One potential concern over lung targeting with the 273-34A antibody is that it induces poor internalization following binding to pulmonary ECs *in vivo*, despite the efficient internalization observed in cultured ECs (L. Huang, personal communication). It remains to be

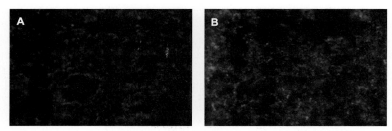

Figure 2.3. Antibody-mediated selective delivery of ODN to pulmonary circulation in mice. Mice received a tail vein injection of 20 μg cy3-ODN as free ODN or 273-34A-targeted lipidic ODN. Mice were sacrificed 15 min later and the lungs were observed for cy3-ODN uptake. (A) Free ODN; (B) 273-34A-targeted lipidic ODN. Blue is Hoescht-stained nuclei and red is cy3-ODN. (See Color Insert.)

examined whether 273-34A induces internalization *in vivo* following incorporation into a lipid vector. Also remaining to be tested is whether antibodies to be developed that are targeted to other epitopes of TM are more efficient than 273-34A in inducing internalization *in vivo*.

c. ACE

ACE (kininase II, CD143), an important regulator of vascular tone and remodeling (Soffer, 1976), is expressed on the luminal surface of ECs of different types of blood vessels (Caldwell *et al.*, 1976). ACE is a type 1 integral membrane protein that is anchored to the plasma membrane near its C terminus. Somatic ACE also exists as a soluble form in plasma, amniotic and cerebrospinal fluids, and seminal plasma (Ehlers and Riordan, 1989). Plasma ACE is probably derived from vascular ECs (Ching *et al.*, 1983) and lacks a transmembrane portion at the C terminus (Hooper *et al.*, 1987). ACE secretase was shown to be a metalloprotease (Hooper *et al.*, 1987; Oppong and Hooper, 1993) that cleaves ACE in a number of tissues. ACE must be anchored in a membrane in order for cleavage by ACE secretase to occur (Ehlers *et al.*, 1996). Little is known about the effects of antibody binding on ACE release. The unique tissue distribution of endothelial ACE, along with preferential expression in lung capillaries (Franke *et al.*, 1997), makes it an almost ideal target for therapy directed toward pulmonary endothelium. It has been shown that binding of mAbs to ACE depends not only on the size of ACE (short, soluble ACE versus long ACE with transmembrane anchor), but also on the availability of the epitope for a particular mAb in the circulation. For example, binding of mAb i1A8 with ACE (somatic two-domain ACE with transmembrane anchor) in solution was the lowest among the 10 studied mAbs to ACE, whereas binding of this antibody with ACE on the surface of CHO-ACE cells was the highest

(Balyasnikova et al., 1999). Thus such Ab is more suitable for targeted delivery of therapeutic agents to pulmonary ECs.

d. ICAM-1

ICAM-1 is a transmembrane glycoprotein constitutively exposed on the luminal surface of ECs (Almenar-Queralt et al., 1995). It is an attractive target because it is upregulated in vascular inflammation, oxidative stress, and thrombosis (Diamond et al., 1990). In addition to acting as a potential targeting agent, antibody blocking of ICAM-1 suppresses leukocyte adhesion to ECs, providing an anti-inflammatory benefit to the effects of drugs (Broide et al., 1998). ICAM-1 is internalized by ECs by cell adhesion molecule (CAM)-mediated endocytosis (Muro et al., 2003). ICAM-1 binding triggers signaling via protein kinase C, Src family kinases, and Rho-dependent kinase and also involves dynamin and amiloride-sensitive Na^+/H^+ exchangers, leading to rapid reorganization of the actin cytoskeleton and formation of endocytotic compartments (Muro et al., 2003). It has been shown that a large fraction of ICAM-1 dissociated from internalized anti-ICAM-1 and recycled to the EC surface, providing a potential prolonged therapeutic effect (Muro et al., 2004).

2. Peptide phage display

In vivo phage display is a powerful tool to study tissue- and organ-specific vascular addresses. Using this approach, peptides capable of tissue-specific homing can be identified by performing a selection for that trait in vivo (Pasqualini and Ruoslahti, 1996). A phage display library may be injected and then the bound phage collected for identification. The GFE-1 phage (CGFECVRQC-PERC) isolated from the $CX_3CX_3CX_3C$ library, where C is cysteine and X is any amino acid, was found to home to the lung vasculature 35-fold greater than other organ systems (Rajotte and Ruoslahti, 1999). The receptor of this peptide was identified as membrane dipeptidase, a cell surface zinc metallo-protease involved in the metabolism of glutathione, leukotriene D_4, and certain β-lactam antibiotics. Novel peptides LSIPPKA, FQPTPPL, and LTPATAI (White et al., 2001) have also been developed that target the lectin-like oxidized LDL receptor (LOX-1). LOX-1 is normally expressed at low levels on venous and arterial vascular ECs, but is upregulated in dysfunctional endothelium associated with hypertension and atherogenesis (Kataoka et al., 1999). One of the potential advantages of peptide-mediated targeting is its relatively lower immunogenecity compared to antibody. Peptides can also be produced with ease in large quantities. The potential drawback of using peptides as a targeting ligand is their relatively lower affinity. No successful report thus far has been reported of peptide-mediated targeting of therapeutic agents to pulmonary endothelium *in vivo*. It remains to be tested whether the targeting efficiency of peptide can be improved via incorporation into lipid vectors via a multivalency effect.

IV. LIPID VECTORS FOR PULMONARY OLIGONUCLEOTIDE DELIVERY

A. Cationic lipid vectors

Cationic lipid vectors have been widely used for pulmonary gene transfer via the vascular route. It was first reported by Zhu *et al.* (1993). However, it was not until recently that researchers were able to obtain high and reproducible levels of gene expression by IV injection (Hong *et al.*, 1997; Li *et al.*, 1997; Liu *et al.*, 1997; Templeton *et al.*, 1997; Wang *et al.*, 1998). There are several advantages of using cationic liposomes for gene delivery. Cationic liposomes readily form complexes with negatively charged nucleic acids, unlike neutral or anionic liposomes, which require entrapment of nucleic acids within vesicles. Theoretically, there is no size restriction of the nucleic acid and almost 100% of the nucleic acid can be recovered in the complexed form. Second, the liposome/DNA complexes are prepared such that there is a slight excess positive charge enhancing its interaction with negatively charged cell surfaces. Depending on the lipid composition, there may be an endosome disruption mechanism to enhance release into the cytosol following endocytosis. Finally, the complexation of DNA with cationic liposomes may offer some protection from physical forces and enzymatic digestion.

Pulmonary gene transfer via cationic lipids is largely mediated by a passive mechanism of serum-induced aggregation of lipid vectors, although the nonspecific interaction of the cationic lipid vector with the negatively charged endothelium may also play a role (Li *et al.*, 1999). Following IV administration, blood components such as negatively charged serum proteins induce the aggregation of cationic lipid vectors. The large-sized aggregates are mainly entrapped in pulmonary microvasculature, because lung has the first and the largest capillary bed the lipid vector will interact with following systemic administration. We have shown that cationic lipid vectors can also be employed in the systemic delivery of ODN to pulmonary endothelium (Ma *et al.*, 2002b). However, ODN delivery and plasmid DNA delivery differ with respect to the optimized lipid composition of the vectors. For example, gene transfer is the most efficient when delivered by a vector composed of 1,2-dioleoyl-3-trimethylammonium propane (DOTAP) and a helper lipid, cholesterol (Li *et al.*, 1998; Song *et al.*, 1997). In contrast, a lipid vector composed of DOTAP alone is most efficient in mediating the delivery of ODN to pulmonary endothelium (Ma *et al.*, 2002b). One possible reason for this may be the structural differences between plasmid DNA and ODN. Plasmid DNA is a large molecule with a size of ~5 kb or greater. The mixing of plasmid DNA with cationic liposomes will lead to a substantial condensation of the plasmid along with lipid rearrangement. In contrast, ODN are much smaller molecules and little, if any, condensation is expected. In the absence of condensation the affinity between ODN and cationic liposomes will be largely determined by the charge density on the liposome surface. Inclusion of any neutral lipid into the liposomes will decrease the surface charge density and therefore decrease their interaction with ODN. The decreased affinity of cationic liposomes for ODN will result in a rapid dissociation of ODN from lipid vectors following exposure to serum and, thus, inefficient pulmonary uptake (Ma et al., 2002b).

Intravenous delivery of an ODN without a CpG motif via a cationic lipid vector is associated with a minimal induction of proinflammatory cytokine response and other hematologic toxicities (Ma et al., 2002b). Immunofluorescence staining of BrdU-labeled ODN suggested efficient accumulation of ODN in the alveolar capillary region. Transmission electron microscopy of immunogold localization of BrdU-labeled ODN confirmed that pulmonary ECs were indeed targeted by the vector (Ma et al., 2002b). Pretreatment of mice with an ODN against ICAM-1 formulated in a lipid vector significantly decreased ICAM-1 expression in the lung following lipopolysaccharide (LPS) challenge. One of the advantages of ODN delivery using cationic lipid vectors is its simplicity. However, the nonspecific interaction of cationic lipid vector with serum proteins and blood cells may limit the amount of ODN that can be delivered without causing significant toxicity.

B. Neutral lipid vector

Neutral liposomes, in general, are more biocompatible, less toxic, and potentially more stable in plasma than cationic lipid vectors. Furthermore, neutral vectors can be rendered tissue specific following conjugation with a specific ligand. Neutral liposome-based drugs have a high clinical profile and are currently used for the treatment of cancers and infectious diseases (Allen, 1998; Herbrecht et al., 2003; Hussein, 2003). However, as a gene or ODN delivery vehicle conventional liposomes suffer from low encapsulation efficiency due to the limited interior size of liposomes and the poor interaction of ODN with neutral lipids. Stuart and colleagues (2004) described a method by which 80–100% of input ODN were entrapped in a neutral lipid vector that was stable in human plasma. This involves the formation in and extraction of the cationic lipid/ODN complex from an organic solvent. The cationic lipid/ODN complex was then mixed with other neutral lipids and ODN-containing lipid vesicles were obtained by a reverse phase evaporation method (Stuart et al., 2004).

A much simpler method was developed by Semple and colleagues (2001), which utilizes an ionizable aminolipid (1,2-dioleoyl-3-dimethylammonium propane, DODAP) and an ethanol-containing buffer system. DODAP is an ionizable cationic lipid that has a pK_a of 6.6 in lipid bilayer systems (Semple *et al.*, 2001). At acidic pH values (i.e. pH 4.0), this lipid is positively charged and helps improve ODN entrapment via electrostatic interactions. DODAP/ODN complexes can interact with other lipids [distearoylphosphatidylcholine (DSPC), cholesterol, and *N*-palmitoylsphingo-sine-1-[succinyl-(methoxypoly (ethylene-glycol)2000] (PEG-CerC$_{16}$)] in an ethanol-containing buffer, and ODN-containing lipid vesicles are formed upon the removal of the ethanol by dialysis. Subsequent adjustment of the external pH to neutral pH values results in a neutral surface charge on the resulting particles. This method led to a significant increase in ODN entrapment efficiency (60–80%) with a final lipid/ODN ratio of 0.15–0.25 (w/w). This formulation has also been shown to be long circulating in the blood following systemic administration. These ODN-containing lipid vesicles, however, are inefficient in interacting with cells. Furthermore, they lack cell type specificity.

We examined whether a neutral lipid vector system as described previously can be modified with an endothelium-specific antibody to achieve targeted delivery of ODN to pulmonary ECs in intact mice. The endothelium-specific antibody to TM or PECAM-1 was coupled to ODN-encapsulated lipid vesicles via a chemical reaction with *p*-nitrophenylcarbonyl (pNP)-PEG-phosphoethanolamine (PE), which was introduced into the surface of the vesicles. A 60–80% of ODN entrapment efficiency was routinely achieved. ODN were protected from nuclease-mediated degradation. These particles are neutral or slightly anionic, and delivery of ODN into cultured ECs via this vector is largely dependent on the presence of an endothelium-specific antibody that was introduced. Furthermore, the cellular uptake of targeted lipidic ODN is significantly inhibited in the presence of excess amounts of free antibody, suggesting that intracellular delivery of ODN via the lipid vector is selectively mediated by the specific antibody. Intravenous administration of the targeted lipidic ODN led to substantial accumulation of ODN in the pulmonary microcirculation (Fig. 2.3B). In contrast, little ODN was found in lungs with either free ODN or ODN formulated in a lipid vector that was conjugated with a control IgG (Fig. 2.3A). Importantly, IV injection of this novel neutral lipidic vector containing ODN did not affect either (a) pulmonary or systemic hemodynamics or (b) blood chemistry in fully catheterized intact mice. It has been shown more recently that this vector can also be modified for targeted delivery of siRNA to pulmonary ECs. Studies are currently underway to examine its potential in the targeted delivery of therapeutic ODN for the treatment of pulmonary diseases.

V. CONCLUSION

ODN show great promise both as new therapeutics and as a research tools for gene function studies. Pulmonary endothelium represents an attractive site for the targeted delivery of therapeutic ODN because of its large surface area and a first passage effect, which allow a sufficient dose to be delivered to pulmonary circulation for the treatment of a number of lung diseases. The success of their application is largely dependent on the development of a delivery vehicle that is (a) capable of mediating efficient and selective intracellular delivery; (b) inert in inducing an immune response and in inducing no or little tissue or hemato-logical toxicities; and (c) causes minimal changes in normal pulmonary physiol-ogy. The recent success in developing a neutral lipid vector for targeted delivery of ODN to pulmonary endothelium represents a significant step toward achiev-ing that goal. Future studies include characterization of the *in vivo* delivery efficiency, particularly in an animal model of pulmonary disease, and more comprehensive studies on the potential toxicity of lipidic ODN. Development of small molecule-based targeting ligands will also help overcome the problem of immunogenecity. These studies are currently ongoing in this laboratory.

Acknowledgments

This work was supported by National Institutes of Health Grants HL-63080 and HL-68688 (to S. Li) and HL-32154 and GM-53789 (to B. Pitt) and by American Heart Association Grant 026540U (to S. Li).

References

Allen, T. M. (1998). Liposomal drug formulations: Rationale for development and what we can expect for the future. *Drugs* **56,** 747–756.

Almenar-Queralt, A., Duperray, A., Miles, L., Felez, J., and Altieri, D. (1995). Apical topography and modulation of ICAM-1 expression on activated endothelium. *Am. J. Pathol.* **147,** 1278–1288.

Amarzguioui, M., Holen, T., Babaie, E., and Prydz, H. (2003). Tolerance for mutations and chemical modifications in a siRNA. *Nucleic Acids Res.* **31,** 589–595.

Balyasnikova, I. V., Gavriljuk, V. D., McDonald, T. D., Berkowitz, R., Miletich, D. J., and Danilov, S. M. (1999). Antibody-mediated lung endothelium targeting: *In vitro* model of lung endothelium targeting using a cell line expressing angiotensin-converting enzyme. *Tumor Targeting* **4,** 70–83.

Benimetskaya, L., Berton, M., Kolbanovsky, A., Benimetsky, S., and Stein, C. A. (1997). Formation of a G-tetrad and higher order structures correlates with biological activity of the RelA (NF-kappaB p65) 'antisense' oligodeoxynucleotide. *Nucleic Acids Res.* **25,** 2648–2656.

Betts, L., Josey, J. A., Veal, J. M., and Jordan, S. R. (1995). A nucleic acid triple helix formed by a peptide nucleic acid-DNA complex. *Science* **270,** 1838–1841.

Borgatti, M., Finotti, A., Romanelli, A., Saviano, M., Bianchi, N., Lampronti, I., Lambertini, E., Penolazzi, L., Nastruzzi, C., Mischiati, C., Piva, R., Pedone, C., and Gambari, R. (2004). Peptide nucleic acids (PNA)-DNA chimeras targeting transcription factors as a tool to modify gene expression. *Curr. Drug Targets* 5(8), 735–744.

Bridge, A. J., Pebernard, S., Ducraux, A., Nicoulaz, A. L., and Iggo, R. (2003). Induction of an interferon response by RNAi vectors in mammalian cells. *Nature Genet.* **34**, 263–264.

Broide, D. H., Sullivan, S., Gifford, T., and Sriramarao, P. (1998). Inhibition of pulmonary eosinophilia in P-selectin- and ICAM-1-deficient mice. *Am. J. Respir. Cell Mol. Biol.* **18**(2), 218–225.

Brummelkamp, T. R., Bernards, R., and Agami, R. (2002). A system for stable expression of short interfering RNAs in mammalian cells. *Science* **296,** 550–553.

Burgess, T. L., Fisher, E. F., Ross, S. L., Bready, J. V., Qian, Y. X., Bayewitch, L. A., Cohen, A. M., Herrera, C. J., Hu, S. S., Kramer, T. B., Lott, F. D., Martin, F. H., Pierce, G. F., Simonet, L., and Farrell, C. L. (1995). The antiproliferative activity of c-myb and c-myc antisense oligonucleotides in smooth muscle cells is caused by a nonantisense mechanism. *Proc. Natl. Acad. Sci. USA* **92,** 4051–4055.

Caldwell, P. R., Seegal, B. C., Hsu, K. C., Das, M., and Soffer, R. L. (1976). Angiotensin-converting enzyme: Vascular endothelial localization. *Science* **191,** 1050–1051.

Campbell, A. I., Zhao, Y., Sandhu, R., and Stewart, D. J. (2001). Cell-based gene transfer of vascular endothelial growth factor attenuates monocrotaline-induced pulmonary hypertension. *Circulation* **104,** 2242–2248.

Campbell, J. M., Bacon, T. A., and Wickstrom, E. (1990). Oligodeoxynucleoside phosphotothioate stability in subcellular extracts, culture media, sera and cerebrospinal fluid. *J. Biochem. Biophys. Methods* **20,** 259–267.

Ching, S. F., Hayes, L. W., and Slakey, L. L. (1983). Angiotensin-converting enzyme in cultured endothelial cells: Synthesis, degradation, and transfer to culture medium. *Arteriosclerosis* **3,** 581–588.

Christofidou-Solomidou, M., Kennel, S., Scherpereel, A., Wiewrodt, R., Solomides, C. C., Pietra, G. G., Murciano, J. C., Shah, S. A., Ischiropoulos, H., Albelda, S. M., and Muzykantov, V. R. (2002). Vascular immunotargeting of glucose oxidase to the endothelial antigens induces distinct forms of oxidant acute lung injury: Targeting to thrombomodulin, but not to PECAM-1, causes pulmonary thrombosis and neutrophil transmigration. *Am. J. Pathol.* **160**(3), 1155–1169.

Christofidou-Solomidou, M., Scherpereel, A., Wiewrodt, R., Ng, K., Sweitzer, T., Arguiri, E., Shuvaev, V., Solomides, C. C., Albelda, S. M., and Muzykantov, V. R. (2003). PECAM-directed delivery of catalase to endothelium protects against pulmonary vascular oxidative stress. *Am. J. Physiol. Lung Cell Mol. Physiol.* **285**(2), L283–L292.

Cole-Strauss, A., Yoon, K., Xiang, Y., Byrne, B. C., Rice, M. C., Gryn, J., Holloman, W. K., and Kmiec, E. B. (1996). Correction of the mutation responsible for sickle cell anemia by an RNA-DNA oligonucleotide. *Science* **273,** 1386–1389.

Crooke, S. T. (2000). Progress in antisense technology: The end of the beginning. *Methods Enzymol.* **313,** 3–45.

Danilov, S. M., Gavrilyuk, V. D., Franke, F. E., Pauls, K., Harshaw, D. W., McDonald, T. D., Miletich, D. J., and Muzykantov, V. R. (2001). Lung uptake of antibodies to endothelial antigens: Key determinants of vascular immunotargeting. *Am. J. Physiol. Lung Cell Mol. Physiol.* **280**(6), L1335–L1347.

Diamond, M. S., Staunton, D. E., de Fougerolles, A. R., Stacker, S. A., Garcia-Aguilar, J., Hibbs, M. L., and Springer, T. A. (1990). ICAM-1 (CD54): A counter-receptor for Mac-1 (CD11b/CD18). *J. Cell Biol.* **111**(6 Pt 2), 3129–3139.

Egholm, M., Buchardt, O., Christensen, L., Behrens, C., Freier, S. M., Driver, D. A., Berg, R. H., Kim, S. K., Norden, B., and Nielsen, P. E. (1993). PNA hybridizes to complementary oligonucleotides obeying the Watson-Crick hydrogen bonding rules. *Nature* **365,** 556–568.

Ehlers, M. R., and Riordan, J. F. (1989). Angiotensin-converting enzyme: New concepts concerning its biological role. *Biochemistry* **28,** 5311–5318.

Ehlers, M. R., Schwager, S. L., Scholle, R. R., Manji, G. A., Brandt, W. F., and Riordan, J. F. (1996). Proteolytic release of membrane-bound angiotensin-converting enzyme: Role of the juxtamembrane stalk sequence. *Biochemistry* **25,** 95449–95459.

Elbashir, S. M., Harborth, J., Lendeckel, W., Yalcin, A., Weber, K., and Tuschl, T. (2001). Duplexes of 21-nucleotide RNAs mediate RNA interference in cultured mammalian cells. *Nature* **411,** 494–498.

Franke, F., Metzger, R., Bohle, R., Kerkman, L., Alhenc-Gelas, G., and Danilov, S. M. (1997). Angiotesin-I-converting enzyme (CD143) on endothelial cells in normal and in pathological conditions. *In* "Leukocyte Typing VI: White Cell Differentiation Antigens" (T. Kishimoto *et al.*, ed.), pp. 749–751. Garland, New York.

Galbraith, W. M., Hobson, W. C., Giclas, P. C., Schechter, P. J., and Agrawal, S. (1994). Complement activation and hemodynamic changes following intravenous administration of phosphorothioate oligonucleotides in the monkey. *Antisense Res. Dev.* **4,** 201–206.

Grotte, G. (1956). Passage of dextran molecules across the blood-lymph barrier. *Acta Chir. Scand. Suppl.* **211,** 1–84.

Hamilton, S. E., Simmons, C. G., Kathiriya, I. S., and Corey, D. R. (1999). Cellular delivery of peptide nucleic acids and inhibition of human telomerase. *Chem. Biol.* **6,** 343–351.

Hanvey, J. C., Peffer, N. C., Bisi, J. E., Thompson, S. A., Cadilla, R., Josey, J. A., Ricca, D. J., Hassman, C. F., Bonham, M. A., Au, K. G., Carter, S. G., Bruckenstein, D. A., Boyd, A. L., Noble, S. A., and Babiss, L. E. (1992). Antisense and antigene properties of peptide nucleic acids. *Science* **258,** 1481–1484.

Henry, S. P., Giclas, P. C., Leeds, J., Pangburn, M., Auletta, C., Levin, A. A., and Kornbrust, D. J. (1997). Activation of the alternative pathway of complement by a phosphorothioate oligonucleotide: Potential mechanism of action. *J. Pharmacol. Exp. Ther.* **281,** 810–816.

Herbrecht, R., Natarajan-Ame, S., Nivoix, Y., and Letscher-Bru, V. (2003). The lipid formulations of amphotericin B. *Expert Opin. Pharmacother* **4,** 1277–1287.

Ho, S. P., Bao, Y., Lesher, T., Malhorta, R., Ma, L. Y., Fluharty, S. J., and Sakai, R. R. (1998). Mapping of RNA accessible sites for antisense experiments with oligonucleotide libraries. *Nature Biotechnol.* **16,** 59–63.

Ho, S. P., Britton, D. H. O., Stone, B. A., Behrens, D. L., Leffet, L. M., Hobbs, F. W., Millaer, J. A., and Trainor, G. L. (1996). Potent antisense oligonucleotides to the human multidrug resistance-1 mRNA are rationally selected by mapping RNA-accessible sites with oligonucleotide libraries. *Nucleic Acids Res.* **24,** 1901–1907.

Hong, K., Zheng, W., Baker, A., and Papahadjopoulos, D. (1997). Stabilization of cationic liposome-plasmid DNA complexes by polyamines and poly(ethylene glycol)-phospholipid conjugates for efficient *in vivo* gene delivery. *FEBS Lett.* **400,** 233–237.

Hooper, N. M., Keen, J., Pappin, D. J., and Turner, A. J. (1987). Pig kidney angiotensin converting enzyme. *Biochem. J* **247,** 85–93.

Hughes, B. J., Kennel, S., Lee, R., and Huang, L. (1989). Monoclonal antibody targeting of liposomes to mouse lung *in vivo*. *Cancer Res.* **1549**(22), 6214–6220.

Hussein, M. (2003). Pegylated liposomal doxorubicin, vincristine, and reduced-dose dexamethasone as first-line therapy for multiple myeloma. *Clin. Lymphoma* **4**(Suppl. 1), S18–S22.

Kataoka, H., Kume, N., Miyamoto, S., Minami, M., Moriwaki, H., Murase, T., Sawamura, T., Masaki, T., Hashimoto, N., and Kita, T. (1999). Expression of lectinlike oxidized low-density lipoprotein receptor-1 in human atherosclerotic lesions. *Circulation* **99**(24), 3110–3117.

Kennel, S. J., Lankford, T., Hughes, B., and Hotchkiss, J. A. (1988). Quantitation of a murine lung endothelial cell protein, P112, with a double monoclonal antibody assay. *Lab. Invest.* **59**(5), 692–701.

Kmiec, E. B. (1999). Targeted gene repair. *Gene Ther.* **6,** 1–3.

Krieg, A. (1999). Mechanisms and applications of immune stimulatory CpG oligodeoxynucleotides. *Biochim. Biophys. Acta* **1489**, 107–116.

Krieg, A., Matson, S., and Fisher, E. (1996). Oligodeoxynucleotide modifications determine the magnitude of B cell stimulation by CpG motifs. *Antisense Nucleic Acid Drug Dev.* **6**, 133–139.

Kurreck, J., Wyszko, E., Gillen, C., and Erdmann, V. A. (2002). Design of antisense oligonucleotides stabilized by locked nucleic acids. *Nucleic Acids Res.* **30**, 1911–1918.

Larsen, H. J., Bentin, T., and Nielsen, P. E. (1999). Antisense properties of peptide nucleic acid. *Biochim. Biophys. Acta* **1489**(1), 159–166.

Lewis, D. L., Hagstrom, J. E., Loomis, A. G., Wol, J. A., and Herweijer, H. (2002). Efficient delivery of siRNA for inhibition of gene expression in postnatal mice. *Nature Genet.* **32**, 107–108.

Li, S., and Huang, L. (1997). In vivo gene transfer via intravenous administration of cationic lipid-protamine-DNA (LPD) complexes. *Gene Ther.* **4**(9), 891–900.

Li, S., and Ma, Z. (2001). Non-viral gene therapy. *Curr. Gene Ther.* **2**(2), 201–226.

Li, S., Rizzo, M. A., Bhattacharya, S., and Huang, L. (1998). Characterization of cationic lipid-protamine-DNA (LPD) complexes for intravenous gene delivery. *Gene Ther.* **5**, 930–937.

Li, S., Tan, Y., Viroonchatapan, E., Pitt, B. R., and Huang, L. (2000). Targeted gene delivery to pulmonary endothelium by anti-PECAM antibody. *Am. J. Physiol. Lung Cell Mol. Physiol.* **278**(3), L504–L511.

Li, S., Tseng, W. C., Stolz, D. B., Wu, S. P., Watkins, S. C., and Huang, L. (1999). Dynamic changes in the characteristics of cationic lipidic vectors after exposure to mouse serum: Implications for intravenous lipofection. *Gene Ther.* **6**(4), 585–594.

Liu, Y., Mounkes, L. C., Liggitt, H. D., Brown, C. S., Solodin, I., Heath, T. D., and Debs, R. J. (1997). Factors influencing the efficiency of cationic liposome-mediated intravenous gene delivery. *Nature Biotechnol.* **15**(2), 167–173.

Ma, Z., Mi, Z., Wilson, A., Alber, S., Robbins, P. D., Watkins, S., Pitt, B., and Li, S. (2002a). Redirecting adenovirus to pulmonary endothelium by cationic liposomes. *Gene Ther.* **9**, 176–182.

Ma, Z., Zhang, J., Alber, S., Dileo, J., Negishi, Y., Stolz, D., Watkins, S., Huang, L., Pitt, B., and Li, S. (2002b). Lipid-mediated delivery of oligonucleotide to pulmonary endothelium. *Am. J. Respir. Cell Mol. Biol.* **27**(2), 151–159.

Manoharan, M. (1999). 2′-carbohydrate modifications in antisense oligonucleotide therapy: Importance of conformation, configuration and conjugation. *Biochim. Biophys. Acta* **1489**(1), 117–130.

Maruyama, K., Holmberg, E., Kennel, S. J., Klibanov, A., Torchilin, V. P., and Huang, L. (1990). Characterization of in vivo immunoliposome targeting to pulmonary endothelium. *J. Pharm. Sci.* **79**(11), 978–984.

McCarey, A. P., Meuse, L., Pham, T.-T. T., Conklin, D. S., Hannon, G. J., and Kay, M. A. (2002). RNA interference in adult mice. *Nature* **418**, 38–39.

McIntosh, D. P., Tan, X. Y., Oh, P., and Schnitzer, J. E. (2002). Targeting endothelium and its dynamic caveolae for tissue-specific transcytosis in vivo: A pathway to overcome cell barriers to drug and gene delivery. *Proc. Natl. Acad. Sci. USA* **99**(4), 1996–2001.

McManus, M. T., and Sharp, P. A. (2002). Gene silencing in mammals by small interfering RNAs. *Nature Rev.* **3**, 737–747.

Mukherjee, S., Ghosh, R. N., and Maxfield, F. R. (1997). Endocytosis. *Physiol. Rev.* **77**, 759–803.

Muro, S., Cui, X., Gajewski, C., Murciano, J. C., Muzykantov, V. R., and Koval, M. (2003). Slow intracellular trafficking of catalase nanoparticles targeted to ICAM-1 protects endothelial cells from oxidative stress. *Am. J. Physiol. Cell Physiol.* **285**, C1339–C1347.

Muro, S., Gajewski, C., Koval, M., and Muzykantov, V. R. (2005). ICAM-1 recycling in endothelial cells: A novel pathway for sustained intracellular delivery and prolonged effects of drugs. *Blood* **105**(2), 650–658.

Muro, S., Wiewrodt, R., Thomas, A., Koniaris, L., Albelda, S. M., Muzykantov, V. R., and Koval, M. (2003). A novel endocytic pathway induced by clustering endothelial ICAM-1 or PECAM-1. *J. Cell Sci.* **116**(Pt. 8), 1599–1609.

Muzykantov, V., Christofidou-Solomidou, M., Balyasnikova, I., Harshaw, D., Schultz, L., Fisher, A., and Albelda, S. (1999). Streptavidin facilitates internalization and pulmonary targeting of anti-endothelial antibody (PECAM-1): A strategy for intraendothelial drug delivery. *Proc. Natl. Acad. Sci. USA* **96**, 2379–2384.

Nagaya, N., Kangawa, K., Kanda, M., Uematsu, M., Horio, T., Fukuyama, N., Hino, J., Harada-Shiba, M., Okumura, H., Tabata, Y., Mochizuki, N., Chiba, Y., Nishioka, K., Miyatake, K., Asahara, T., Hara, H., and Mori, H. (2003). Hybrid cell-gene therapy for pulmonary hypertension based on phagocytosing action of endothelial progenitor cells. *Circulation* **108**, 889–895.

Newman, P. J. (1997). The biology of PECAM-1. *J. Clin. Invest.* **99**, 3–7.

Nielsen, P. E. (1999). Antisense properties of peptide nucleic acid. *Methods Enzymol.* **313**, 56–164.

Nielsen, P. E., Egholm, R. H., Berg, R. H., and Buchardt, O. (1991). Sequence-selective recognition of DNA by strand displacement with a thymine-substituted polyamide. *Science* **254**, 1497–1500.

Oppong, S. Y., and Hooper, N. M. (1993). Characterization of a secretase activity which releases angiotensin-converting enzyme from the membrane. *Biochem. J.* **292**, 597–603.

Pappenheimer, R. J., Renkin, E. M., and Borrero, L. M. (1951). Filtration, diffusion and molecular sieving through peripheral capillary membranes: A contribution to the pore theory of capillary permeability. *Am. J. Physiol.* **167**, 13–21.

Pasqualini, R., and Ruoslahti, E. (1996). Organ targeting *in vivo* using phage display peptide libraries. *Nature* **380**(6572), 364–366.

Phillips, M. I., and Zhang, Y. C. (2000). Basic principles of using antisense oligonucleotides *in vivo*. *Methods Enzymol.* **313**, 46–56.

Pooga, M., Sommets, U., Hallbrink, M., Valkna, A., Saar, K., Rezaei, K., Kahl, U., Hao, J.-X., Wiesenfeld-Hallin, Z., Hokfelt, T., Bartfai, T., and Langel, U. (1998). Cell penetrating PNA constructs regulate galanin receptor levels and modify pain transmission *in vivo*. *Nature Biotechnol.* **16**, 857–861.

Predescu, D., and Palade, G. E. (1993). Plasmalemmal vesicles represent the large pore system of continuous microvascular endothelium. *Am. J. Physiol. Heart Circ. Physiol.* **265**, H725–H733.

Predescu, D., Vogel, S. M., and Malik, A. B. (2004). Functional and morphological studies of protein transcytosis in continuous endothelia. *Am. J. Physiol. Lung Cell Mol. Physiol.* **287**, L895–L901.

Rajotte, D., and Ruoslahti, E. (1999). Membrane dipeptidase is the receptor for a lung-targeting peptide identified by *in vivo* phage display. *J. Biol. Chem* **274**(17), 11593–11598.

Renkin, E. M. (1985). Capillary transport of molecules: Pores and other endothelial pathways. *J. Appl. Physiol.* **58**, 315–325.

Reynolds, A., Leake, D., Boese, Q., Scaringe, S., Marshall, W. S., and Khorva, A. (2004). Rational siRNA design for RNAi. *Nature Biotechnol.* **22**, 326–330.

Reynolds, P. N., Zinn, K. R., Gavrilyuk, V. D., Balyasnikova, I. V., Rogers, B. E., Buchsbaum, D. J., Wang, M. H., Miletich, D. J., Grizzle, W. E., Douglas, J. T., Danilov, S. M., and Curiel, D. T. (2000). A targetable, injectable adenoviral vector for selective gene delivery to pulmonary endothelium *in vivo*. *Mol. Ther.* **2**, 562–578.

Rippe, B., and Haraldsson, B. (1987). How are macromolecules transported across the capillary wall? *News Physiol. Sci.* **2**, 1135–1138.

Scacheri, P. C., Rozenblatt-Rosen, O., Caplen, N. J., Wolfsberg, T. G., Umayam, L., Lee, J. C., Hughes, C. M., Shanmugam, K. S., Bhattacharjee, A., Meyerson, M., and Collins, F. S. (2004). Short interfering RNAs can induce unexpected and divergent changes in the levels of untargeted proteins in mammalian cells. *Proc. Natl. Acad. Sci. USA* **101**, 1892–1897.

Schnitzer, J. E., and Oh, P. (1994). Albondin-mediated capillary permeability to albumin: Differential role of receptors in endothelial transcytosis and endocytosis of native and modified albumins. J. Biol. Chem. **269**, 6072–6082.

Schnitzer, J. E., and Oh, P. (1996). Aquaporin-1 in plasma membrane and caveolae provides mercury-sensitive water channels across lung endothelium. Am. J. Physiol. **270**, H416–H422.

Semple, S. C., Klimuk, S. K., Harasym, T. O., Dos Santos, N., Ansell, S. M., Wong, K. F., Mauer, N., Stark, H., Cullis, P. R., Hope, M. J., and Scherrer, P. (2001). Efficient encapsulation of antisense oligonucleotides in lipid vesicles using ionizable aminolipids: Formation of novel small multilamellar vesicle structures. Biochim. Biophys. Acta **1510**(1–2), 152–166.

Siflinger-Birnboim, A., Del Vecchio, P. J., Cooper, J. A., Blumenstock, F. A., Shepard, J. M., and Malik, A. B. (1987). Molecular sieving characteristics of the cultured endothelial monolayer. J. Cell. Physiol. **132**, 111–117.

Simionescu, M. (1991). Lung endothelium: Structure-function correlates. In "The Lung: Scientific Foundations" (R. G. Crystal and J. B. West, eds.), pp. 301–331. Raven, New York.

Simionescu, M., Simionescu, N., and Palade, G. E. (1982). Differentiated microdomains on the luminal surface of capillary endothelium: Distribution of lectin receptors. J. Cell Biol. **94**(2), 406–413.

Sledz, C. A., Holko, M., de Veer, M. J., Silverman, R. H., and Williams, B. R. (2003). Activation of the interferon system by short-interfering RNAs. Nature Cell Biol. **5**, 834–839.

Soffer, R. L. (1976). Angiotensin-converting enzyme and the regulation of vasoactive peptides. Annu. Rev. Biochem. **45**, 73–94.

Song, Y. K., Liu, F., Chu, S., and Liu, D. (1997). Characterization of cationic liposome-mediated gene transfer in vivo by intravenous administration. Hum. Gene Ther. **8**(13), 1585–1594.

Stuart, D. D., Semple, S. C., and Allen, T. M. (2004). High efficiency entrapment of antisense oligonucleotides in liposomes. Methods Enzymol. **387**, 171–188.

Taylor, A. E., and Granger, D. N. (1984). Exchange of macromolecules across the microcirculation. In "Handbook of Physiology: The Cardiovascular System," Vol. IV, pp. 467–520. Bethesda, MD. Am. Physiol. Sot.

Templeton, N. S., Lasic, D. D., Frederik, P. M., Strey, H. H., Roberts, D. D., and Pavlakis, G. N. (1997). Improved DNA: Liposome complexes for increased systemic delivery and gene expression. Nature Biotechnol. **15**, 647–652.

Tuschl, T. (2002). Expanding small RNA interference. Nature Biotechnol. **20**, 446–448.

Vaporician, A. A., Delisser, H. M., Yam, H. C., Mendiguren, I., Thom, S. R., Jones, M. L., Ward, P. A., and Albelda, S. M. (1993). Platelet-endothelial cell adhesion molecule-1 (PECAM-1) is involved in neutrophil recruitment in vivo. Science **262**, 580–582.

Wallace, T. L., Bazemore, S. A., Kornbrust, D. J., and Cossum, P. A. (1996). Single-dose hemodynamic toxicity and pharmacokinetics of a partial phosphorothioate anti-HIV oligonucleotide (AR177) after intravenous infusion to cynomolgus monkeys. J. Pharmacol. Exp. Ther. **278**, 1306–1312.

Wang, J., Guo, X., Xu, Y., Barron, L., and Szoka, F. C., Jr. (1998). Synthesis and characterization of long chain alkyl acyl carnitine esters: Potentially biodegradable cationic lipids for use in gene delivery. J. Med. Chem. **41**, 2207–2215.

West, J. B., and Dolley, C. T. (1965). Distribution of blood flow and the pressure-flow relations of the whole lung. J. Appl. Physiol. **20**, 175–183.

White, S. J., Nicklin, S. A., Sawamura, T., and Baker, A. H. (2001). Identification of peptides that target the endothelial cell-specific LOX-1 receptor. Hypertension **37**(2 Part 2), 449–455.

Wickstrom, E. (1986). Oligodeoxynucleotide stability in subcellular extracts and culture media. J. Biochem. Biophys. Methods **13**, 97–102.

Yu, J.-Y., DeRuiter, S. L., and Turner, D. L. (2002). RNA interference by expression of short-interfering and hairpin RNAs in mammalian cells. *Proc. Natl. Acad. Sci. USA* **99,** 6047–6052.

Yuan, X., Ma, Z., Zhou, W., Niidome, T., Alber, S., Huang, L., Watkins, S., and Li, S. (2003). Lipid-mediated delivery of peptide nucleic acids to pulmonary endothelium. *Biochem. Biophys. Res. Commun.* **302**(1), 6–11.

Zamaratski, E., Pradeepkumar, P. I., and Chattopadhyaya, J. (2001). A critical survey of the structure-function of the antisense oligo/RNA heteroduplex as substrate for RNase H. *J. Biochem. Biophys. Methods* **48,** 189–208.

Zhu, N., Liggitt, D., Liu, Y., and Debs, R. (1993). Systemic gene expression after intravenous DNA delivery into adult mice. *Science* **261,** 209–211.

Zon, G. (1988). Oligonucleotide analogs as potential chemotherapeuticagents. *Pharm. Res.* **5,** 539–549.

3

Naked DNA for Liver Gene Transfer

Feng Liu and Pradeep Tyagi

Center for Pharmacogenetics, School of Pharmacy
University of Pittsburgh, Pittsburgh, Pennsylvania 15261

ABSTRACT

The majority of acquired and inherited genetic disorders, including most inborn errors of metabolism, are manifested in the liver. Therefore, it is hardly any surprise to see a large number of Medline reports describing gene therapy efforts in preclinical settings directed toward this organ (Inoue *et al.*, 2004; Oka and

0065-2660/05 $35.00
DOI: 10.1016/S0065-2660(05)54003-3

Chen, 2004). Of late, non-viral vectors have garnered a lot of attention from the biomedical research community engaged in liver gene therapy (Gupta *et al.*, 2004). However, the first initiative toward gene transfer to the liver using a non-viral approach was taken by Hickman *et al.* (1994), who applied the technique of naked DNA injection pioneered by Wolff (1990) for skeletal muscle. Direct injection of naked DNA resulted in low, variable and localized gene expression in the rat liver. Consequently, several developments reported in the literature since then aimed to improve hepatic gene expression by employing both surgical and nonsurgical methods. These developments include the exploitation of the unique vasculature of liver as well as the use of electric and mechanical force as an adjunct to the systemic administration of the naked plasmid gene. This chapter focuses on these developments reported from various laboratories, including ours. In addition, the underlying mechanism responsible for the dramatic increase in gene expression using these latest approaches for non-viral gene transfer to the liver is also discussed. © 2005, Elsevier Inc.

I. INTRODUCTION

The liver is the primary organ, which is responsible for controlling the anabolic and catabolic fates of carbohydrates and lipids in the human body. The protein synthesis apparatus of this versatile organ also plays a crucial role in maintaining a steady amount of the active unbound fraction of steroidal hormones in the body, as most of the carrier proteins for transporting these hormones in the serum are synthesized in liver. Parenchymal liver cells known as hepatocytes perform a host of posttranslational modifications that may be required for the activity of certain gene products. Due to the multiplicity in the roles played by the liver, diseases affecting the liver not only affect intrinsic hepatic functions, but the function of other organs is frequently compromised as well. Numerous diseases such as hemophilia and metabolic diseases such as phenylketonuria (PKU), familial hypercholesterolemia, and other enzyme deficiencies are due to defective proteins synthesized in the liver as a result of a single gene defect (Herweijer and Wolff, 2003; VandenDriessche *et al.*, 2003). Therefore, the liver represents an ideal target organ and hepatocytes as important target cells for gene therapy against many diseases such as liver cancer, viral hepatitis, and allograft rejection (Cheng and Smith, 2003). The aim of such therapy is to provide a functional copy for the gene of interest, incorporated into a viral, plasmid, or other genetic element, for expression in hepatic tissues (Marshall *et al.*, 2002). Liver gene transfer must be successful in transfecting a large popula; tion of hepatocytes for proteins that remain within the cells to correct the underlying genetic defect responsible for the disease. In contrast, if the protein encoded by transgene is secreted by the cell, then transfection of a limited number of cells is required, as protein secreted from transfected cells can then accomplish paracrine and endocrine modulation of bioactivity in nearby and distant cells.

It is true for most organs that viral vectors have the ability to transduce a large number of cells. This was the main reason for the selection of viral vectors for liver gene transfer (Kaleko et al., 1991; Stoll et al., 2001). Recombinant viral vectors derived from human adenovirus have been the vectors of popular choice for laboratory studies and clinical trials (Chuah et al., 2004; Ferber, 2001; Jaffe et al., 1992). Many routes for gene transfer into the liver have been attempted using viral vectors (Ferry et al., 1991; Jung et al., 2001). However, a few drawbacks of the viral vectors have been identified in the past decade (Liu et al., 1993). First, despite being replication deficient, there are persistent concerns about their safety, including the possibility of recombination with endogenous viruses, which could produce a replication competent (infectious) variant (High, 2004). Second, many vectors induce immune responses against the intrinsic viral antigens, which can cause serious toxicity to the patient and render repetitive treatment problematic (Schiedner et al., 2003).

While it is possible that these problems will be solved eventually, the unfavorable characteristics of viral vectors have motivated the efforts to develop non-viral vectors. Non-viral vectors such as cationic liposomes, polymers, and naked DNA are being developed under the assumption that they will overcome problems associated with viral gene delivery. Non-viral systems can be classified into two categories: synthetic vectors, including cationic lipids and polymers; and naked plasmid DNA. In the case of synthetic vectors, the highest gene expression, in most cases, is observed in the lung after intravenous or intraportal injection of DNA–cationic liposome (or polymers) complexes. This is mainly because an embolization mechanism is behind their ability to transfect the lung endothelial cells (Liu et al., 1999). When the positively charged DNA complexes are injected into the blood circulation, they form aggregates with the negatively charged serum proteins. Embolization results when the aggregates are trapped in the lung capillaries. The increased retention time of plasmid DNA due to this mechanism correlates with higher gene expression in the lung (Templeton and Lasic, 1999). These observations pointed toward the difficulty that may be encountered in delivering a gene to the liver using simple synthetic vectors (Takakura et al., 2001).

Consequently, many efforts employed a receptor-mediated fashion for the genes to home onto hepatocytes. For instance, plasmid DNA was complexed to polylysine-bearing targeting ligands such as galactose groups for in vitro transfection of hepatocytes (Smith and Wu, 1999). DNA is able to bind to the polylysine bearing vectors through electrostatic interactions (Shi et al., 2001). The application of this approach for the in vivo setup is unlikely as there is potential for immunogenicity of the complexes used for in vitro study and serum might disrupt the forces involved in the binding of DNA to its vector (Shi *et al.*, 2001). This inherent pitfall in synthetic vectors has led to interest in other possible alternatives (Schmitz *et al.*, 2002).

A phenomenon first reported by Wolff (1990) saw gene expression in rodent muscle after a simple injection of naked plasmid DNA into muscle. It was certainly a groundbreaking discovery as gene expression was shown using only naked DNA free of any recombinant virus or DNA complexed with polymers and lipids. The distinct histology of skeletal muscle probably rendered some contribution for successful gene expression observed by Wolff *et al.* (1990). Multiple nuclei present in each skeletal muscle cell might provide several targets per cell for nuclear gene transfer, and an elaborate transport system for calcium ions from outside the cell could also assist gene transfer. Significant protein expression was observed following injection of either pure DNA coding for β-galactosidase or RNA coding for CAT into the quadriceps muscles of mice. Two days later, a slightly higher CAT activity was determined with DNA injection compared to injection of RNA. Roughly one-third of the muscle cells near the injection site where DNA was injected a week before were positive for histochemical stain of LacZ.

Direct injection of either DNA or RNA coding for firefly luciferase in escalating doses showed a linear increase in gene expression in murine skeletal muscle. The peak luciferase activity was determined to be 18 h after RNA injection and nadir was reached 42 h later. Naked DNA injection into the muscle sustained luciferase expression for 60 days after injection. The 24-h half-life of both luciferase protein and its *in vitro* RNA transcript further supported that sustained transgene expression was responsible for luciferase activity in muscle for 60 days. The luciferase gene exists as a nonintegrated circular form in the skeletal muscle after injection, but a low level of chromosomal integration could not be ruled out from the Southern blot analysis for the exogenous DNA. The sensitivity toward endonuclease digestion of the transgene revealed that injected DNA does not replicate itself after injection.

The mechanism of DNA entry into muscle was suggested to involve the intrinsic regenerative ability of skeletal muscle. The technique has been described in detail elsewhere (Wolff *et al.*, 1990). The transgene expression in muscle following a simple injection of plasmid DNA with a syringe has immensely influenced many aspects of gene therapy research, including liver gene transfer. The technique of naked DNA injection for gene transfer has drawn increasing attention from the biomedical community due to its safety, simplicity, and repeatability.

II. LIVER GENE TRANSFER BY NAKED DNA INJECTION

The exciting success of naked gene transfer to the skeletal muscle inspired the application of this relatively simple technique for gene transfer to other organs (Miao *et al.*, 2001). Naked gene transfer to the muscle was a simple technique that could be done without any surgery. However, surgery was an obvious prerequisite for application of this simple technique for liver gene transfer (Stecenko and Brigham, 2003). Hickman *et al.* (1994) were the first to report success in achieving gene expression in liver after direct injection of DNA into hepatic parenchyma of cat and rats. The animal liver was exposed following ventral midline incision prior to injection of the human α_1-antitrypsin (hAAT) gene or reporter genes coding for LacZ and luciferase gene. Forty eight hours later, serial sections of the isolated animal liver were made for X-gal staining. The sections were also used for making liver homogenates to determine their luciferase activity.

The hepatic gene expression was localized near the injection site with not much expression in the more distant tissue. The pattern of X-gal staining in liver was nearly identical to that seen in skeletal muscle following direct injection. As reported for skeletal muscle, liver gene transfer also exhibited a dose-dependent increase in gene expression. However, in direct contrast to skeletal muscle, an increase in the number of injections caused a linear increase in the luciferase activity of liver. This increase in activity can be explained by the increase in the number of cells directly exposed to DNA after a liver is injected multiple times. Compared to skeletal muscle, the unique vasculature of liver allows administration of much higher volumes for injection. Therefore, efficiency in gene expression was optimized by varying the volume of injection. It was determined that the force of direct injection into the liver has to be strong enough for visually blanching the surrounding tissue without distending the organ.

The study demonstrated a simple economical method of gene delivery to the liver that can be used as a tool in assessing the efficacy of a new DNA construct. However, inflammation occurring at the surgical site adversely affected gene expression following the direct injection of DNA in the liver. Therefore, Malone *et al.* (1994) applied the rationale that suppression of inflammation would be favorable for gene expression. As expected, a daily subcutaneous injection of dexamethasone at a dose of 1 mg/kg for 2 days following direct injection of plasmid DNA drastically increased luciferase expression in livers of cats and rats (Malone *et al.* 1994). Results suggest that dexamethasone operates on twin mechanisms for enhancing gene expression following direct injection: one is transcriptional modulation and the other suppresses the adverse effect of inflammation on gene expression.

III. LIVER GENE TRANSFER BY ELECTROPORATION

A. Electroporation coupled with direct injection

The use of electric pulses for cell electropermeabilization (also called cell electroporation) has been used since the mid-1980s to introduce foreign DNA into prokaryotic and eukaryotic cells *in vitro* (Neumann *et al.*, 1982; Wong and

Neumann, 1982). Electroporation leads to a temporary increase in membrane permeability, allowing DNA in the extracellular space to enter the cell rapidly, up to 100 times faster than accounted for by simple diffusion, probably involving electrophoresis (Smith *et al.*, 2004). Pulsed electrical fields are applied by using an electroporation device, typically delivering pulses of 20- to 60-ms duration, thus creating a shallow electromagnetic field for precise gene delivery along the electrode. Electrotransfer drastically increases the efficiency of im gene transfer, resulting in a two to four logarithmic increase in gene expression and reduces variation (Mir *et al.*, 1999). In addition to muscle, various other organs have been transfected successfully using this approach, including skin, male and female germ cells, artery, gut, kidney, retinal ganglion cells, cornea, spinal cord, joint synovium, and brain (Golzio *et al.*, 2004; Maruyama *et al.*, 2001; Yamazaki *et al.*, 1998). However, the prominent limitations of this technique are a high mortality of cells after high-voltage exposure and difficulties in optimization. Optimization of this approach depends on a number of parameters, such as target organ accessibility, cell turnover, microelectrode design, electric pulsing protocols, and the physiological response to the transferred gene. Similar to the gene gun approach discussed later, electroporation does not induce an immune response. It is possible to achieve excellent levels of cell permeabilization that are compatible with cell survival using optimal electrical parameters.

Use of electroporation to assist liver gene transfer with direct injection was reported for the first time by Heller *et al.* (1996), a technique better known as electroinjection. A reporter gene was injected at a single location just beneath the liver surface in small volume (0.1 ml). Normal saline was used as a vehicle for the reporter gene and not cell medium. Needle electrodes were inserted in liver after an interval of 1.5 min lapsed after direct injection. The geometry of the rotating electrical field from inserted electrodes formed an annular pattern around the DNA injection site. Forty-eight hours after the procedure the expression of luciferase and β-galactosidase in the electroinjected liver tissue was determined. Optimal gene expression was achieved at field strengths between 1 and 1.5 kV/cm. Electroporation increased the *in vivo* expression efficiency of the injected DNA by as much as 30% as measured by flow cytometry. In another report published by Suzuki *et al.* (1998), the vehicle for reporter gene was replaced with 0.1 ml of phosphate-buffered saline instead of normal saline. Immediately after direct injection of DNA into the liver, electric pulses were administered by tweezer-type electrodes. The tweezer disks were placed around the lobe of the liver that received DNA injection. The study demonstrated that the expression level of green fluorescent protein (GFP) increased with an increase in the number of electric pulses from one to four. The optimum transfection efficiency was found to be at electric pulses having a voltage of 50 V, and prolonged duration of electric pulses had a favorable effect on the GFP signal.

This technique was employed to study the nutritional regulation of hepatic transgene expression following gene transfer to the liver by Muramatsu *et al.* (2001). Electric pulses were only applied to the left lobe of murine liver immediately after direct injection of DNA. A liver-specific promoter for the phosphoenolpyruvate carboxykinase (PEPCK) gene was fused upstream of the luciferase gene to drive its expression so that transgene expression was restricted to liver alone. A 13-fold induction of luciferase activity in the liver was observed with the liver-type PEPCK promoter upon fasting mice. However, neither the same PEPCK promoter nor the SV 40 promoter was able to induce gene expression upon fasting mice. Fasting leads to hypoglycemia, which triggers the activation of glycogenolysis following elevation of cAMP by glucagon and other mediators. Therefore, the homeostatic condition and hormonal milieu created by fasting in mouse can also be generated by cAMP administration to mice. This fact may explain the increased expression in the liver of fed mice for the gene driven by PEPCK following administration of cAMP. Moreover, a hundredth-fold increase in reporter gene expression was observed when dry ice was used prior to the application of pulses to offset the heat damage generated by electrodes at the injection site. Luciferase expression in the liver of live intact mouse was measured *in situ* by bioluminescence imaging of the mouse body. A suitable promoter and controlled nutrition might be able to achieve tissue-specific expression in the liver (Muramatsu *et al.*, 2001).

A similar approach of liver gene transfer was used in the rat to examine the protective effect of proapoptotic genes against hepatocellular carcinomas (HCC) (Baba *et al.*, 2001). HCC was induced in Sprague–Dawley rats by feeding them water containing 175 mg/liter N-nitrosomorpholine (NNM) for 8 weeks. The bcl-xs plasmid (pCR3.1-rat bcl-xs cDNA) or pCR3.1 encapsulated in cationic empty liposomes at a dose of 80 μg plasmid/kg body weight was injected directly into the left lobe of the liver following surgical incision at 1, 4, and 7 weeks on rats injected previously with NMM. One minute after direct injection of plasmid, pincette electrodes having circular poles of 1 cm in diameter were used to apply eight electric pulses to the left lobe of rat livers, with each pulse lasting 50 ms at the field strength of 70–100 V/cm.

Overexpression of the bcl-xs protein in liver tissues was probably responsible for the higher apoptotic index of hepatocytes adjacent to hepatic lesions in the livers of rats injected with NMM. The progression of neoplastic nodules was decreased by the gene transfer of *bcl-xs* plasmids. The hepatocellular carcinoma was completely inhibited on both sides of electrode contact as well as sites 0.5–1.0 cm away from the electrode. Moreover, CAT mRNA of the reporter gene was expressed in higher amounts at 0.5–1.0 cm away from the sites of contact for electrodes on the liver, suggesting that gene expression crossed the 1 cm diameter edges of made by electrodes. The recurrence of hepatocellular carcinoma might be prevented by the transfer of apoptotic genes to liver.

B. Electroporation coupled with injection via portal blood vessels of liver

The technique of direct injection coupled with electroporation was successful in achieving the gene expression only to a limited area around the injected site in the liver. Moreover, the needle used for direct injection could cause mechanical tissue damage and bleeding. In a modified method, plasmid DNA was infused via the portal vein to avoid tissue damage from the needle and to obtain diffuse gene expression in the liver (Matsuno et al., 2003). The portal vein of rat liver was exposed surgically and the infusion of plasmid DNA preceded electroporation. A dramatic improvement in gene expression rendered by electroporation with other methods also encouraged its application in the present technique.

The transgene expression of GFP in the liver began to appear 24 h after infusion, which was sustained for 3 weeks, and peak expression was seen around 72 h after transfection. The contact sites for electrodes on the liver lobe revealed very mild tissue damage on histological analysis. Using this improved method, rat liver showed a dose-dependent gene expression of the human HGF gene. However, higher gene expression did not correlate with the serum levels of HGF. Moreover, the survival of all dimethylnitrosamine-treated rats was prolonged by gene transfer of 500 μg HGF gene into the rat liver; the intensity of fibrosis in the liver was also reduced.

In another study, the aforementioned method was further modified by employing the strategy of blood flow occlusion for liver gene transfer to 7-week-old male Shionogi–Wistar rats. Conditions for gene transfer into the liver using this technique were optimized by Koboyashi et al. (2003a), which involved three steps: portal vein infusion, electroporation of the caudal lobe of the liver, and occlusion of afferent vessels in the right and caudal lobes of the liver. A tapered 3Fr. catheter was inserted through a branch of the colonic vein of the anesthetized rat and a catheter tip was secured into the portal vein. Fluorescein isothiocyanate (FITC)-labeled oligodeoxynucleotides (ODN) were used in the study to delineate the site of gene transfer and compare its distribution with the expression pattern of reporter genes. FITC-labeled ODN and reporter plasmid were injected slowly through the secured catheter immediately after clamping the inferior and superior vena cava of rat and the Glisson capsule lying between the branches of the middle and left lobes and right and caudal lobes. While infusion continued, electric pulses were applied only to the caudal lobe at the rate of 1 pulse/s with a pulse duration of 100 ms; the right lobe did not receive any electric pulses. Two sets of three square wave electric pulses, with each set having opposite polarity and constant voltage, were applied with a pair of stainless steel electrodes with round pincers 10 mm in diameter and 4–5 mm apart.

The study observed diffuse distribution of fluorescently labeled ODN in the hepatocytes of the Glisson capsule and around central veins in caudal lobes of the liver. GFP expression was also observed in the Glisson capsule, in addition to the presence of labelled ODN, but sites around central veins lacked GFP expression. It is possible that the plasmid solution was diluted in the region of central veins to create the observed difference in the distribution of ODN and expression pattern of GFP. Transfection efficiency of this new method was estimated from luciferase activity. The plasmid concentration and voltage of the electric pulse had an effect on luciferase activity. However, the volume of plasmid solution used for injection did not correlate with luciferase activity. The study showed a drastic improvement in luciferase activity due to electroporation, and only localized expression in the absence of electroporation.

The method, once optimized, was used to demonstrate successful *ex vivo* gene transfection into liver grafts (Koboyashi *et al.*, 2003b). The male Shionogi–Wistar rats used as donors and recipients for rat liver transplantation in this study were 2 weeks older than the rats used previously for the optimization study (Koboyashi *et al.*, 2003a). The liver graft was maintained at 4°C in saline solution during all the subsequent procedures. The afferent blood vessels at the portal triad between left and middle liver lobes and caudal and right liver lobes were clamped. A tapered 3Fr. catheter was inserted into the portal vein and clamped the end of the portal vein.

The catheter was used for the slow injection of plasmid solutions coding for luciferase and GFP into the liver graft and all clamps were removed after injection. The study examined gene transfer efficacy using different cold preservation solutions used routinely for liver graft, namely cold lactated Ringer's solution, University of Wisconsin solution, and histidine–tryptophan–ketoglutarate solution. These cold preservation solutions were used as electroporator solutions in the study. Luciferase activity in the *ex vivo* situation was not only influenced by the current and voltage of the electric pulse, but also by the type of preservation solution. Moreover, luciferase activity decreased with an increase in plasmid solution osmotic pressure. Luciferase activity was localized only in lobes of the liver graft injected with plasmid. Luciferase activity was highest 24 h after injection and it decreased gradually thereafter. GFP fluorescence was observed predominantly in perivascular cells, including hepatocytes. The study describes an efficient non-viral method for *ex vivo* gene transfer into rat liver grafts.

C. Electroporation coupled with intravenous injection

The gene expression was expected to be more widespread following the systemic administration of plasmid by tail vein injection than against local injection. However, despite the maximum uptake of DNA in liver following intravenous injection of naked plasmid DNA, there is hardly any detectable gene expression in the liver (Nishikawa *et al.*, 2003). Among many factors responsible for the fast clearance of DNA, one is the rapid elimination of DNA from the blood circulation as a result of DNA uptake by liver nonparenchymal cells such as Kupffer and endothelial cells (Koboyashi *et al.*, 2001). Degradation of DNA by nucleases in the blood could be another factor responsible for the afore-mentioned outcome, although clearance by degradation has been estimated to be only one-eighth of the clearance by hepatic uptake. Methods of liver gene transfer employing direct injection of DNA into the liver (Hickman *et al.*, 1994), gene gun (Kuriyama *et al.*, 2000), and electroporation (Koboyashi *et al.*, 2003) were able to avoid these barriers through local regional delivery.

Systemic administration has been successful in achieving liver gene transfer after injecting a large volume by the hydrodynamics-based method (discussed later). Liu and Huang (2002) sought to improve liver gene transfer with electroporation by injecting conservative volumes via the tail vein of the mouse. We tried systemic administration of 80 μg of DNA in 0.1 ml of normal saline via the tail vein of mice. Immediately following tail vein injection, eight electric pulses having a field strength of 250 V/cm with each pulse having a duration of 20 ms were delivered to the median lobe of the mouse liver. All sections of the liver displayed high levels of luciferase activity following an intravenous injection of plasmid. In contrast, the high level of gene expression was restricted to only a single section of liver following direct injection in combination with electroporation. Using the technique of direct injection into the liver, the gene expression was reduced by approximately 20 to 1000 times less in other sections of the liver. Therefore, in our study, systemic delivery of DNA produced a much broader distribution of cells expressing the reporter gene as evident from β-gal staining of liver compared to direct injection into the liver. This was the first report describing the positive effect of electric gene transfer to the liver on hepatic gene expression following the systemic adminis-tration of plasmid DNA. Contrary to the report by Muramatsu et al. (2001) on tissue-specific gene expression by direct injection, our study suggested that electroporation can accomplish localized gene expression even after the systemic administration of plasmid.

IV. LIVER GENE TRANSFER BY THE GENE GUN

The technique of the gene gun is fast, simple, and versatile enough to be applied for gene transfer to visceral tissues such as liver. However, surgical intervention is necessary for liver gene transfer because gene guns can only deliver genetic material to a limited area and very small depths when applied from the surface (Cheng *et al.*, 1993). The requirement of surgery is common for a strategy employing either electroporation or the gene gun. The gene gun technique involves coating plasmid DNA onto gold or tungsten microcarriers, which are then accelerated toward a stopping plate by either an electric discharge or a pressurized helium pulse (Heiser, 1994; Sanford et al., 1993). The microcarriers then penetrate into the cytoplasm of the target cells. Multiple genes can be delivered simply by coating more than one plasmid onto the same carrier. Because the spheres are composed of relatively inert materials such as gold, no specific or nonspecific immune response is invoked by gene transfer. The gene gun was assessed as a delivery technique for RNA by measuring the transfection of three reporter genes, firefly luciferase, human growth hormone, and human α_1-antitrypsin in rat liver and mouse epidermis, after bombarding their respective RNA transcripts (Qiu et al., 1996). In this study, particle-mediated gene delivery was used to counter the instability of RNA transcripts.

Various designs of gene gun have been used for in vitro applications and it is now a widely used technique for introducing DNA into animal cells. An electric charge was used to accelerate DNA-coated gold particle to achieve liver gene transfer (Cheng et al., 1993). A different design of hand-held type gun known as the Helios gene gun system accelerated DNA-coated gold particles by pressurized helium gas to supersonic velocities. The surface of rat liver was exposed by a transverse skin incision to bombard plasmid DNA coding for LacZ (Yoshida et al., 1997). Radioactive DNA was used to study the distribution of DNA in the organ after each shot. The general distribution of radioactive DNA in rat liver slices was similar to the expression pattern of β-galactosidase. Penetration of DNA into the rat liver was dependent on the helium pressure used in each shot. Hemorrhage and fissure in the rat liver were observed with the use of higher pressures. In contrast to earlier designs, the Helios gun operated in the absence of any vacuum.

The same group of researchers also used the Helios gene gun system to achieve gene transfer in rat liver allografts (Nakamura et al., 2003). The objective of this study was similar to another study investigating gene transfer to a liver allograft by employing electroporation (Koboyashi et al., 2003b). In this study, liver gene transfer to the donor livers was accomplished by either ex vivo bombardment or in vivo bombardment of the liver before donation or after a 30% hepatectomy. Reporter genes coding for firefly luciferase and GFP were transfected to evaluate the efficiency of gene transfer. A 30% partial hepatectomy in rats favored luciferase activity to a sevenfold increase relative to normal rats. Gene expression was further augmented after immunosuppression with a regimen of tacrolimus (FK506). The study further examined the feasibility of ex vivo gene gun-mediated vaccination of rat liver allografts before transplantation for inducing immunity against infection and malignancy. A much stronger primary humoral response was seen after local DNA vaccination of liver by the gene gun using an expression plasmid encoding the hepatitis B virus surface (HBs) antigen compared to skin immunization alone. Moreover, antibody production against the HBs antigen with an immunized partial liver graft was enhanced significantly by tacrolimus. Immune suppression can be combined with *ex vivo* liver gene transfer using the gene gun for evolving a potential strategy in the future for preventing infectious disease associated with liver transplantation.

The gene gun technique has been used to decipher gene regulation in several studies. It was used to identify tissue-specific elements in the $5'$-upstream promoter region for the ovalbumin gene in the oviduct and liver of laying hens (Muramatsu et al., 1998). The CAT reporter gene containing varying lengths of sequence for the truncated ovalbumin promoter was bombarded with the gene gun to transfect selected organs. CAT gene expression was independent of the length of truncated ovalbumin gene promoters fused to the CAT gene in the oviduct of laying hens. However, there was a decrease in the CAT activity measured in the liver when the sequence from -3200 to 8 bp of the ovalbumin promoter region was fused upstream of the CAT gene. There was an increase in the activity of CAT in liver when the upstream region of the CAT gene was fused with either the sequence from -2800 to 8 bp or a shorter length of the ovalbumin promoter region. It is probable that the promoter region between -3200 and -2800 bp of the ovalbumin gene is responsible for a tissue-specific suppression in transcription of the ovalbumin gene only in the liver, but not in the oviduct of laying hens. The Helios gene gun system was also used to study regulation of the cytochrome P450 gene in rat liver by Muangmoonchai et al. (2002). Reporter gene constructs for CYP2B1 having either the SV40 promoter or the minimal promoter containing the phenobarbital response element were delivered to the rat liver by particle bombardment. Although the SV40 promoter was eight times stronger than a minimal promoter, but a similar level of induction was produced in both promoters by phenobarbitone.

The efficiency of particle-mediated gene transfer by the gene gun depends on the type of gun and target organ. Existing designs of gene guns produce gene expression levels that are typically 20–30 times greater than those observed following naked DNA injection (Kuriyama et al., 2000; Pertmer et al., 1995). Another design of gene gun used for liver gene transfer employed a hammering bullet to discharge microprojectiles from a vibration plate coated with 0.2-mg gold particles containing 0.8 μg LacZ expression plasmid (Kuriyama et al., 2000). Ventral midline incision in mouse exposed the liver to the gene gun and gold particles coated with DNA were bombarded onto the liver surface. Compared to simple plasmid injection or electroporation-mediated gene transfer to liver, the bombarded liver tissue showed a broad and random distribution of profound LacZ expression.

The relatively nontraumatic injection of spheres by the gene gun allows repeated applications, but the risk of significant cell damage increases with repeat applications. Unfortunately, the use of gene guns for gene therapy has been limited to genetic vaccinations and applications where secreted proteins are produced, such as cancer therapy (Oshikawa *et al.*, 1999). It has been shown that for equivalent immune responses after vaccination, the gene gun required 100 times less DNA than im injection (Fynan *et al.*, 1993). The duration of gene expression following this therapy was shown to be longer than 60 days when used for skeletal muscle in muscle (Yang *et al.*, 1990). This technology has been used to develop a high-throughput screen for searching preerythrocytic vaccine target antigens among newly annotated malaria genes. The approach aims to find new malaria parasite proteins for which targeting by the immune system will be lethal to the parasite (Haddad *et al.*, 2004).

V. LIVER GENE TRANSFER BY INJECTION VIA PORTAL AND EFFERENT VESSELS

Successful gene expression was restricted to only a limited area with strategies employing direct injection, which prompted the search for alternative routes of gene delivery. Using a surgical approach similar to that employed for direct injection, a hypertonic DNA solution along with an anticoagulant heparin was injected into the portal blood supply of mouse liver via the portal vein (Budker *et al.*, 1996). It was observed that occlusion of blood outflow through the hepatic vein was a crucial factor for efficient gene expression ($\sim$1% of hepatocytes) in mice.

Zhang and colleagues (1997) extended the findings of a previous study of portal vein injection done on mice by injecting DNA against the direction of blood flow from liver in the hepatic vein and retrograde to bile flow in the bile duct of mice and higher mammals such as dog and rats. Retrograde injection counter to the flow of blood and bile increased the efficiency of transgene expression dramatically ($\sim$10% of hepatocytes) as indicated by β-gal expression. There was a clear correlation between the level of luciferase expression and the hydrostatic pressure in the liver with peak gene expression seen at intrahepatic pressures of 40 mm Hg or greater. The reporter gene expression in the liver was distributed evenly following DNA injection into any of three hepatic vessels, i.e., portal vein, hepatic vein, and bile duct, compared to localized gene expression following direct injection into the liver as discussed in previous sections. Limitations of this technique are, however, that the blood flow must be blocked temporarily in order to establish the pressure necessary for DNA transfer to occur and good surgical skills are essential to obtain consistent results.

Liver gene transfer using the method of portal vein injection with a transient occlusion of blood flow was examined for the treatment of acute intermittent porphyria (AIP) by Johansson *et al.* (2004). The mouse model used in the study was a transgenic mouse manifesting AIP due to deficiency of porphobilinogen deaminase (PBGD), an enzyme involved in heme biosyntheis in mouse liver. Plasmid DNA encoding PBGD was injected into the portal vein of a transgenic mouse followed by transient occlusion of blood flow by the hepatic vein. Although luciferase expression was increased in liver 24 h after injection, there was hardly any increase in PBGD activity. Polymerase chain reation also failed to detect any increase in the message for PBGD after transfection.

A novel method of liver surface instillation was used for naked DNA transfer to the liver in mouse by Hirayama et al. (2003). This method appears similar to intraperitoneal injection, but involves insertion of a polyethernylon catheter through the abdominal wall to be retained between right and left medial lobes of the liver. DNA solution instilled over the liver surface is less than 0.1 ml, and gene expression can be made lobe selective, which was, however, very low compared to other methods of gene transfer of naked DNA to the liver.

VI. LIVER GENE TRANSFER BY SYSTEMIC ADMINISTRATION

A. Liver gene transfer by the hydrodynamics method

In 1999, the field of hepatic transfection saw a radical transformation with a relatively simple and efficient technique (transfecting approximately 40% hepatocytes) that can be done without the need for any surgical procedures to achieve liver gene transfer (Liu et al., 1999; Zhang et al., 1999). This new, nonsurgical method of pressure-mediated gene transfer is better known as hydrodynamics-based gene transfection (Liu et al., 1999).

A large volume of DNA solution is administered as a rapid bolus into the tail vein of a mouse and the process of DNA solution flushing rapidly into the liver in a large volume creates high intrahepatic pressure (Liu and Knapp, 2001; Song et al., 2002). This high pressure not only stretches the sinusoids from the blood side to allow the transit of DNA across the fenestrated sinusoids for gaining direct access to the hepatocytes, but is also capable of generating a pressure gradient across the plasma membrane of the hepatocytes (higher outside and lower inside). Transient membrane defects (holes, membrane rupture, etc.) across the plasma membrane are created by a pressure gradient between microvilli to drive DNA molecules into the cells (Liu and Knapp, 2001). The crucial role played by fluid dynamics in hepatic gene expression was recognized by the advent of the hydrodynamics method (Koboyashi et al., 2001). Gene expression in the liver was shown in this method without the need for any surgical incision (McCafferty et al., 2002). As discussed in prior sections of this chapter, surgical incision is a must for the success of most methods seeking naked DNA transfer to the liver.

B. Retention time-mediated liver gene transfer

Notwithstanding the intrinsic toxicity in the liver caused by hydrodynamic gene transfer, the method was certainly successful in demonstrating gene expression in the liver following systemic administration of DNA. Systemic administration of DNA for gene transfer to the liver has been the subject of ongoing focus in our laboratory. In a study reported earlier, liver gene transfer was achieved in the absence of any physical force and without any viral or non-viral carrier system (Liu and Huang, 2001). A mere increase in the retention time spent in liver for venous blood carrying the injected DNA was successful in achieving hepatic gene expression. Prior to a small volume tail vein injection of plasmid DNA, mouse liver was exposed under anesthesia for applying either a metal clip at the vena cava (VC) or a combination of portal vein and hepatic artery (PV + HA) for the occlusion of blood flow through the liver for a few seconds. As shown in Fig. 3.1, immediately after injection, the blood flow through liver was transiently blocked in the afferent and efferent blood vessels of the liver. A single injection of 80 μg of plasmid DNA was able to produce up to 560 pg of luciferase protein

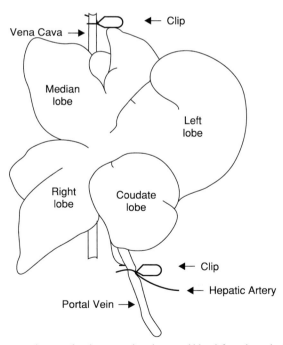

Figure 3.1. Schematic diagram for the sites of occlusion of blood flow through the liver (Liu and Huang, 2001).

per milligram of extracted protein from the liver. Clamping the blood vessels was found to be essential for detecting gene expression in the liver, and the time interval between injection and clamping influenced the efficiency of liver gene transfer However, the duration and volume of injection did not influence gene expression. A significant level of luciferase protein was obtained when the clamping time was as short as 1 s for occlusion of blood inflow via PV + HA and blood outflow via VC. Gene expression did not improve any further with an increased duration of occlusion ($P > 0.5$).

In the literature, there are two probable mechanisms reported for explaining the weak binding of plasmid DNA to the surface of hepatic cells: it is either a receptor-mediated or a nonspecific binding/uptake process (Budker et al., 2000; Kawabata et al., 1995). The clamping experiments of our study do rule out the role of a pressure-dependent mechanism in retention time-mediated gene transfer. We propose another hypothesis for gene transfer by our method of clamping: that DNA binding is not strong enough and can be easily dissociated and washed away by the blood flow in normal physiological conditions. Consequently, plasmid DNA is unlikely to yield any transfection even after establishing a steady-state plasma concentration of DNA in the blood. However, a temporary block in the blood flow through the liver drastically reduces the washing force of blood flow in addition to slowing down the circulating speed of DNA as well as increasing the retention time of DNA in the liver. In such circumstances, systemically administered DNA is likely to have an improved chance of being taken up and expressed by the hepatocytes.

C. Liver gene transfer by mechanical massage

In an innovation reported from our laboratory, a simple application of mechanical force to the liver in the form of manual massage (MML) was successful in achieving naked gene transfer to the liver. Performing MML on an anesthetized mouse without any surgery resulted in efficient hepatic expression of genes in the liver (Liu and Huang, 2002b). As illustrated in Fig. 3.2, inhalation of isoflurane anesthetized the mice prior to an intravenous (tail vein) injection of 80 μg plasmid DNA in 100 μl saline. Immediately after injection, a manual massage of the liver was done by placing both thumbs over the upper abdomen of the mouse directly underneath the rib cage and pushing the liver against the back of the animal for 1 s. The procedure was repeated four times within 4 s. Six hours later, the mouse liver was removed for measuring reporter gene expression. The firefly luciferase gene was used for making a quantitative comparison of gene expression using MML with respect to other methods. The levels of luciferase expression in the liver using MML were not comparable to those observed after hydrodynamic delivery, but were found to be lower by more than one order of magnitude. In contrast, the toxicity profile of mouse liver after MML was much

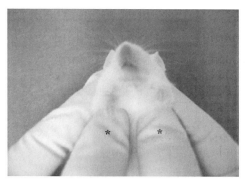

Figure 3.2. Gene transfer to the mouse liver by manually massaging the liver. The procedure is done by placing both thumbs (indicated by stars) over the abdomen and pushing the liver against the back of the animal four times for 1 s each (Liu and Huang, 2002). (See Color Insert.)

better than that seen after hydrodynamic delivery. This easy-to-use gene transfer method does not require sophisticated equipment and expensive reagents and appears to be nontraumatic. The lower levels of protein expression following gene transfer of a therapeutic gene by MML into mouse liver were assessed as a therapeutic option for liver disease.

Fulminant hepatic failure (FHF) is one such liver disease whose mortality rate in humans is very high, approximately 35–45%. We tested our method in the most commonly used animal model of FHF, which is endotoxin-induced lethal hepatic failure in mice. Injection of bacterial endotoxin lipopolysaccharide (LPS) into mice in combination with GalN induces a failure of mouse liver far more severe than that seen in humans. The hepatic gene transfer of hepatocyte growth factor (HGF) plasmid DNA in our study by MML was successful in preventing FHF, which led to a dramatically increased survival in mice (Liu and Huang, 2002). As shown in Fig. 3.3, all of the untreated mice or mice treated with empty plasmid plus MML died within 20 h after the injection of LPS and GalN, in which eight out of nine in each group died within 8 h. In contrast, seven out of nine mice transfected with the HGF plasmid using MML survived until the experiment was terminated at 7 days. These data suggest that HGF gene transfer by MML strongly inhibits hepatic failure and the lethality induced by LPS and GalN (Kosai et al., 1999).

Liu et al. (2004) described the underlying mechanism for liver gene transfer by mechanical massage. The process of MML involves applying pressure to the abdomen, and we investigated the changes in blood pressure of mice during MML. A sensor inserted into the inferior vena cava of a mouse notices a brief rise in blood pressure of the mouse while MML was applied. However, the rise in blood pressure dropped quickly to normal baseline once the thumbs were released from the abdomen. The sudden drop suggested a transient nature of the

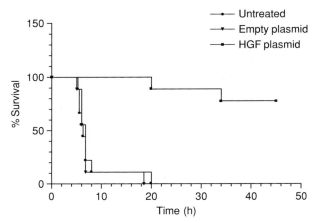

Figure 3.3. Survival of mice after LPS and GalN injection with no treatment, injection of pNGVL-3 empty plasmid, and pCMV-HGF plasmid. $n = 9$ in each group (Liu and Huang, 2002).

pressure rise from MML. An increase in venous blood pressure during MML correlated with a linear increase in gene expression, which argues for the role of pressure-mediated gene delivery in gene transfer by MML. Most likely a transient permeablized state of hepatocytes is created by the pressure exerted by MML, which allows DNA to be taken up by cells. Moreover, other experiments conducted in our laboratory employing MML for liver gene transfer further supported our inference about the transient state of hepatocyte membrane. It was observed that liver transfection could be achieved even when MML preceded DNA injection into the tail vein. The role of a scavenger receptor in MML gene transfer was also ruled out by our studies.

VII. CONCLUSION

Most metabolic diseases are caused by defects in the metabolic pathways or by the absence of specific proteins normally produced and secreted by the liver. Gene therapy could serve as an ideal therapeutic option for the treatment of these metabolic disorders, as most of them are usually caused by single gene defects. Systemic delivery of naked plasmid DNA for gene transfer into the liver does hold solid promise for the treatment of diseases. To be clinically relevant, a plasmid vector has to be persistently expressed, not to mention easy to use, nontoxic, and has the ability to be targeted. Expression cassettes can be used for liver-specific gene expression. This is clearly one of the most important challenges in non-viral gene therapy today. Efforts are in progress in various laboratories to address this issue.

References

Baba, M., Iishi, H., and Tatsuta, M. (2001). Transfer of bcl-xs plasmid is effective in preventing and inhibiting rat hepatocellular carcinoma induced by N-nitrosomorpholine. *Gene Ther.* **8,** 1149–1156.

Budker, V., Zhang, G., Knechtle, S., and Wolff, J. A. (1996). Naked DNA delivered intraportally expresses efficiently in hepatocytes. *Gene Ther.* **3,** 593–598.

Budker, V., Budker, T., Zhang, G., Subbotin, V., Loomis, A., and Wolff, J. A. (2000). Hypothesis: Naked plasmid DNA is taken up by cells *in vivo* by a receptor-mediated process. *J. Gene. Med.* **2** (2), 76.

Cheng, L., Ziegelhoffer, P. R., and Yang, N. S. (1993). *In vivo* promoter activity and transgene expression in mammalian somatic tissues evaluated by using particle bombardment. *Proc. Natl. Acad. Sci. USA* **90,** 4455–4459.

Cheng, S. H., and Smith, A. E. (2003). Gene therapy progress and prospects: Gene therapy of lysosomal storage disorders. *Gene Ther.* **10,** 275–281.

Chuah, M. K., Collen, D., and VandenDriessche, T. (2004). Clinical gene transfer studies for hemophilia. A. *Semin. Thromb. Hemost.* **30,** 249–256.

Ferber, D. (2001). Gene therapy, safer and virus-free. *Science* **294,** 1638–1642.

Ferry, N., Duplessis, O., Houssin, D., Danos, O., and Heard, J. M. (1991). Retroviral-mediated gene transfer into hepatocytes *in vivo. Proc. Natl. Acad. Sci. USA* **88,** 8377–8381.

Fynan, E. F., Websser, R. G., Fuller, D. H., Haynes, J. R., Santoro, J. C., and Robinson, H. L. (1993). DNA vaccines: Protective immunizations by parenteral, mucosal, and gene-gun inoculations. *Proc. Natl. Acad. Sci. USA* **90,** 11478–11482.

Golzio, M., Rols, M. P., and Teissie, J. (2004). *In vitro* and *in vivo* electric field-mediated permeabilization, gene transfer, and expression. *Methods* **33,** 26–35.

Gupta, M., Jansen, E. E., Senephansiri, H., Jakobs, C., Snead, O. C., Grompe, M., and Gibson, K. M. (2004). Liver-directed adenoviral gene transfer in murine succinate semialdehyde dehydrogenase deficiency. *Mol. Ther.* **9,** 527–539.

Haddad, D., Bilcikova, E., Witney, A. A., Carlton, J. M., White, C. E., Blair, P. L., Chattopadhyay, R., Russell, J., Abot, E., Charoenvit, Y., Aguiar, J. C., Carucci, D. J., and Weiss, W. R. (2004). Novel antigen identification method for discovery of protective malaria antigens by rapid testing of DNA vaccines encoding exons from the parasite genome. *Infect. Immun.* **72,** 1594–1602.

Heiser, W. C. (1994). Gene transfer into mammalian cells by particle bombardment. *Anal. Biochem.* **217,** 185–196.

Heller, R., Jaroszeski, M., Atkin, A., Moradpour, D., Gilbert, R., Wands, J., and Nicolau, C. (1996). *In vivo* gene electroinjection and expression in rat liver. *FEBS Lett.* **389,** 225–228.

Herweijer, H., and Wolff, J. A. (2003). Progress and prospects: Naked DNA gene transfer and therapy. *Gene Ther.* **10,** 453–458.

Hickman, M. A., Malone, R. W., Lehmann-Bruinsma, K., Sih, T. R., Knoell, D., Szoka, F. C., Walzem, R., Carlson, D. M., and Powell, J. S. (1994). Gene expression following direct injection of DNA into liver. *Hum. Gene Ther.* **5,** 1477–1483.

High, K. A. (2004). Clinical gene transfer studies for hemophilia B. *Semin. Thromb. Hemost.* **30,** 257–267.

Hirayama, R., Kawakami, S., Nishida, K., Nakashima, M., Sasaki, H., Sakeda, T., and Nakamura, J. (2003). Development of the liver- and lobe-selective non-viral gene transfer following the instillation of naked plasmid DNA using catheter on the liver surface in mice. *Pharm. Res.* **20,** 328–332.

Inoue, S., Hakamata, Y., Kaneko, M., and Kobayashi, E. (2004). Gene therapy for organ grafts using rapid injection of naked DNA: Application to the rat liver. *Transplantation* **77,** 997–1003.

Jaffe, H. A.,Danel, C.,Longenecker, G.,Metzger, M.,Setoguchi, Y.,Rosenfeld, M. A.,Gant, T. W., et al. (1992). Adenovirus-mediated *in vivo* gene transfer and expression in normal rat liver. *Nature Genet.* **1**, 372–378.

Johansson, A., Nowak, G., Moller, C., and Harper, P. (2004). Non-viral delivery of the porphobilinogen deaminase cDNA into a mouse model of acute intermittent porphyria. *Mol. Genet. Metab.* **82**, 20–26.

Jung, S. C., Han, I. P., Limaye, A., Xu, R., Gelderman, M. P., Zerfas, P., and Tirumalai, K., et al. (2001). Adeno-associated viral vector-mediated gene transfer results in long-term enzymatic and functional correction in multiple organs of Fabry mice. *Proc. Natl. Acad. Sci. USA* **98**, 2676–2681.

Kaleko, M., Garcia, J. V., and Miller, A. D. (1991). Persistent gene expression after retroviral gene transfer into liver cells *in vivo*. *Hum. Gene Ther.* **2**, 27–32.

Kawabata, K., Takakura, Y., and Hashida, M. (1995). The fate of plasmid DNA after intravenous injection in mice: Involvement of scavenger receptors in its hepatic uptake. *Pharm. Res.* **12**, 825–830.

Koboyashi, N., Kuramoto, T., Yamaoka, K., Hashida, M., and Takakura, Y. (2001). Hepatic uptake and gene expression mechanisms following intravenous administration of plasmid DNA by conventional and hydrodynamics-based procedures. *J. Pharmacol. Exp. Ther.* **297**, 853–860.

Koboyashi, S., Dono, K., Takahara, S., Isaka, Y., Imai, E., Zhenhui, L., Nagano, H., Tomoaki, K., Umeshita, K., Nakamori, S., Sakon, M., and Monden, M. (2003a). Gene transfer into the liver by plasmid injection into the portal vein combined with electroporation. *J. Gene Med.* **5**, 201–208.

Koboyashi, S., Dono, K., Tanaka, T., Takahara, S., Isaka, Y., Imai, E., Nagano, H., Tomoaki, K., Umeshita, K., Nakamori, S., Sakon, M., and Monden, M. (2003b). Electroporation-mediated *ex vivo* gene transfer into graft not requiring injection pressure in orthotopic liver transplantation. *J. Gene Med.* **5**, 510–517.

Kosai, K., Matsumoto, K., Funakoshi, H., and Nakamura, T. (1999). Hepatocyte growth factor prevents endotoxin-induced lethal hepatic failure in mice. *Hepatology* **30**, 151–159.

Kuriyama, S., Mitoro, A., Tsujinoue, H., Nakatani, T., Yoshiji, H., Tsujimoto, T., Yamazaki, M., and Fukui, H. (2000). Particle-mediated gene transfer into murine livers using a newly developed gene gun. *Gene Ther.* **7**, 1132–1136.

Liu, D., Knapp, E., and Song, Y. K. (1999). Mechanism of cationic liposome-mediated transfection of the lung endothelium. *In* "Non-viral Vectors for Gene Therapy" (L. Huang, M. C. Hung, and E. Wagner, eds.), pp. 230–267. Academic Press, San Diego.

Liu, D., and Knapp, J. E. (2001). Hydrodynamics-based gene delivery. *Curr. Opin. Mol. Ther.* **3**, 192–197.

Liu, F., and Huang, L. (2001). Improving plasmid DNA-mediated liver gene transfer by prolonging its retention in the hepatic vasculature. *J. Gene Med.* **3**, 569–576.

Liu, F., and Huang, L. (2002a). Electric gene transfer to the liver following systemic administration of plasmid DNA. *Gene Ther.* **9**, 1116–1119.

Liu, F., and Huang, L. (2002b). Noninvasive gene delivery to the liver by mechanical massage. *Hepatology* **35**, 1314–1319.

Liu, F., Qi, H., Huang, L., and Liu, D. (1999). Factors controlling the efficiency of cationic lipid-mediated transfection *in vivo* via intravenous administration. *Gene Ther.* **4**, 517–523.

Liu, F., Song, Y. K., and Liu, D. (1999). Hydrodynamics-based transfection in animals by systemic administration of plasmid DNA. *Gene Ther.* **6**, 1258–1266.

Liu, F., Lei, J., Vollmer, R., and Huang, L. (2004). Mechanism of liver gene transfer by mechanical massage. *Mol. Ther.* **9**(3), 452–457.

Liu, Q., Kay, M. A., Finegold, M., Stratford-Perricaudet, L. D., and Woo, S. L. (1993). Assessment of recombinant adenoviral vectors for hepatic gene therapy. *Hum. Gene Ther.* **4**, 403–409.

Liu, V. W. Y., Falo, L. D., and Huang, L. (2001). Systemic production of IL-12 by naked DNA mediated gene transfer, toxicity and attenuation of transgene expression *in vivo. J. Gene Med.* **3,** 384–393.

Malone, R. W., Hickman, M. A., Lehmann-Bruinsma, K., Sih, T. R., Walzem, R., Carlson, D. M., and Powell, J. S. (1994). Dexamethasone enhancement of gene expression after direct hepatic DNA injection. *J. Biol. Chem.* **269,** 29903–29907.

Marshall, J., McEachern, K. A., Kyros, J. A. C., Nietubski, J. B., Budzinski, T., Ziegler, R. J., Yew, N. S., Sullivan, J., Scaria, A., van Rooijen, N., Barranger, J. A., and Cheng, S. H. (2002). Demonstration of feasibility of *in vivo* gene therapy for Gaucher disease using a chemically induced mouse model. *Mol. Ther.* **6,** 179–189.

Maruyama, H., Ataka, K., Gejyo, F., Higuchi, N., Ito, Y., Hirahara, H., Imazeki, I., Hirata, M., Ichikawa, F., Neichi, T., Kikuchi, H., Sugawa, M., and Miyazaki, J. (2001). Long-term production of erythropoietin after electroporation-mediated transfer of plasmid DNA into the muscles of normal and uremic rats. *Gene Ther.* **8,** 461–468.

Matsuno, Y., Iwata, H., Umeda, Y., Takagi, H., Mori, Y., Kosugi, A., Matsumoto, K., Nakamura, T., and Hirose, H. (2003). Hepatocyte growth factor gene transfer into the liver via the portal vein using electroporation attenuates rat liver cirrhosis. *Gene Ther.* **10,** 1559–1566.

McCafferty, A. P., Meuse, L., Pham, T. T., Conklin, D. S., Hannon, G. J., and Kay, M. A. (2002). RNA interference in adult mice. *Nature* **418,** 38–39.

Miao, C. H., Thompson, A. R., Loeb, K., and Ye, X. (2001). Long-term and therapeutic-level hepatic gene expression of human factor IX after naked plasmid transfer *in vivo. Mol. Ther.* **3,** 947–957.

Mir, L. M., Bureau, M. F., Gehl, J., Rangara, R., Rouy, D, Caillaud, J. M., Delaere, P., Branellec, D., Schwartz, B., and Scherman, D (1999). High-efficiency gene transfer into skeletal muscle mediated by electric pulses. *Proc. Natl. Acad. Sci. USA* **96,** 4262–4267.

Muangmoonchai, R., Wong, S. C., Smirlis, D., Phillips, I. R., and Shephardl, E. A. (2002). Transfection of liver *in vivo* by biolistic particle delivery: Its use in the investigation of cytochrome P450 gene regulation. *Mol. Biotechnol.* **20,** 145–151.

Muramatsu, T., Imai, T., Park, H. M., Watanabe, H., Nakamura, A., and Okumura, J. (1998). Gene gun-mediated *in vivo* analysis of tissue-specific repression of gene transcription driven by the chicken ovalbumin promoter in the liver and oviduct of laying hens. *Mol. Cell. Biochem.* **185,** 27–32.

Muramatsu, T., Ito, N., Tamaoki, N., Oda, H., and Park, H. M. (2001). *In vivo* gene electroporation confers nutritionally-regulated foreign gene expression in the liver. *Int. J. Mol. Med.* **7,** 61–66.

Nakamura, M., Wang, J., Murakami, T., Ajiki, T., Hakamata, Y., Kaneko, T., Takahashi, M., Okamoto, H., Mayumi, M., and Kobayashi, E. (2003). DNA immunization of the grafted liver by particle-mediated gene gun. *Transplantation* **76,** 1369–1375.

Neumann, E., Schaefer-Ridder, M., Wang, Y., and Hofschneider, P. H. (1982). Gene transfer into mouse lyoma cells by electroporation in high electric fields. *EMBO J.* **1,** 841–845.

Nishikawa, M., Nakano, T., Okabe, T., Hamaguchi, N., Yamasaki, Y., Takakura, Y., Yamashita, F., and Hashida, M. (2003). Residualizing indium-111-radiolabel for plasmid DNA and its application to tissue distribution study. *Bioconjug. Chem.* **14,** 955–961.

Oka, K., and Chan, L. (2004). Liver-directed gene therapy for dyslipidemia and diabetes. *Curr. Atheroscler. Rep.* **46,** 203–209.

Oshikawa, K., Shi, F., Rakhmilevich, A. L., Sondel, P. M., Mahvi, D. M., and Yang, N. S. (1999). Synergistic inhibition of tumor growth in a murine mammary adenocarcinoma model by combinational gene therapy using IL-12, pro-IL-18, and IL-1beta converting enzyme cDNA. *Proc. Natl. Acad. Sci. USA* **96,** 13351–13356.

Pertmer, T. M., Eisenbraun, M. D., Mccabe, D., Prayaga, S. K., Fuller, D. H., and Haynes, J. R. (1995). Gene gun-based nucleic acid immunization: Elicitation of humoral and cytotoxic T lymphocyte responses following epidermal delivery of nanogram quantities of DNA. *Vaccine* **13,** 1427–1430.

Qiu, P., Ziegelhoffer, P., Sun, J., and Yang, N. S. (1996). Gene gun delivery of mRNA *in situ* results in efficient transgene expression and genetic immunization. *Gene Ther.* **3,** 262–268.

Sanford, J. C., Smith, F. D., and Russell, J. A. (1993). Optimizing the biolistic process for different biological applications. *Methods Enzymol.* **217,** 483–509.
Schiedner, G., Bloch., W., Hertel, S., Johnston, M., Molojavyi, A., Dries, V., Varga, G., Van Rooijen, N., and Kochanek, S. (2003). A hemodynamic response to intravenous adenovirus vector particles is caused by systemic Kupffer cell-mediated activation of endothelial cells. *Hum. Gene Ther.* **14,** 1631–1641.
Schmitz, V., Qian, C., Ruiz, J., Sangro, B., Melero, I., Mazzolini, G., Narvaiza, I., and Prieto, J. (2002). Gene therapy for liver diseases, recent strategies for treatment of viral hepatitis and liver malignancies. *Gut* **50,** 130–135.
Shi, N., Boado, R. J., and Pardridge, W. M. (2001). Receptor-mediated gene targeting to tissues *in vivo* following intravenous administration of pegylated immunoliposomes. *Pharm. Res.* **18,** 1091–1095.
Smith, K. C., Neu, J. C., and Krassowska, W. (2004). Model of creation and evolution of stable electropores for DNA delivery. *Biophys. J.* **86,** 2813–2826.
Smith, R. M., and Wu, G. Y. (1999). Hepatocyte-directed gene delivery by receptor-mediated endocytosis. *Semin. Liver Dis.* **19,** 83–92.
Song, Y. K., Liu, F., Zhang, G. S., and Liu, D. (2002). Hydrodynamics-based transfection: Simple and efficient method for introducing and expressing transgenes in animals by intravenous injection of DNA. *Method Enzymol.* **346,** 92–105.
Stecenko, A. A., and Brigham, K. L. (2003). Gene therapy progress and prospects: Alpha-1 antitrypsin. *Gene Ther.* **10,** 95–99.
Stoll, S. M., Sclimenti, C. R., Baba, E. J., Meuse, L., Kay, M. A., and Calos, M. P. (2001). Epstein-Barr virus/human vector provides high-level, long-term expression of alpha1-antitrypsin in mice. *Mol. Ther.* **4,** 122–129.
Suzuki, T., Shin, B. C., Fujikura, K., Matsuzaki, T., and Takata, K. (1998). Direct gene transfer into rat liver cells by *in vivo* electroporation. *FEBS Lett.* **425,** 436–440.
Takakura, Y., Nishikawa, M., Yamashita, F., and Hashida, M. (2001). Development of gene drug delivery systems based on pharmacokinetic studies. *Eur. J. Pharm. Sci.* **13,** 71–76.
Templeton, N. S., and Lasic, D. D. (1999). Directions in Liposome gene delivery. *Mol. Biotechnol.* **11**(2), 175–180.
VandenDriessche, T., Collen, D., and Chuah, M. K. (2003). Gene therapy for the hemophilias. *J. Thromb. Haemost.* **1,** 1550–1558.
Wolff, J. A., Malone, R. W., Williams, P., Chong, W., Acsadi, G., Jani, A., and Felgner, P. L. (1990). Direct gene transfer into mouse muscle *in vivo*. *Science* **247,** 1465–1468.
Wong, T. K., and Neumann, E. (1982). Electric field mediated gene transfer. *Biochem. Biophys. Res. Commu.* **107,** 584–587.
Yamazaki, Y., Fujimoto, H., Ando, H., Ohyama, T., Hirota, Y., and Noce, T. (1998). *In vivo* gene transfer to mouse spermatogenic cells by deoxyribonucleic acid injection into seminiferous tubules and subsequent electroporation. *Biol. Reprod.* **59,** 1439–1444.
Yang, N. S., Burkholder, J., Roberts, B., Martinell, B., and Mccabe, D. (1990). *In vivo* and *in vitro* gene transfer to mammalian somatic cells by particle bombardment. *Proc. Natl. Acad. Sci. USA* **87,** 9568–9572.
Yoshida, Y.,Kobayashi, E.,Endo, H.,Hamamoto, T.,Yamanaka, T.,Fujimura, A.,Kagawa, Y., *et al.* (1997). Introduction of DNA into rat liver with a hand-held gene gun, distribution of the expressed enzyme, [^{32}P]DNA, and Ca^{2+} flux. *Biochem. Biophys. Res. Commun.* **234,** 695–700.
Zhang, G., Budker, V., and Wolff, J. A. (1999). High levels of foreign gene expression in hepatocytes after tail vein injections of naked plasmid DNA. *Hum. Gene Ther.* **10,** 1735–1737.
Zhang, G., Vargo, D., Budker, V., Armstrong, N., Knechtle, S., and Wolff, J. A. (1997). Expression of naked plasmid DNA injected into the afferent and efferent vessels of rodent and dog livers. *Hum. Gene Ther.* **8,** 1763–1772.

4

Hydrodynamic Delivery

Mohammed S. Al-Dosari, Joseph E. Knapp, and Dexi Liu
Department of Pharmaceutical Sciences, University of Pittsburgh School of
Pharmacy, Pittsburgh, Pennsylvania 15261

ABSTRACT

Hydrodynamic delivery has emerged as a near-perfect method for intracellular
DNA delivery *in vivo*. For gene delivery to parenchymal cells, only essential
DNA sequences need to be injected via a selected blood vessel, eliminating
safety concerns associated with current viral and synthetic vectors. When
injected into the bloodstream, DNA is capable of reaching cells in the different
tissues accessible to the blood. Hydrodynamic delivery employs the force

0065-2660/05 $35.00
DOI: 10.1016/S0065-2660(05)54004-5

generated by the rapid injection of a large volume of solution into the incompressible blood in the circulation to overcome the physical barriers of endothelium and cell membranes that prevent large and membrane-impermeable compounds from entering parenchymal cells. In addition to the delivery of DNA, this method is useful for the efficient intracellular delivery of RNA, proteins, and other small compounds *in vivo*. This review discusses the development, current application, and clinical potential of hydrodynamic delivery. © 2005, Elsevier Inc.

I. INTRODUCTION

Efforts to introduce gene sequences into tissues of animals to generate sufficient amounts of gene products have led to the development of various methods using principles of biology (e.g., viral vectors), chemistry (e.g., lipids, polymers), and physics (e.g., microinjection, electroporation). Hydrodynamic delivery, one of the newer methods developed in recent years, has become a frequent choice for gene therapy studies. Excellent review articles summarizing recent progress in hydrodynamic delivery have been published elsewhere (Hagstrom, 2003; Hodges and Scheule, 2003; Liu and Knapp, 2001). Mechanistically, hydrodynamic delivery employs the physical force generated by a rapid injection of a large volume of DNA solution into the incompressible blood in circulation to rupture the endothelium and cell membranes to allow DNA to enter parenchymal cells. Practically, hydrodynamic delivery can be used to facilitate the intracellular delivery of DNA, RNA, proteins, polymers, and other membrane-impermeable compounds. The following sections describe the principle, the experimental procedures, and the current applications of hydrodynamic delivery. Future perspectives are provided and reflect our views on further development.

II. PRINCIPLES OF HYDRODYNAMIC DELIVERY

Direct injection of naked DNA into a blood vessel to achieve intracellular gene delivery *in vivo* was first reported by Zhang and colleagues (1997). In their study, transfection of approximately 5% of hepatocytes in mouse liver was achieved by an injection of hypertonic solutions of reporter-containing plasmids into the portal vein, hepatic vein, or bile duct along with occlusion of blood flow (Zhang *et al.*, 1997). This study showed that the level of reporter gene expression was directly proportional to the volume of DNA solution injected. The larger the volume of DNA solution injected into the liver, the higher the level of reporter gene expression. From these results, it was hypothesized that hydrostatic pressure resulting from the injection of DNA solution drives DNA into the hepatocytes where gene expression occurs. However, the experimental procedure developed in this study was not widely adapted by others due to the need for rather challenging surgical procedures required to achieve consistent results.

The principle of hydrodynamic delivery using large volume and rapid injection as the driving force for DNA delivery was established when Liu *et al.* (1999) and Zhang *et al.* (1999) demonstrated highly efficient gene delivery in mice by a rapid injection via the tail vein of DNA solutions in volumes equivalent to 8–10% of the body weight. It was shown that, among the internal organs (heart, lung, liver, kidneys, and spleen) that exhibited significant levels of transgene expression, the liver was the most sensitive to this procedure. Approximately 40% of hepatocytes were transfected on administration of doses as low as 10 μg plasmid DNA per mouse. We coined the term "hydrodynamics-based procedure" for this method to reflect the need for a large volume and high injection speed (Liu *et al.*, 1999). Importantly, this procedure was well tolerated by animals. Both histochemical and serum biochemistry analyses revealed no obvious tissue damage. The transient increase of liver enzymes observed returned to the normal range in 3 days (Kobayashi *et al.*, 2004c; Liu *et al.*, 1999). Because of its simplicity, convenience, high efficiency, and lack of toxicity to animals, the hydrodynamics-based procedure (now more often referred to as hydrodynamic delivery or hydrodynamic injection) has become a frequent choice for the delivery of DNA and RNA (especially siRNA) for gene function and gene therapy studies in mice.

Significant efforts have been made in the past few years to mechanistically dissect the process of hydrodynamic delivery of DNA to hepatocytes in mice. Budker *et al.* (2000) proposed that successful gene transfer involves receptor-mediated DNA internalization into hepatocytes. According to their hypothesis, the function of hydrostatic pressure established by hydrodynamic injection was to enlarge the fenestrae of liver sinusoids to allow large sized plasmids to pass through the capillary walls and reach the hepatocytes (Budker *et al.*, 2000). We have taken a systematic approach to studying the mechanisms underlying the efficient gene transfer to hepatocytes mediated by hydrodynamic injection. We have focused our attention on the impact of hydrodynamic injection on two major organs that are directly involved: the heart, which serves as the primary organ in maintaining blood circulation; and the liver, which exhibits the highest susceptibility to hydrodynamic gene delivery. Employing various techniques, we demonstrated that hydrodynamic injection into the tail vein of a mouse induces a transient decrease in heart rate, a rapid rise of venous pressure in the inferior vena cava, enlargement of liver fenestrae, and generation of pores in heptatocyte membranes (Zhang *et al.*, 2004a). Based on these studies and those of others (Kobayashi *et al.*, 2001; Lecocq *et al.*, 2003), we proposed the following sequence of events that occur in mice after hydrodynamic DNA delivery via the tail vein:

1. Bolus injection of a large volume of DNA solution into the tail vein reduces the heart rate and, at the same time, causes accumulation of injected solution in the inferior vena cava.
2. The solution in the inferior vena cava refluxes into the liver and generates hydrodynamic pressure in the liver.
3. Liver fenestrae in the perivenous region are enlarged, imposing the pressure directly onto the plasma membrane of the hepatocytes.
4. The exterior and interior pressure difference across the hepatocyte membrane pushes the membrane inward and creates membrane defects (pores) when the pressure reaches a certain level.
5. DNA near the membrane defects (pores) enters the cells and, with the declining pressure, the membrane reseals, trapping DNA inside the hepatocytes.
6. With continuous cardiac activity, the body adapts to the volume load and, over time, homeostasis is restored.

The physical nature of the hydrodynamic delivery is also supported by the fact that an array of compounds with no structural similarities have been efficiently delivered by hydrodynamic injection. These structurally different compounds include polyethylene glycol, immunoglobulins (Kobayashi *et al.*, 2001), propidium iodine (Kobayashi *et al.*, 2004c), β-galactosidase protein, Evans blue (Zhang *et al.*, 2004a), genomic RNA (Chang *et al.*, 2001), and siRNA (Giladi *et al.*, 2003; Layzer *et al.*, 2004; Lewis *et al.*, 2002; McCaffrey *et al.*, 2002b, 2003b; Song *et al.*, 2003; Zender *et al.*, 2003). At the cellular level, the mechanism of hydrodynamic delivery is the generation of pores in the plasma membrane, which allow substances that would otherwise be excluded to enter cells. The term "hydroporation" has been proposed to reflect the physical nature and the mechanism of hydrodynamic delivery (Zhang *et al.*, 2004a).

To determine how long the membrane pores stay open after hydrodynamic injection, we and others have employed a strategy of sequential injections in which a hydrodynamic injection of saline is followed by a conventional injection of a small volume of solution containing a reporter plasmid or a membrane-impermeable fluorescence marker (Kobayashi *et al.*, 2004c; Zhang *et al.*, 2004a). In mice, we found that a significant number of membrane pores were resealed within a short time (<1 min) after the hydrodynamic injection, although minimal reporter gene expression was seen in the liver when the second injection was administered 10 min after the hydrodynamic treatment (Zhang *et al.*, 2004a). The long-lasting effect of hydrodynamic injection on membrane permeability of this small number of cells is not currently understood.

We also employed scanning and transmission electron microscopy to examine the structural impact of hydrodynamic injection on mouse livers.

Significant enlargement of the fenestrae is evident in the perivenuous region but not in the periportal region. This observation is in agreement with the observation of transgene expression clustered around the central vein (Zhang et al., 2000). Heterogeneous impact of hydrodynamic injection at the cellular level is also evident. While normally appearing cells were observed to be the significant majority, hepatocytes that appeared to have suffered total release of their contents were seen in some liver sections. At the cellular level, disruption of the hepatocyte membrane structure was identifiable in mouse liver tissue fixed immediately after hydrodynamic injection.

III. HYDRODYNAMICS-BASED PROCEDURES

Since the initial reports of the hydrodynamics-based procedure, modifications in experimental details have been made to satisfy the needs of various applications. Table 4.1 summarizes some of the optimized hydrodynamics-based procedures developed so far. In rodents, optimal gene delivery through tail vein injection requires a volume equivalent to approximately 10% body weight. A 27-gauge needle is used most commonly for tail vein injection in mice (Zhang et al., 1999, 2000), whereas a 22-gauge needle appears appropriate for rats (Maruyama et al., 2002). Other sized needles can also be used depending on the injection volume, the injection speed, and the type of blood vessels receiving the injection. With a fixed volume, as the injection speed is increased, the force generated against the endothelium and cell membrane is increased and a higher gene transfer efficiency results. However, it is important to point out that too rapid an injection rate and/or the use of an excessive injection volume can result in significant tissue damage, even animal death. When a local injection is used, occlusion of blood flow to the area can make a significant difference in gene transfer efficiency over that obtained in the absence of such occlusion (Hagstrom et al., 2004; Zhang et al., 2001). While saline is the most commonly used vehicle for hydrodynamic procedures (Liu et al., 1999), Ringers solution (Chabicovsky et al., 2003; Dagnaes-Hansen et al., 2002; Hanawa et al., 2004; Holst et al., 2001; Jiang et al., 2001; Slattum et al., 2003; Vorup-Jensen et al., 2001) and phosphate-buffered saline (Giladi et al., 2003; Imagawa et al., 2002; Layzer et al., 2004; McCaffrey et al., 2002a, 2003a; Miao et al., 2003; Ye et al., 2004) have also been employed. DNA dosages employed range from 0.1 to 10 mg/kg depending on the application. An optimal dose for rats is 3 mg/kg (Maruyama et al., 2002), 7 mg/kg for rabbits (Eastman et al., 2002), and 0.5–2.5 mg/kg for mice (Liu et al., 1999). Hagstrom et al. (2004) demonstrated that the administration of permeability enhancers prior to hydrodynamic injection can effectively decrease the volume required for optimal gene delivery. Interestingly, it has been shown that chloroquine, an endocytosis-blocking agent, was able to

Table 4.1. Hydrodynamics-Based Procedures for Intracellular Delivery

Target tissue	Injection site	Animal species	Needle size	Volume/injection time	Refs.
Kidney	Renal vein	Rats	24G	1.0 ml/5 s	Maruyama et al. (2002)
Liver	Tail vein	Mice (18–20 g)	27G	1.6–1.8 ml/5–7 s	Liu et al. (1999)
Liver	Tail vein	Rats (250 g)	22G	25 ml/15 s	Maruyama et al. (2002)
Muscle[a]	Iliac artery or tail vein	Mice (4–6 weeks old)	25G	2 ml/5 s	Liang et al. (2004)
Muscle[a]	GSV[b]	Rats	(?)	3 ml/15–30 s	Hagstrom et al. (2004)
Muscle[a]	GSV	Dog (9.5 kg)	20G	36 ml/18 s	Hagstrom et al. (2004)
Muscle[a]	GSV	Mouse	(?)	0.6 ml/7.5 s	Hagstrom et al. (2004)
Muscle[a]	GSV	Monkey (4.2–8.8 kg)	22G	40–100 ml/20–50 s	Hagstrom et al. (2004)

[a]Blood flow was restrained.
[b]GSV, great saphenous vein; papaverine solution was injected at the same site 2 min prior to DNA injection.

enhance and sustain exogenous gene expression when coinjected systemically with the transgene (Chen, H. Y. *et al.*, 2004).

IV. APPLICATIONS OF HYDRODYNAMIC DELIVERY

As the simplest and one of the most effective methods for *in vivo* intracellular delivery, hydrodynamic delivery has been widely employed for the delivery of DNA (plasmids, DNA fragments, bacterial artificial chromosomes) (Magin-Lachmann *et al.*, 2004), RNA (single or double stranded, synthetic or genomic RNA) (Chang *et al.*, 2001; Giladi *et al.*, 2003; Kobayashi *et al.*, 2004b; Layzer *et al.*, 2004; McCaffrey *et al.*, 2002a,b), oligonucleotides (McCaffrey *et al.*, 2003a), proteins (antibodies, enzymes) (Kobayashi *et al.*, 2001; Zhang *et al.*, 2004a), polymers (Kobayashi *et al.*, 2001), and small compounds (Kobayashi *et al.*, 2004c; Zhang *et al.*, 2004a). Compounds with or without charge, large or small, can all be delivered efficiently. Currently, the most common applications of hydrodynamic delivery are, however, in DNA and RNA delivery.

A. Applications in gene therapy studies

Although additional work is needed before hydrodynamic delivery can be used in the clinic, this method has been widely used for gene therapy studies in mice. Genes that have been evaluated for their therapeutic activities include cytokine genes (Brady *et al.*, 2004; Chen, H. W. *et al.*, 2003; Higuchi *et al.*, 2003; Hong *et al.*, 2003; Itokawa *et al.*, 2004; Jiang *et al.*, 2001, 2003; Kishida *et al.*, 2003; Kobayashi *et al.*, 2002; Lui *et al.*, 2002; Wang *et al.*, 2001, Wang, G. *et al.*, 2003; Zabala *et al.*, 2004), genes encoding growth factors (Dagnaes-Hansen *et al.*, 2002; Dai *et al.*, 2002; Sondergaard *et al.*, 2003; Yang *et al.*, 2001a,b), and those associated with diseases of genetic (Dai *et al.*, 2003; Ehrhardt *et al.*, 2003; He *et al.*, 2003, 2004; Imagawa *et al.*, 2002; Miao *et al.*, 2003, 2004; Notley *et al.*, 2002; Razzini *et al.*, 2004; Vorup-Jensen *et al.*, 2001; Wang, J. H. *et al.*, 2003; Wen *et al.*, 2004; Wu and Hui, 2004; Yasutomi *et al.*, 2003; Ye *et al.*, 2003, 2004) and viral origin (Cui *et al.*, 2003). The activity of the secretory flt3 ligand in enhancing the number of dendritic and natural killer cells in mice has been demonstrated (Fong and Hui, 2002; He *et al.*, 2000). More recent work demonstrated the effective delivery of the dystrophin gene into muscle cells using hydrodynamic delivery principles (Liang *et al.*, 2004; Zhang *et al.*, 2004b). In practice, the therapeutic activity of any candidate gene, especially those encoding secretory proteins, can be studied in mice by application of hydrodynamic delivery methodology.

B. Applications in gene function analysis

While the transfer of genes for therapeutic purposes is an important application of gene therapy, an additional important application of hydrodynamic delivery is in the *in vivo* transfection of cells for gene function studies (Chabicovsky *et al.*, 2003; Holm *et al.*, 2003; Holst *et al.*, 2001; Rossmanith *et al.*, 2002). Similar to the procedures for transfecting cells in culture, hydrodynamic delivery could be viewed as a method of transfection in whole animals. Thus, hydrodynamic delivery can be used to define the physiological function of a gene in the context of a whole organism. Using the strategy of gain of function or loss of function, one can introduce a transgene in the form of plasmid or bacteria artificial chromosome (BAC) into an animal and evaluate the biochemical, cellular, or physiological changes that are manifested as the consequence of expression of the introduced transgene. Similarly, a siRNA specific to an mRNA of a gene can be injected hydrodynamically into animals to reduce the gene product level in order to establish the relationship between the downregulation of the target gene and the biochemical or physiological changes that occur in the animal. To sustain the physiological effects of up- or downregulation of selected gene expression, one can repeat the procedure of hydrodynamic injection on the same animals. Repeat application is a unique advantage of hydrodynamic delivery compared to viral or non-viral vectors.

With the completion of human and other genome projects, determining the function of genomic sequences becomes one of the focal points in biomedical research. Toward this end, hydrodynamic delivery provides a powerful means to allow for the study of the genomic sequences in animals through *in vivo* transfection. Because BAC is the primary technology employed for mapping the entire human genome, most genes can be recovered intact as BAC clones from BAC libraries. Thus, delivery of BACs carrying complete gene sequences in their genomic setting presents the possibility of gene expression under the control of its genomic arrangement in sequence. Such expression is not achieved easily with current viral or synthetic vectors due to the size limitation imposed in their packaging or to their low delivery efficiency. Magin-Lachmann *et al.* (2004) demonstrated an effective delivery of a BAC (157 kbp) in mice using hydrodynamic delivery.

C. Applications in the study of regulatory DNA sequences

With well-established molecular biology techniques, cloning of a DNA fragment into a reporter-containing plasmid can be accomplished easily. Application of hydrodynamic delivery provides a convenient means to study DNA sequences with potential functions in regulating gene expression in mice. For example, the activity of a promoter, enhancer, suppressor, insulator, intron, or 5′- and 3′-flanking sequences in general can be evaluated in whole animals by hydrodynamic delivery of reporter plasmids with the test sequence inserted at the appropriate site (Alino et al., 2003; Chen et al., 2001, Chen, Z. Y. et al., 2003, 2004; Gehrke et al., 2003; Kameda et al., 2003; Kramer et al., 2003; Miao et al., 2001; Xu et al., 2001; Yant and Kay, 2003; Yew et al., 2001, 2002; Zhang et al., 2000). By hydrodynamic delivery of a plasmid construct containing a reporter gene in the presence of regulatory elements, one can also assess the activities of the regulatory sequence in maintaining persistence of transgene expression. For example, the function of the apoE locus control region, various promoters, intron sequences, viral sequences such as Epstein–Barr virus nuclear antigen gene, and the *oriP* sequence in driving persistent transgene expression have been evaluated using hydrodynamic delivery in mice (Cui et al., 2001; Sclimenti et al., 2003; Stoll et al., 2001). In addition, hydrodynamic delivery has also been used for studies involving integrase (Olivares et al., 2002) and transposon-based recombination (Montini et al., 2002; Nakai et al., 2003; Yant et al., 2000, 2002). Practically any transfection that can be accomplished in cell culture can now be realized in animals.

D. Applications in establishing animal models

Viruses or other infectious agents causing human diseases cannot be propagated in animals because of their species specificity, which is partially due to the lack of human receptors on animal cells that are capable of facilitating binding of the virus and injection of the viral genome into the target cells. Hydrodynamic delivery of the viral genome directly into cells in whole animals without involving the intervention of any viral proteins provides a powerful means for establishing animal models to study human viral diseases. Two examples of such applications include establishment of an animal model of human hepatitis B (Chang et al., 2003; McCaffrey et al., 2003b; Suzuki et al., 2003; Yang et al., 2002) and delta (Bordier et al., 2003; Chang et al., 2001) virus infection. Similar applications for other human diseases are anticipated in the near future.

Hydrodynamic delivery has also been employed for the establishment of a rapid real-time assay that measures the induced transcription of luciferase genes under the control of the 5′-flanking region of CYP genes in transfected mouse hepatocytes *in vivo* (Schuetz et al., 2002; Tirona et al., 2003; Zhang et al., 2003). By means of quantitative whole body imaging of bioluminescence generated by luciferase in transfected animals, Schuetz et al. (2002) and Zhang et al. (2003) were able to measure the effects on transcription as it occurred when hydrodynamically transfected mice were administered pharmaceuticals capable of inducing CYP gene expression. Not only does this system provide biological conditions for the most informative studies of gene regulation, but it also substantially reduces the cost and number of animals required for gene expression analysis.

E. Applications in protein production

Taking advantage of the high capacity of the liver for protein synthesis, hydro-dynamic delivery of gene expression cassettes into hepatocytes can turn a mouse liver into a protein production site. For certain genes, their gene products are retained within the target cells and protein purification will have to begin with a liver homogenate. For genes encoding secretory proteins, the gene products are secreted into the bloodstream and can be isolated directly from the blood. Plasmid constructs with a poly-His Tag sequence attached to the end of the transgene can be used for efficient and convenient purification employing an affinity column. An obvious advantage of protein production in animals as compared to bacterial systems is that proteins made in animals are properly folded and glycosylated.

F. Hydrodynamic delivery as a general means for intracellular delivery *in vivo*

Due to its physical and nonspecific nature, hydrodynamic delivery can be used for intracellular delivery of any membrane-impermeable molecules. Recent work has demonstrated the successful delivery of DNA, RNA, proteins, small com-pounds, fluorescence-labeled beads (Kobayashi *et al.*, 2004a), and even Lenti viral particles (Condiotti *et al.*, 2004). In principle, the only limitation to the use of hydrodynamic delivery is the toxicity of the substance injected. Because a large injection volume is required, the amount of compound to be used is, of necessity, usually large. The estimated amount of compound delivered to the liver cells following the hydrodynamics-based procedure is less than 5%, leaving more than 95% of the injected dose staying in circulation before being removed or degraded.

V. CONCLUSION AND FUTURE PERSPECTIVES

Similar to many physical methods such as microinjection, particle bombard-ment, and electroporation, hydrodynamic delivery employs a physical force to permeabilize plasma membranes and allow substances to reach the cytoplasm. Different from these device-based methods, hydrodynamic delivery does not require sophisticated equipment. Other than a needle and syringe, saline and the test substance are the only additional materials needed. Since the establish-ment of the hydrodynamics-based procedure in 1999, many laboratories have adapted this method for their routine use in studies requiring intracellular delivery of membrane-impermeable compounds. Compared to viral and synthet-ic vectors, hydrodynamic delivery offers great convenience for the *in vivo* intracellular delivery of DNA, RNA, protein, or other membrane-impermeable compounds.

Despite the fact that the physical nature of intrahepatic gene delivery following hydrodynamic injection has been firmly established, gene transfer into cells in other organs has not been fully explored. In principle, successful application of hydrodynamic delivery is dependent on two factors: the structure of endothelium and surrounding parenchymal cells. Although a vein or artery often serves as the site for hydrodynamic injection due to their larger diameter, intracellular delivery actually takes place primarily in the capillaries. Two types of capillaries, continuous and fenestrated, exist. In a continuous capillary, the endothelium is a complete lining that permits the diffusion of water, small solutes, and lipid-soluble materials into the surrounding interstitial fluid, but prevents direct access of large molecules. In contrast, fenestrated capillaries contain pores that span the endothelial lining and permit the rapid exchange of large molecular weight blood components, such as serum proteins, between blood and interstitial fluids. Evidence has been presented to support that hydrodynamic delivery is applicable to either continuous or fenestrated capillaries (Hagstrom et al., 2004; Liu et al., 1999). In general, delivery of DNA or other substances to cells through continuous capillaries requires higher hydrodynamic pressure as compared to capillaries with fenestrae. Other than hydrodynamic injection via tail vein in mouse, systematic studies are needed, however, to establish experimental conditions fine-tuned for continuous and fenestrated capillaries, respectively, for new applications. For localized delivery, occlusion of blood flow or transient clamping of relevant blood vessels helps in reducing the injection volume required, and therefore the invasiveness of the procedure. It is foreseeable in the near future that optimized procedures will be established for hydrodynamic delivery to various tissues.

The high efficiency and simplicity of hydrodynamic delivery have raised interest in the gene therapy community toward the potential application of hydrodynamic delivery to human subjects. The current focus has centered on the development of a modified procedure involving a significant reduction in the volume of the injected gene-containing solution. Although it remains a research tool in small animals, largely mice, some encouraging results have been reported in the literature demonstrating hydrodynamic delivery of reporter genes in large animals (Hagstrom et al., 2004; Zhang et al., 2001). For example, Hagstrom et al. (2004) showed that highly efficient gene transfer into muscle tissue in dogs and primates can be achieved by a procedure involving injection of plasmid DNA or siRNA into a distal vein of a limb that is transiently isolated by a tourniquet or blood pressure cuff (Hagstrom et al., 2004). Similarly, Eastman and colleagues (2002) had devised several methods for gene delivery to the isolated rabbit liver using minimally invasive catheter-based techniques. Employing balloon catheters, they demonstrated in rabbits that the volume and rates of injection are within acceptable limits for rapid bolus delivery in humans. Unfortunately, a lack of appropriate catheters to satisfy the needs of the injection has limited further development. The bottleneck along this direction appears to be the lack of catheters specifically engineered to reduce the leakage around the occlusion balloons. Once these technical issues are resolved, hydrodynamic delivery can be further developed to become a useful clinical procedure for the amelioration of disease.

References

Alino, S. F., Crespo, A., and Dasi, F. (2003). Long-term therapeutic levels of human alpha-1 antitrypsin in plasma after hydrodynamic injection of non-viral DNA. *Gene Ther.* **10,** 1672–1679.

Bordier, B. B., Ohkanda, J., Liu, P., Lee, S. Y., Salazar, F. H., Marion, P. L., Ohashi, K., Meuse, L., Kay, M. A., Casey, J. L., Sebti, S. M., Hamilton, A. D., and Glenn, J. S. (2003). In vivo antiviral efficacy of prenylation inhibitors against hepatitis delta virus. *J. Clin. Invest.* **112,** 407–414.

Brady, J., Hayakawa, Y., Smyth, M. J., and Nutt, S. L. (2004). IL-21 induces the functional maturation of murine NK cells. *J. Immunol.* **172,** 2048–2058.

Budker, V., Budker, T., Zhang, G. F., Subbotin, V., Loomis, A., and Wolff, J. A. (2000). Hypothesis: Naked plasmid DNA is taken up by cells in vivo by a receptor-mediated process. *J. Gene Med.* **2,** 76–88.

Chabicovsky, M., Herkner, K., and Rossmanith, W. (2003). Overexpression of activin beta(C) or activin beta(E) in the mouse liver inhibits regenerative deoxyribonucleic acid synthesis of hepatic cells. *Endocrinology* **144,** 3497–3504.

Chang, J. H., Sigal, L. J., Lerro, A., and Taylor, J. (2001). Replication of the human hepatitis delta virus genome is initiated in mouse hepatocytes following intravenous injection of naked DNA or RNA sequences. *J. Virol.* **75,** 3469–3473.

Chang, W. W., Su, I. J., Lai, M. D., Chang, W. T., Huang, W. Y., and Lei, H. Y. (2003). The role of inducible nitric oxide synthase in a murine acute hepatitis B virus (HBV) infection model induced by hydrodynamics-based in vivo transfection of HBV-DNA. *J. Hepatol.* **39,** 834–842.

Chen, H. W., Lee, Y. P., Chung, Y. F., Shih, Y. C., Tsai, J. P., Tao, M. H., and Ting, C. C. (2003). Inducing long-term survival with lasting anti-tumor immunity in treating B cell lymphoma by a combined dendritic cell-based and hydrodynamic plasmid-encoding IL-12 gene therapy. *Int. Immunol.* **15,** 427–435.

Chen, H. Y., Zhu, H. Z., Lu, B., Xu, X., Yao, J. H., Shen, Q., and Xue, J. L. (2004). Enhancement of naked FIX minigene expression by chloroquine in mice. *Acta Pharmacol. Sin.* **25,** 570–575.

Chen, Z. Y., He, C. Y., Ehrhardt, A., and Kay, M. A. (2003). Minicircle DNA vectors devoid of bacterial DNA result in persistent and high-level transgene expression in vivo. *Mol. Ther.* **8,** 495–500.

Chen, Z. Y., He, C. Y., Meuse, L., and Kay, M. A. (2004). Silencing of episomal transgene expression by plasmid bacterial DNA elements in vivo. *Gene Ther.* **11,** 856–864.

Chen, Z. Y., Yant, S. R., He, C. Y., Meuse, L., Shen, S., and Kay, M. A. (2001). Linear DNAs concatemerize in vivo and result in sustained transgene expression in mouse liver. *Mol. Ther.* **3,** 403–410.

Condiotti, R., Curran, M. A., Nolan, G. P., Giladi, H., Ketzinel-Gilad, M., Gross, E., and Galun, E. (2004). Prolonged liver-specific transgene expression by a non-primate lentiviral vector. *Biochem. Biophys. Res. Commun.* **320,** 998–1006.

Cui, F. D., Asada, H., Kishida, T., Itokawa, Y., Nakaya, T., Ueda, Y., Yamagishi, H., Gojo, S., Kita, M., Imanishi, J., and Mazda, O. (2003). Intravascular naked DNA vaccine encoding glycoprotein

B induces protective humoral and cellular immunity against herpes simplex virus type 1 infection in mice. *Gene Ther.* **10,** 2059–2066.

Cui, F. D., Kishida, T., Ohashi, S., Asada, H., Yasutomi, K., Satoh, E., Kubo, T., Fushiki, S., Imanishi, J., and Mazda, O. (2001). Highly efficient gene transfer into murine liver achieved by intravenous administration of naked Epstein-Barr virus (EBV)-based plasmid vectors. *Gene Ther.* **8,** 1508–1513.

Dagnaes-Hansen, F., Holst, H. U., Sondergaard, M., Vorup-Jensen, T., Flyvbjerg, A., Jensen, U. B., and Jensen, T. G. (2002). Physiological effects of human growth hormone produced after hydrodynamic gene transfer of a plasmid vector containing the human ubiquitin promotor. *J. Mol. Med.* **80,** 665–670.

Dai, C. S., Li, Y. J., Yang, J. W., and Liu, Y. H. (2003). Hepatocyte growth factor preserves beta cell mass and mitigates hyperglycemia in streptozotocin-induced diabetic mice. *J. Biol. Chem.* **278,** 27080–27087.

Dai, C. S., Yang, J. W., and Liu, Y. H. (2002). Single injection of naked plasmid encoding hepatocyte growth factor prevents cell death and ameliorates acute renal failure in mice. *J. Am. Soc. Nephrol.* **13,** 411–422.

Eastman, S. J., Baskin, K. M., Hodges, B. L., Chu, Q. M., Gates, A., Dreusicke, R., Anderson, S., and Scheule, R. K. (2002). Development of catheter-based procedures for transducing the isolated rabbit liver with plasmid DNA. *Hum. Gene Ther.* **13,** 2065–2077.

Ehrhardt, A., Peng, P. D., Xu, H., Meuse, L., and Kay, M. A. (2003). Optimization of cis-acting elements for gene expression from non-viral vectors *in vivo. Hum. Gene Ther.* **14,** 215–225.

Fong, C. L., and Hui, K. M. (2002). Generation of potent and specific cellular immune responses via *in vivo* stimulation of dendritic cells by pNGVL3-hFLex plasmid DNA and immunogenic peptides. *Gene Ther.* **9,** 1127–1138.

Gehrke, S., Jerome, V., and Muller, R. (2003). Chimeric transcriptional control units for improved liver-specific transgene expression. *Gene* **322,** 137–143.

Giladi, H., Ketzinel-Gilad, M., Rivkin, L., Felig, Y., Nussbaum, O., and Galun, E. (2003). Small interfering RNA inhibits hepatitis B virus replication in mice. *Mol. Ther.* **8,** 769–776.

Hagstrom, J. E. (2003). Plasmid-based gene delivery to target tissues *in vivo*: The intravascular approach. *Curr. Opin. Mol. Ther.* **5,** 338–344.

Hagstrom, J. E., Hegge, J., Zhang, G., Noble, M., Budker, V., Lewis, D. L., Herweijer, H., and Wolff, J. A. (2004). A facile non-viral method for delivering genes and siRNAs to skeletal muscle of mammalian limbs. *Mol. Ther.* **10,** 386–398.

Hanawa, H., Watanabe, R., Hayashi, M., Yoshida, T., Abe, S., Komura, S., Liu, H., Elnaggar, R., Chang, H., Okura, Y., Kato, K., Kodama, M., Maruyama, H., Miyazaki, J., and Aizawa, Y. (2004). A novel method to assay proteins in blood plasma after intravenous injection of plasmid DNA. *Tohoku J. Exp. Med.* **202,** 155–161.

He, C. X., Feng, D. M., Wu, W. J., Ding, Y. F., Chen, L., Chen, H. M., Yao, J. H., Shen, Q., Lu, D. R., and Xue, J. L. (2003). Efficient expression of human factor IX cDNA in liver mediated by hydrodynamics-based plasmid administration. *Chin. Sci. Bull.* **48,** 790–795.

He, C. X., Shi, D., Wu, W. J., Ding, Y. F., Feng, D. M., Lu, B., Chen, H. M., Yao, J. H., Shen, Q., Lu, D. R., and Xue, J. L. (2004). Insulin expression in livers of diabetic mice mediated by hydrodynamics-based administration. *World J. Gastroenterol.* **10,** 567–572.

He, Y. K., Pimenov, A. A., Nayak, J. V., Plowey, J., Falo, L. D., and Huang, L. (2000). Intravenous injection of naked DNA encoding secreted flt3 ligand dramatically increases the number of dendritic cells and natural killer cells *in vivo. Hum. Gene Ther.* **11,** 547–554.

Higuchi, N., Maruyama, H., Kuroda, T., Kameda, S., Iino, N., Kawachi, H., Nishikawa, Y., Hanawa, H., Tahara, H., Miyazaki, J., and Gejyo, F. (2003). Hydrodynamics-based delivery of the viral interleukin-10 gene suppresses experimental crescentic glomerulonephritis in Wistar-Kyoto rats. *Gene Ther.* **10,** 1297–1310.

Hodges, B. L., and Scheule, R. K. (2003). Hydrodynamic delivery of DNA. *Expert Opin. Biol. Ther.* **3**, 911–918.

Holm, D. A., Dagnaes-Hansen, F., Simonsen, H., Gregersen, N., Bolund, L., Jensen, T. G., and Corydon, T. J. (2003). Expression of short-chain acyl-COA dehydrogenase (SCAD) proteins in the liver of SCAD deficient mice after hydrodynamic gene transfer. *Mol. Genet. Metab.* **78**, 250–258.

Holst, H. U., Dagnaes-Hansen, F., Corydon, T. J., Andreasen, P. H., Jorgensen, M. M., Kolvraa, S., Bolund, L., and Jensen, T. G. (2001). LDL receptor-GFP fusion proteins: New tools for the characterisation of disease-causing mutations in the LDL receptor gene. *Eur. J. Hum. Genet.* **9**, 815–822.

Hong, I. C., Mullen, P. M., Precht, A. F., Khanna, A., Li, M., Behling, C., Lopez, V. F., Chiou, H. C., Moss, R. B., and Hart, M. E. (2003). Non-viral human IL-10 gene expression reduces acute rejection in heterotopic auxiliary liver transplantation in rats. *Microsurgery* **23**, 432–436.

Imagawa, T., Watanabe, S., Katakura, S., Boivin, G. P., and Hirsch, R. (2002). Gene transfer of a fibronectin peptide inhibits leukocyte recruitment and suppresses inflammation in mouse collagen-induced arthritis. *Arthritis Rheum.* **46**, 1102–1108.

Itokawa, Y., Mazda, O., Ueda, Y., Kishida, T., Asada, H., Cui, F. D., Fuji, N., Fujiwara, H., Shin-Ya, M., Yasutomi, K., Imanishi, J., and Yamagishi, H. (2004). Interleukin-12 genetic administration suppressed metastatic liver tumor unsusceptible to CTL. *Biochem. Biophys. Res. Commun.* **314**, 1072–1079.

Jiang, J. J., Yamato, E., and Miyazaki, J. (2001). Intravenous delivery of naked plasmid DNA for *in vivo* cytokine expression. *Biochem. Biophys. Res. Commun.* **289**, 1088–1092.

Jiang, J. J., Yamato, E., and Miyazaki, J. (2003). Long-term control of food intake and body weight by hydrodynamics-based delivery of plasmid DNA encoding leptin or CNTF. *J. Gene Med.* **5**, 977–983.

Kameda, S., Maruyama, H., Nakamura, G., Iino, N., Nishikawa, Y., Miyazaki, J., and Gejyo, F. (2003). Hydrodynamics-based transfer of PCR-amplified DNA fragments into rat liver. *Biochem. Biophys. Res. Commun.* **309**, 929–936.

Kishida, T., Asada, H., Itokawa, Y., Cui, F. D., Shin-Ya, M., Gojo, S., Yasutomi, K., Ueda, Y., Yamagishi, H., Imanishi, J., and Mazda, O. (2003). Interleukin (IL)-21 and IL-15 genetic transfer synergistically augments therapeutic antitumor immunity and promotes regression of metastatic lymphoma. *Mol. Ther.* **8**, 552–558.

Kobayashi, N., Hirata, K., Chen, S., Kawase, A., Nishikawa, M., and Takakura, Y. (2004a). Hepatic delivery of particulates in the submicron range by a hydrodynamics-based procedure: Implications for particulate gene delivery systems. *J. Gene Med.* **6**, 455–463.

Kobayashi, N., Kuramoto, T., Chen, S., Watanabe, Y., and Takakura, Y. (2002). Therapeutic effect of intravenous interferon gene delivery with naked plasmid DNA in murine metastasis models. *Mol. Ther.* **6**, 737–744.

Kobayashi, N., Kuramoto, T., Yamaoka, K., Hashida, M., and Takakura, Y. (2001). Hepatic uptake and gene expression mechanisms following intravenous administration of plasmid DNA by conventional and hydrodynamics-based procedures. *J. Pharmacol. Exp. Ther.* **297**, 853–860.

Kobayashi, N., Matsui, Y., Kawase, A., Hirata, K., Miyagishi, M., Taira, K., Nishikawa, M., and Takakura, Y. (2004b). Vector-based *in vivo* RNA interference: Dose- and time-dependent suppression of transgene expression. *J. Pharmacol. Exp. Ther.* **308**, 688–693.

Kobayashi, N., Nishikawa, M., Hirata, K., and Takakura, Y. (2004c). Hyrodynamics-based procedure involves transient hyperpermeability in the hepatic cellular membrane: Implication of a nonspecific process in efficient intracellular gene delivery. *J. Gene Med.* **6**, 584–592.

Kramer, M. G., Barajas, M., Razquin, N., Berraondo, P., Rodrigo, M., Wu, C., Qian, C., Fortes, P., and Prieto, J. (2003). *In vitro* and *in vivo* comparative study of chimeric liver-specific promoters. *Mol. Ther.* **7**, 375–385.

Layzer, J. M., McCaffrey, A. P., Tanner, A. K., Huang, Z., Kay, M. A., and Sullenger, B. A. (2004). *In vivo* activity of nuclease-resistant siRNAs. *RNA* **10,** 766–771.

Lecocq, M., Andrianaivo, F., Warnier, M. T., Wattiaux-De Coninck, S., Wattiaux, R., and Jadot, M. (2003). Uptake by mouse liver and intracellular fate of plasmid DNA after a rapid tail vein injection of a small or a large volume. *J. Gene Med.* **5,** 142–156.

Lewis, D. L., Hagstrom, J. E., Loomis, A. G., Wolff, J. A., and Herweijer, H. (2002). Efficient delivery of siRNA for inhibition of gene expression in postnatal mice. *Nature Genet.* **32,** 107–108.

Liang, K. W., Nishikawa, M., Liu, F., Sun, B., Ye, Q., and Huang, L. (2004). Restoration of dystrophin expression in mdx mice by intravascular injection of naked DNA containing full-length dystrophin cDNA. *Gene Ther.* **11,** 901–908.

Liu, D. X., and Knapp, J. E. (2001). Hydrodynamics-based gene delivery. *Curr. Opin. Mol. Ther.* **3,** 192–197.

Liu, F., Song, Y., and Liu, D. (1999). Hydrodynamics-based transfection in animals by systemic administration of plasmid DNA. *Gene Ther.* **6,** 1258–1266.

Lui, V. W. Y., He, Y. K., Falo, L., and Huang, L. (2002). Systemic administration of naked DNA encoding interleukin 12 for the treatment of human papillomavirus DNA-positive tumor. *Hum. Gene Ther.* **13,** 177–185.

Magin-Lachmann, C., Kotzamanis, G., D'Aiuto, L., Cooke, H., Huxley, C., and Wagner, E. (2004). *In vitro* and *in vivo* delivery of intact BAC DNA: Comparison of different methods. *J. Gene Med.* **6,** 195–209.

Maruyama, H., Higuchi, N., Nishikawa, Y., Kameda, S., Iino, N., Kazama, J. J., Takahashi, N., Sugawa, M., Hanawa, H., Tada, N., Miyazaki, J., and Gejyo, F. (2002). High-level expression of naked DNA delivered to rat liver via tail vein injection. *J. Gene Med.* **4,** 333–341.

McCaffrey, A. P., Hashi, K., Meuse, L., Shen, S. L., Lancaster, A. M., Lukavsky, P. J., Sarnow, P., and Kay, M. A. (2002a). Determinants of hepatitis C translational initiation *in vitro*, in cultured cells and mice. *Mol. Ther.* **5,** 676–684.

McCaffrey, A. P., Meuse, L., Karimi, M., Contag, C. H., and Kay, M. A. (2003a). A potent and specific morpholino antisense inhibitor of hepatitis C translation in mice. *Hepatology* **38,** 503–508.

McCaffrey, A. P., Meuse, L., Pham, T. T. T., Conklin, D. S., Hannon, G. J., and Kay, M. A. (2002b). Gene expression: RNA interference in adult mice. *Nature* **418,** 38–39.

McCaffrey, A. P., Nakai, H., Pandey, K., Huang, Z., Salazar, F. H., Xu, H., Wieland, S. F., Marion, P. L., and Kay, M. A. (2003b). Inhibition of hepatitis B virus in mice by RNA interference. *Nature Biotechnol.* **21,** 639–644.

Miao, C. H., Thompson, A. R., Loeb, K., and Ye, X. (2001). Long-term and therapeutic-level hepatic gene expression of human factor IX after naked plasmid transfer *in vivo*. *Mol. Ther.* **3,** 947–957.

Miao, C. H., Ye, X., and Thompson, A. R. (2003). High-level factor VIII gene expression *in vivo* achieved by non-viral liver-specific gene therapy vectors. *Hum. Gene Ther.* **14,** 1297–1305.

Miao, H. Z., Sirachainan, N., Palmer, L., Kucab, P., Cunningham, M. A., Kaufman, R. J., and Pipe, S. W. (2004). Bioengineering of coagulation factor VIR for improved secretion. *Blood* **103,** 3412–3419.

Montini, E., Held, P. K., Noll, M., Morcinek, N., Al-Dhalimy, M., Finegold, M., Yant, S. R., Kay, M. A., and Grompe, M. (2002). *In vivo* correction of murine tyrosinemia type I by DNA-mediated transposition. *Mol. Ther.* **6,** 759–769.

Nakai, H., Montini, E., Fuess, S., Storm, T. A., Meuse, L., Finegold, M., Grompe, M., and Kay, M. A. (2003). Helper-independent and AAV-ITR-independent chromosomal integration of double-stranded linear DNA vectors in mice. *Mol. Ther.* **7,** 101–111.

Notley, C., Killoran, A., Cameron, C., Wynd, K., Hough, C., and Lillicrap, D. (2002). The canine factor VIII 3′-untranslated region and a concatemeric hepatocyte nuclear factor 1 regulatory element enhance factor VIII transgene expression *in vivo*. *Hum. Gene Ther.* **13,** 1583–1593.

Olivares, E. C., Hollis, R. P., Chalberg, T. W., Meuse, L., Kay, M. A., and Calos, M. P. (2002). Site-specific genomic integration produces therapeutic Factor IX levels in mice. *Nature Biotechnol.* **20,** 1124–1128.

Razzini, G., Parise, F., Calebiro, D., Battini, R., Bagni, B., Corazzari, T., Tarugi, P., Angelelli, C., Molinari, S., Falqui, L., and Ferrari, S. (2004). Low-density lipoprotein (LDL) receptor/transferrin fusion protein: *In vivo* production and functional evaluation as a potential therapeutic tool for lowering plasma LDL cholesterol. *Hum. Gene Ther.* **15,** 533–541.

Rossmanith, W., Chabicovsky, M., Herkner, K., and Schulte-Hermann, R. (2002). Cellular gene dose and kinetics of gene expression in mouse livers transfected by high-volume tail-vein injection of naked DNA. *DNA Cell Biol.* **21,** 847–853.

Schuetz, E., Lan, L. B., Yasuda, K., Kim, R., Kocarek, T. A., Schuetz, J., and Strom, S. (2002). Development of a real-time *in vivo* transcription assay: Application reveals pregnane X receptor-mediated induction of CYP3A4 by cancer chemotherapeutic agents. *Mol. Pharmacol.* **62,** 439–445.

Sclimenti, C. R., Neviaser, A. S., Baba, E. J., Meuse, L., Kay, M. A., and Calos, M. P. (2003). Epstein-Barr virus vectors provide prolonged robust factor IX expression in mice. *Biotechnol. Prog.* **19,** 144–151.

Slattum, P. S., Loomis, A. G., Machnik, K. J., Watt, M. A., Duzeski, J. L., Budker, V. G., Wolff, J. A., and Hagstrom, J. E. (2003). Efficient *in vitro* and *in vivo* expression of covalently modified plasmid DNA. *Mol. Ther.* **8,** 255–263.

Sondergaard, M., Dagnaes-Hansen, F., Flyvbjerg, A., and Jensen, T. G. (2003). Normalization of growth in hypophysectomized mice using hydrodynamic transfer of the human growth hormone gene. *Am. J. Physiol. Endocrinol. Metab.* **285,** E427–E432.

Song, E., Lee, S. K., Wang, J., Ince, N., Ouyang, N., Min, J., Chen, J., Shankar, P., and Lieberman, J. (2003). RNA interference targeting Fas protects mice from fulminant hepatitis. *Nature Med.* **9,** 347–351.

Stoll, S. M., Sclimenti, C. R., Baba, E. J., Meuse, L., Kay, M. A., and Calos, M. P. (2001). Epstein-Barr virus/human vector provides high-level, long-term expression of alpha(1)-antitrypsin in mice. *Mol. Ther.* **4,** 122–129.

Suzuki, T., Takehara, T., Ohkawa, K., Ishida, H., Jinushi, M., Miyagi, T., Sasaki, Y., and Hayashi, N. (2003). Intravenous injection of naked plasmid DNA encoding hepatitis B virus (HBV) produces HBV and induces humoral immune response in mice. *Biochem. Biophys. Res. Commun.* **300,** 784–788.

Tirona, R. G., Lee, W., Leake, B. F., Lan, L. B., Cline, C. B., Lamba, V., Parviz, F., Duncan, S. A., Inoue, Y., Gonzalez, F. J., Schuetz, E. G., and Kim, R. B. (2003). The orphan nuclear receptor HNF4alpha determines PXR- and CAR-mediated xenobiotic induction of CYP3A4. *Nature Med.* **9,** 220–224.

Vorup-Jensen, T., Jensen, U. B., Liu, H., Kawasaki, T., Uemura, K., Thiel, S., Dagnaes-Hansen, F., and Jensen, T. G. (2001). Tail-vein injection of mannan-binding lectin DNA leads to high expression levels of multimeric protein in liver. *Mol. Ther.* **3,** 867–874.

Wang, G., Tschoi, M., Spolski, R., Lou, Y. Y., Ozaki, K., Feng, C. G., Kim, G., Leonard, W. J., and Hwu, P. (2003). *In vivo* antitumor activity of interleukin 21 mediated by natural killer cells. *Cancer Res.* **63,** 9016–9022.

Wang, J. H., Yao, M. Z., Zhang, Z. L., Gu, J. F., Zhang, Y. H., Li, B. H., Sun, L. Y., and Liu, X. Y. (2003). Enhanced suicide gene therapy by chimeric tumor-specific promoter based on HSF1 transcriptional regulation. *FEBS Lett.* **546,** 315–320.

Wang, Z., Qiu, S. J., Ye, S. L., Tang, Z. Y., and Xiao, X. (2001). Combined IL-12 and GM-CSF gene therapy for murine hepatocellular carcinoma. *Cancer Gene Ther.* **8,** 751–758.

Wen, J. H., Matsumoto, K., Taniura, N., Tomioka, D., and Nakamura, T. (2004). Hepatic gene expression of NK4, an HGF-antagonist/angiogenesis inhibitor, suppresses liver metastasis and invasive growth of colon cancer in mice. *Cancer Gene Ther.* **11,** 419–430.

Wu, X. F., and Hui, K. M. (2004). Induction of potent TRAIL-mediated tumoricidal activity by hFLEX/furin/TRAIL recombinant DNA construct. *Mol. Ther.* **9,** 674–681.

Xu, Z. L., Mizuguchi, H., Ishii-Watabe, A., Uchida, E., Mayumi, T., and Hayakawa, T. (2001). Optimization of transcriptional regulatory elements for constructing plasmid vectors. *Gene* **272,** 149–156.

Yang, J., Dai, C., and Liu, Y. (2001a). Systemic administration of naked plasmid encoding hepato-cyte growth factor ameliorates chronic renal fibrosis in mice. *Gene Ther.* **8,** 1470–1479.

Yang, J. W., Chen, S. P., Huang, L., Michalopoulos, G. K., and Liu, Y. H. (2001b). Sustained expression of naked plasmid DNA encoding hepatocyte growth factor in mice promotes liver and overall body growth. *Hepatology* **33,** 848–859.

Yang, P. L., Althage, A., Chung, J., and Chisari, F. V. (2002). Hydrodynamic injection of viral DNA: A mouse model of acute hepatitis B virus infection. *Proc. Natl. Acad. Sci. USA* **99,** 13825–13830.

Yant, S. R., Ehrhardt, A., Mikkelsen, J. G., Meuse, L., Pham, T., and Kay, M. A. (2002). Transposition from a gutless adeno-transposon vector stabilizes transgene expression *in vivo*. *Nature Biotechnol.* **20,** 999–1005.

Yant, S. R., and Kay, M. A. (2003). Nonhomologous-end-joining factors regulate DNA repair fidelity during Sleeping Beauty element transposition in mammalian cells. *Mol. Cell. Biol.* **23,** 8505–8518.

Yant, S. R., Meuse, L., Chiu, W., Ivics, Z., Izsvak, Z., and Kay, M. A. (2000). Somatic integration and long-term transgene expression in normal and haemophilic mice using a DNA transposon system. *Nature Genet.* **25,** 35–41.

Yasutomi, K., Itokawa, Y., Asada, H., Kishida, T., Cui, F. D., Ohashi, S., Gojo, S., Ueda, Y., Kubo, T., Yamagishi, H., Imanishi, J., Takeuchi, T., and Mazda, O. (2003). Intravascular insulin gene delivery as potential therapeutic intervention in diabetes mellitus. *Biochem. Biophys. Res. Commun.* **310,** 897–903.

Ye, P. Q., Thompson, A. R., Sarkar, R., Shen, Z. P., Lillicrap, D. P., Kaufman, R. J., Ochs, H. D., Rawlings, D. J., and Miao, C. H. (2004). Naked DNA transfer of factor VIII induced transgene-specific, species-independent immune response in hemophilia A mice. *Mol. Ther.* **10,** 117–126.

Ye, X., Loeb, K. R., Stafford, D. W., Thompson, A. R., and Miao, C. H. (2003). Complete and sustained phenotypic correction of hemophilia B in mice following hepatic gene transfer of a high-expressing human factor IX plasmid. *J. Thromb. Haemost.* **1,** 103–111.

Yew, N. S., Przybylska, M., Ziegler, R. J., Liu, D. P., and Cheng, S. H. (2001). High and sustained transgene expression *in vivo* from plasmid vectors containing a hybrid ubiquitin promoter. *Mol. Ther.* **4,** 75–82.

Yew, N. S., Zhao, H. M., Przybylska, M., Wu, I. H., Tousignant, J. D., Scheule, R. K., and Cheng, S. H. (2002). CpG-depleted plasmid DNA vectors with enhanced safety and long-term gene expression *in vivo*. *Mol. Ther.* **5,** 731–738.

Zabala, M., Wang, L., Hernandez-Alcoceba, R., Hillen, W. G., Cheng, Q. A., Prieto, J., and Kramer, M. G. (2004). Optimization of the Tet-on system to regulate interleukin 12 expression in the liver for the treatment of hepatic tumors. *Cancer Res.* **64,** 2799–2804.

Zender, L., Hutker, S., Liedtke, C., Tillmann, H. L., Zender, S., Mundt, B., Waltemathe, M., Gosling, T., Flemming, P., Malek, N. P., Trautwein, C., Manns, M. P., Kuhnel, F., and Kubicka, S. (2003). Caspase 8 small interfering RNA prevents acute liver failure in mice. *Proc. Natl. Acad. Sci. USA* **100,** 7797–7802.

Zhang, G., Budker, V., and Wolff, J. A. (1999). High levels of foreign gene expression in hepatocytes after tail vein injections of naked plasmid DNA. *Hum. Gene Ther.* **10,** 1735–1737.

Zhang, G., Gao, X., Song, Y. K., Vollmer, R., Stolz, D. B., Gasiorowski, J. Z., Dean, D. A., and Liu, D. (2004a). Hydroporation as the mechanism of hydrodynamic delivery. *Gene Ther.* **11,** 675–682.

Zhang, G., Ludtke, J. J., Thioudellet, C., Kleinpeter, P., Antoniou, M., Herweijer, H., Braun, S., and Wolff, J. A. (2004b). Intraarterial delivery of naked plasmid DNA expressing full-length mouse dystrophin in the mdx mouse model of duchenne muscular dystrophy. *Hum. Gene Ther.* **15,** 770–782.

Zhang, G., Song, Y. K., and Liu, D. (2000). Long-term expression of human alpha 1-antitrypsin gene in mouse liver achieved by intravenous administration of plasmid DNA using a hydrodynamics-based procedure. *Gene Ther.* **7,** 1344–1349.

Zhang, G., Vargo, D., Budker, V., Armstrong, N., Knechtle, S., and Wolff, J. A. (1997). Expression of naked plasmid DNA injected into the afferent and efferent vessels of rodent and dog livers. *Hum. Gene Ther.* **8,** 1763–1772.

Zhang, G. F., Budker, V., Williams, P., Subbotin, V., and Wolff, J. A. (2001). Efficient expression of naked DNA delivered intraarterially to limb muscles of nonhuman primates. *Hum. Gene Ther.* **12,** 427–438.

Zhang, W. S., Purchio, A., Chen, K., Burns, S. M., Contag, C. H., and Contag, P. R. (2003). *In vivo* activation of the human CYP3A4 promoter in mouse liver and regulation by pregnane X receptors. *Biochem. Pharmacol.* **65,** 1889–1896.

5

Electric Pulse-Mediated Gene Delivery to Various Animal Tissues

Lluis M. Mir,* Pernille H. Moller,[†] Franck André,* and Julie Gehl[†]

*Laboratory of Vectorology and Gene Transfer, UMR 8121 CNRS
Institut Gustave-Roussy, F-94805 Villejuif Cédex, France
[†]Department of Oncology 54B1, Herlev University Hospital
University of Copenhagen, DK-2730 Herlev, Denmark

0065-2660/05 $35.00
DOI: 10.1016/S0065-2660(05)54005-7

ABSTRACT

Electroporation designates the use of electric pulses to transiently permeabilize the cell membrane. It has been shown that DNA can be transferred to cells through a combined effect of electric pulses causing (1) permeabilization of the cell membrane and (2) an electrophoretic effect on DNA, leading the polyanionic molecule to move toward or across the destabilized membrane. This process is now referred to as DNA electrotransfer or electro gene transfer (EGT). Several studies have shown that EGT can be highly efficient, with low variability both *in vitro* and *in vivo*. Furthermore, the area transfected is restricted by the placement of the electrodes, and is thus highly controllable. This has led to an increasing use of the technology to transfer reporter or therapeutic genes to various tissues, as evidenced from the large amount of data accumulated on this new approach for non-viral gene therapy, termed electrogenetherapy (EGT as well). By transfecting cells with a long lifetime, such as muscle fibers, a very long-term expression of genes can be obtained. A great variety of tissues have been transfected successfully, from muscle as the most extensively used, to both soft (e.g., spleen) and hard tissue (e.g., cartilage). It has been shown that therapeutic levels of systemically circulating proteins can be obtained, opening possibilities for using EGT therapeutically. This chapter describes the various aspects of *in vivo* gene delivery by means of electric pulses, from important issues in methodology to updated results concerning the electrotransfer of reporter and therapeutic genes to different tissues.   © 2005, Elsevier Inc.

I. INTRODUCTION

Electro gene transfer (EGT) opens some interesting possibilities. Indeed the use of naked DNA for gene transfer eliminates the limitations and risks associated with the use of viruses (e.g., limited coding sequence length in the case of adeno-associated virus, insertional mutagenesis with retrovirus, and immunological responses to the adenoviruses). EGT can be performed in both *in vitro* and *in vivo* systems, as well as in the clinical setting. Furthermore, the tissue to be transfected can be selected by placement of the electrodes. Gene expression can be transient (e.g., when skin is transfected) or long term (e.g., when muscle is transfected). This offers a different scope of possibilities, as compared to other non-viral gene transfer techniques.

II. BASIC PRINCIPLES OF ELECTRO GENE TRANSFER

Neumann *et al.* (1982) published the first pioneering study showing that DNA could be introduced into living cells by means of electric pulses using a custom-made chamber. It took 2 years before a second paper describing successful DNA transfer to eukaryotic cells (Potter *et al.*, 1984) could follow up the results. Potter *et al.* (1984) used a classical generator for protein and DNA gel electrophoresis, creating a short circuit through the cell and DNA suspension, which caused the delivery of an exponentially decaying electric pulse to the cells. Use of the classical electrophoresis generator opened up the possibility so many other groups could try this approach. In 1985 Teissié developed the first square wave pulse generator, where both pulse length and pulse amplitude can be controlled independently (Zerbib *et al.*, 1985).

A. Electroporation/electropermeabilization principles

The exposure of living cells to short and intense electric pulses induces position-dependent changes in the transmembrane potential difference. These changes are well described by the equation of Schwann, indicating that the value of the induced change is proportional to the cell radius and the scalar value of the external electric field (Fig. 5.1). The induced change will superimpose on the resting transmembrane potential. When the transmembrane potential difference net value (the sum of the vectorial values of the induced and resting potential differences) is greater than 0.2 to 0.4 V, transient permeation structures are generated at the cell membrane level because the membrane structure cannot resist the electrocompressive forces caused by this potential difference. Electropermeabilization is thus a threshold phenomenon, imposed by the need to overpass a threshold value of the transmembrane potential difference.

The structure of the transient permeation structures is not yet elucidated. Some models have proposed the generation of "electropores" (described as "holes" in the membranes), but cell electropermeabilization can be totally reversible and the theory hardly explains the resealing of these "electropores." Recent modelization by molecular dynamics (Tieleman *et al.*, 2003) has suggested that under transmembrane potential differences much larger than those necessary to obtain "physiological" reversible cell electropermeabilization, pores could indeed be generated. Still, the reversible structure remains undefined. The most interesting property of cell electropermeabilization is that cell permeabilization allows direct delivery of nonpermeant molecules inside the cell, bypassing the normal internalization routes.

B. Cell electroporation and DNA electrophoresis *in vivo*

Experiments have shown that tissues remain permeabilized for several minutes after electric pulses delivery (Gehl *et al.*, 2002; Mir *et al.*, 1999; Satkauskas *et al.*, 2002). However, for efficient transfer, DNA must be injected before the electric pulse delivery. Thus permeabilization of cells is not sufficient for efficient EGT even though intense electric fields are required to overpass the cell

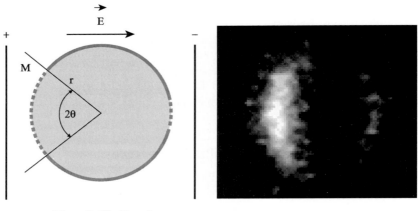

$$\Delta V_M = f\ g(\lambda)\,r\ E\ cos\theta$$

Figure 5.1. Effect of an external electric field applied on a living cell. (Left) The external electric field induces a change (ΔV_M) in the resting transmembrane voltage of the cell. The value of the induced change depends on the shape of the cell (f) and the conductivity of the media $g(\lambda)$. At point **M** on the cell surface, it is also proportional to the cell radius (r), the scalar value of the external electric field (E), and the cosinus of the angle θ (polar coordinate of point **M**). As the resting transmembrane potential is negative on the inside respective to the outside, the first part of the membrane that will be permeabilized is the pole facing the positive electrode. The higher the pulse amplitude, the larger the area of the membrane permeabilized, whereas longer pulses will cause greater perturbation of the area permeabilized. (Right) The positive electrode should be imagined in the left of the picture and the negative electrode on the right. One 20-ms pulse of 1 kV/cm was delivered to a CHO cell suspended in medium containing propidium iodide. The image was recorded less than 40ms after the pulses using a rapid ultralow-light camera, as described by Gabriel and Teissie (1997). This experiment was performed by B. Gabriel, who kindly permitted reproduction.

permeabilization threshold (Gehl and Mir, 1999; Gehl *et al.*, 1999). Moreover, efficient EGT requires sufficiently long pulses. Hence, the mechanism of EGT cannot be explained just by cell electropermeabilization and DNA diffusion through the permeabilized cell membrane.

One of the earliest and most exhaustive series of experiments explaining the mechanism of EGT were performed in muscle using trains of identical pulses (Mir *et al.*, 1999). The authors found a 200-fold increase in expression compared to naked DNA injection and a large decrease in the variability of expression, as well as a long-term, high level of expression, which remained stable for 9 months (Mir *et al.*, 1998). Subsequently, most studies have used trains of identical pulses for the transfer of therapeutic genes into muscles (Table 5.2) or for EGT in other tissues (Tables 5.3 and 5.4). However, the role of electric pulses in EGT can be studied by combining different pulses. Instead of delivering trains of identical pulses, exposing the cells to combinations of high-voltage and low-voltage pulses would elucidate the role of the pulses in EGT. Experiments were set up exposing muscles to combinations of one short high-voltage (HV) pulse followed by one or several long low-voltage (LV) pulses. The HV pulse usually consists of 100 μs at 800 V/cm as the field strength has been adapted to skeletal muscle permeabilization. Indeed, using eight such pulses at 1-Hz repetition frequency, Gehl et al. (1999) showed that this field strength was the highest one could deliver to muscle fibers without provoking irreversible electropermeabilization. The LV pulses were nonpermeabilizing pulses of 100 ms at 80 V/cm, with a field intensity below the threshold for reversible permeabilization in mice skeletal muscle (Bureau et al., 2000).

The efficacy of several combinations of pulses (1 HV alone, 1 HV followed by 1 LV, or 1 HV followed by 4 LV) was compared (Satkauskas et al., 2002). It was shown that the duration of the permeabilized state of the muscle fibers was the same for the three combinations tested. However, the authors found that transfection was only achieved if at least one LV was delivered after the HV. They also reported that the efficacy of a single LV could be observed even if this LV was delivered up to 100 s after the HV, while the gap between the HV and the LV could be extended to at least 3000 s when 4 LV were delivered. The authors drew the conclusion that the permeabilizing HV pulse was mandatory but that the efficacy of the procedure was brought by the LV pulses. Other arguments have contributed to point out that the long LV pulses act on DNA, provoking its electrophoretic displacement (Zaharoff et al., 2002).

Thus the electric pulses have two roles: "electroporation" of the target cells and electrophoretic transport of the DNA "toward and/or across" the cell membrane. Target cell electropermeabilization is mandatory, but the electrophoretic component of the electric pulses is actually instrumental in DNA electrotransfer efficacy.

A new pulse generator (Cliniporator, IGEA s.r.l., Carpi, Italy) (Fig. 5.2A), able to deliver these combinations of pulses, has been developed within the Cliniporator project (QLK3-1999-00484) of the 5th Framework Program of the European Union. In the skeletal muscle, tumor, and skin, the high-voltage plus low-voltage pulse combination has led to a further increase of the expression of the luciferase coding plasmid with almost no histological modification.

C. Tissue reactions to *in vivo* electric pulse delivery: vascular effects

One particular issue about electroporation *in vivo* is that the electric pulses have been shown to cause vascular effects. These effects appear as a transient hypoperfusion of not only the area defined by the electrodes, but also distally

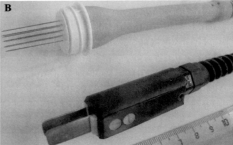

Figure 5.2. The Cliniporator and electrodes. (A) This new pulse generator (Cliniporator, IGEA s.r.l. Carpi, Italy) has been developed within the Cliniporator project (QLK3-1999-00484) of the 5th Framework Program of the European Union. It delivers combinations of HV and LV pulses. (B) Examples of plate and needle electrodes.

thereof (Gehl *et al.*, 2002). The mechanism behind this has been proposed to be a reflexory vasoconstriction of afferent arterioles mediated by the sympathetic nervous system. The effect is most pronounced when high voltage pulses are used (Gehl *et al.*, 2002). In the case of high-voltage pulses applied to deliver chemotherapy to tumors, this is used advantageously. The vascular effect implies that just at the time when the cell is permeabilized, the drug is withheld within the electroporated area by the so-called "vascular lock."

For the pulses used standardly for EGT, the vascular effects are either absent or minimal and should not affect gene transfer and/or expression (Gehl *et al.*, 2002). In fact, it has been shown that a brief ischemia might actually promote the expression of transferred genes (Takeshita *et al.*, 1996; Tsurumi *et al.*, 1996). Actually, the vascular effects of electric pulses may be used to define the optimal conditions for gene transfer to the skeletal muscle, as the expression of transferred genes is abolished when the tissue starts to be irreversibly permeabilized. Indeed this is also the time when vascular effects become more prominent (Gehl *et al.*, 2002). Therefore, a quick examination of blood flow in the muscle after the delivery of electric pulses may reveal what the upper limit for pulse strength for gene transfer is.

III. ELECTRODES FOR *IN VIVO* GENE DELIVERY

Electrodes for *in vivo* gene delivery are a problem to be solved, but at the same time are an option for making choices that benefit the particular aim of the study. In some cases, a great advantage of EGT is that the transfer will take place only in the volume where the local electric field intensity is larger than the threshold value. This volume is delineated by the electrode geometry and the ratio of the applied voltage to the electrode distance. Then, one can choose to transfer DNA to a particular area, leaving a similar area untreated as a control. Field distribution has been calculated for various electrode types (Gehl *et al.*, 1999; Miklavcic *et al.*, 2000). However, there are essentially two types of electrodes (see Fig. 5.2B).

A. Plate electrodes

These electrodes can be either purchased or simply constructed. The corners of the plates should be rounded to avoid "hot spots" on the corners of the electrodes. The material can be, for example, stainless steel. One very important feature is that the plates must be absolutely parallel to avoid inhomogeneities in the field distribution. Plate electrodes can be used either directly in the tissue, as described for liver lobe (Liu and Huang, 2002; Suzuki *et al.*, 1998) or muscle (Ostapenko *et al.*, 2004), or, after depilation and application of conductive gel, directly over the skin of a limb where the muscle underneath is to be transfected (Mir *et al.*, 1999).

B. Needle electrodes

There are many options when it comes to needle electrodes. The most simple is just two opposing needles. In this system, the electric field will be quite heterogeneous. In one study by Mathiesen (1999), this can be seen as an area without gene transfer (field too low), an area with gene transfer (field adequate), and an area with damaged, nonexpressing tissue (field too high). More complex needle arrays include the parallel array, where opposing rows of four needles are used to try and achieve a more even field distribution, as seen with plate electrodes (Gehl *et al.*, 1999). In another system, needles are used in opposing pairs in a hexagonal array where the field is rotating. This also evens the field distribution somewhat (Gilbert *et al.*, 1997). A more complex electrode type with a center needle surrounded by a ring of six needles allows one to submit large tissue volumes by letting all the needles pulse in pairs (Ramirez *et al.*, 1998). Moreover, Liu and Huang (2002a) have used needles of the syringes used to inject DNA as electrodes, ensuring that the high-intensity electric field is confined to the site of injection.

C. Electrode type and various tissues

For muscle tissue, the most extensively used technique is plate electrodes applied on depilated skin (with conducting gel) over the muscle to be transfected. This technique is extremely simple; in fact, gene transfer is performed in the order of seconds. Even when only a small muscle, such as the tibialis cranialis of mice weighing some 50–80 mg is transfected, systemic levels of protein can be detected (Martel-Renoir *et al.*, 2003; Trochon-Joseph *et al.*, 2004). In tumor tissue, plate electrodes have been used most frequently (Cemazar *et al.*, 2003; Trochon-Joseph *et al.*, 2004; Yu *et al.*, 2004). It needs to be emphasized that the distance between the plates must be compatible with good contact of the tumor to both plates so that an adequate field is obtained. For big tumors, needle electrodes are preferable to plate electrodes. In liver, both needle (Ramirez *et al.*, 1998) and plate (tweezers) (Liu and Huang, 2002; Suzuki *et al.*, 1998) electrodes have been used. For the latter, the liver lobe is just placed between two plates during a laparotomy.

IV. PRECLINICAL STUDIES

Several non-viral gene transfer methods have been developed (see Table 5.1). However, the one very big advantage of electric pulse-mediated gene delivery is that it is well suited for *in vivo* gene delivery to very different tissues—both for use in animal research and, in principle, for use in clinical trials. In addition, EGT has also been used to transfect fungal (Robinson and Sharon, 1999) and plant tissues (Dekeyser *et al.*, 1990). An increasing amount of literature is accumulating on the use of EGT for various animal tissues, as listed in Tables 5.2–5.4. Many studies have focused on EGT to muscle as this particular tissue type has the distinct advantages of being (1) easily accessible, (2) with long term, and high-level, expression of transferred genes, and (3) a large cell factory for the production of secreted factors. Table 5.2 lists published studies on transfection with reporter genes and therapeutic genes to muscle tissue. Another tissue of particular interest is tumor tissue. Gene transfer to tumor tissue is interesting for the development of an anticancer gene therapy treatment modality. Table 5.3 lists publications concerning EGT of reporter genes or therapeutic genes in tumor tissue. The therapeutic genes mainly involve im-muno-modulatory molecules or cytokines (Andre and Mir, 2004). Tumors are difficult to transfect mainly due to the high heterogeneity in the tissue. This is reflected in the many different parameters used (Table 5.3).

One particular feature of electroporation is its applicability to a wide variety of tissues. Indeed various tissues (listed in Table 5.4) have been electro-transferred successfully. Because the transfection procedure itself is so short, intraoperative transfection can be foreseen, expanding the range of potential targets. One particular tissue or part of a tissue can be targeted by placement of the electrodes, and thus placing the electrodes around, for example, one testis will leave the untreated testis as an internal control.

A. Preclinical studies using reporter genes

Reporter genes, e.g., firefly luciferase (luc), green fluorescent protein (GFP), or β-galactosidase (β-gal), have been used to elucidate the potential of different tissues to be transfected, to find the optimal parameters for the individual tissue, and to characterize the level and duration of expression. Levels of expression are best evaluated with quantitative reporter genes such as luc or β-gal; of these genes, studies have shown a two to three log increase in gene expression with respect to naked DNA injection alone (Mir et al., 1999). The highest levels of expression are obtained in tissues with an innate ability to produce high levels of proteins, e.g., liver and muscle. Heller and colleagues (1996) were the first to electrotransfer reporter gene DNA in vivo using a train of short pulses (100 μs) to the liver. In 1998, four groups consistently demonstrated very good transfection levels using trains of long pulses (5 to 50 ms): Rols et al. (1998) in tumors, Suzuki et al. (1998) in the liver, and Aihara and Miyazaki (1998) and Mir et al. (1999) in the skeletal muscle. Since then, a variety of tissues have been electrotransfected successfully, among them skin (Babiuk et al., 2003), kidney (Tsujie et al., 2004), bladder (Iwashita et al., 2004), spleen (Tupin et al., 2003), testis (Widlak et al., 2003), ovaries (Sato et al., 2003), lung (Dean et al., 2003), brain (Kondoh et al., 2000), spinal cord (Lin et al., 2002a), cornea (Blair-Parks et al., 2004), retina (Dezawa et al., 2002; Matsuda and Cepko, 2004), vessels (Martin et al., 2000; Matsumoto et al., 2001), cartilage (Grossin et al., 2003), synovium (Ohashi et al., 2002), and tendon (Jayankura et al., 2003) (see Table 5.4 for a complete list of tissues and the appended references). The majority of studies used a train of several (3 to 10) pulses of millisecond to tens of millisecond duration, resulting in a highly significant increase in the level of reporter gene expression compared to naked DNA injection in the target tissue. Short pulses were used by Heller et al. (1996). Since then, they have almost never been used, as this group itself compared short and long pulses and demonstrated that the level of expression was higher, and duration longer, when long pulses were delivered to the tissues (Lucas and Heller, 2001).

The duration of gene expression seems to be closely related to the turnover rate of the cells. In cells with a long lifetime, long-term expression of a gene can be obtained as, for example, in muscles (Mir et al., 1998) where gene expression has been detected 9 months after transfection, or tendon (Jayankura et al., 2003) and cartilage (Grossin et al., 2003) where gene expression was detected for several months. In fast-dividing tissues, e.g., skin (Dujardin et al.,

Table 5.1. Non-viral Transfection Methods

	Injection of naked DNA (Wolff et al., 1990)	Hydrodynamic injection (Liu et al., 1999; Zhang et al., 2004)	Biolistic (ballistic) methods (Furth et al., 1992; Godon et al., 1993; Walther et al., 2002)	Sonoporation (Wang et al., 1998)	Lipofection	DNA electrotransfer
Mechanism	Simple injection with naked DNA	Very rapid intravenous/intraarterial injection of a large volume of DNA solution	Jet injection of DNA at high speed creating an ultrafine stream of high-pressure fluid that penetrates the skin or gene gun injection using DNA-coated gold particles	Focused ultrasounds to permeate cells	Lipids w/wo lipoprotein to form DNA carrying liposomes that will fuse with cell membranes	Short electric pulses that permeabilize cells and electrophoretically drive DNA into the cells
Advantages	Very simple	Very efficient transfection of liver cells after tail vein or intraportal injection, or in clamped muscles after intraarterial delivery	Good transfection in cutaneous/subcutaneous layers	Still in early stages of development	Good transfection in vitro, can be cell type specific through receptor-mediated uptake	High efficiency, performed in both in vitro and in vivo systems, delineation of the transfected volume by placement of the electrodes

Disadvantages	Very low efficiency, high variability	With tail injection the method is dangerous for the animal as it results in transient heart failure, blocks fluid distribution in the body, and provokes overpressure in the liver	No effect in deeper lying tissues	Still in early stages of development	Poor transfection efficiency *in vivo*; introduction of foreign molecules into the cell membranes	At least local anesthesia required
Perspectives for clinical use	Possible in humans	Possible in humans in clamped organs	Possible in humans but only for cutaneous/subcutaneous targets	Still in early stages of development	Not foreseen in humans	Possible in humans, in various tissues

Table 5.2. *In Vivo* Delivery of Genes into Skeletal Muscle by Means of DNA Electrotransfer

Gene	Disease	Animal	Ref.
Green fluorescent protein	Reporter gene	Mouse	Faurie et al. (2004)
Firefly luciferase	Reporter gene	Mouse, rat, rabbit	Bureau et al. (2000); Hoover and Kalhovde (2000); Mennuni et al. (2004); Mir et al. (1999); Selby et al. (2000); Vicat et al. (2000)
β-Galactosidase	Reporter gene	Mouse, rat	Aihara and Miyazaki (1998); Mathiesen (1999); Selby et al. (2000)
Secreted alkaline phosphatase	Reporter gene	Mouse, rabbit	Bettan et al. (2000a); Mennuni et al. (2004)
Interleukin (IL)-4	Diabetes	Mouse	Horiki et al. (2003)
IL-10	Ischemia, myocarditis, arthritis, bronchopulmonary hyperreactivity	Mouse	Deleuze et al. (2002, 2004); Nakano et al. (2001); Perez et al. (2002); Saidenberg-Kermanach et al. (2003); Silvestre et al. (2000); Watanabe et al. (2001)
IL-12	Cancer, immunotherapy	Mouse	Lee et al. (2003a); Lucas and Heller (2001); Lucas et al. (2002)
IL-18	Atherosclerosis, ischemia	Mouse, rat	Mallat et al. (2001, 2002)
Interferon-α	Cancer, vaccination	Mouse	Aurisicchio et al. (2001) Li et al. (2001); Zhang et al. (2003)
Granulocyte–macrophage colony-stimulating factor	Cancer, *in vivo* generation of dendritic cells	Mouse	Perez et al. (2002); Peretz et al. (2002); Tanaka et al. (2004)
B7.1	Cancer, immunotherapy	Mouse	Isaka et al. (2004)

Tissue inhibitor of metalloproteinase	Cancer	Mouse	Jiang et al. (2001)
Metargidine	Cancer	Mouse	Trochon-Joseph et al. (2004)
Endostatin and derivates (K1-5, K1-3-HAS)	Cancer	Mouse	Cichon et al. (2002); Martel-Renoir et al. (2003)
Platelet-activating factor (PAF-AH)	Atherosclerosis	Mouse	Hase et al. (2002)
Vascular endothelium growth factor A and B	Ischemia	Mouse	Silvestre et al. (2003)
Erythropoietin (EPO), and dimeric EPO fusion proteins	Anemia, β-thalassemia	Mouse, rat, rabbit	Ataka et al. (2003); Dalle et al. (2001); Kreiss et al. (1999); Maruyama et al. (2000, 2001a); Mennuni et al. (2004); Nordstrom (2003); Payen et al. (2001); Rizzuto et al. (1999, 2000); Samakoglu et al. (2001); Terada et al. (2001)
Insulin precursors	Diabetes	Mouse	Croze and Prud'homme (2003); Martinenghi et al. (2002); Wang et al. (2003); Yin and Tang (2001)
Insulin growth factor-1	Diabetes, muscle regeneration	Mouse	Rabinovsky and Draghia-Akli (2004); Takahashi et al. (2003)
CD152	Diabetes	Mouse	Prud'homme et al. (2002)
Carcinogenic embryonic antigen	Diabetes	Mouse	Prud'homme et al. (2002)
Hepatocyte growth factor	Kidney regeneration, liver regeneration, dilated cardiomyopathy, lung fibrosis	Mouse, rat, hamster	Komamura et al. (2004); Riera et al. (2004); Tanaka et al. (2002); Umeda et al. (2004); Xue et al. (2002)
Gastrin	Gastric disorder	Mouse, rat	Yasui et al. (2001)
Neurothropin 3	Neuropathy	Mouse	Pradat et al. (2001)

(Continues)

Table 5.2. (*Continued*)

Gene	Disease	Animal	Ref.
Recombinant human thrombopoietin	Thrombocytopenia	Mouse	Zang et al. (2001a,b)
Factor IX	Haemophilia	Mouse, dog	Fewell et al. (2001)
Flt-3 ligand	In vivo generation of dendritic cells	Mouse	Peretz et al. (2002)
Iduronate-2-sulphatase	Mucopolysaccharidosis	Mouse	Tomanin et al. (2002)
Dystrophin, minidystrophin	Myodystrophy	Mouse	Ferrer et al. (2004); Gollins et al. (2003); Murakami et al. (2003); Vilquin et al. (2001)
Cardiothrophin	Neuron degeneration	Mouse	Lesbordes et al. (2002)
Proopiomelanocortin	Arthritis	Mouse, rat	Chuang et al. (2004)
Bone morphogenetic protein (BMP-4)	Bone formation	Mouse	Kishimoto et al. (2002)
Basic fibroblast growth factor	Ischemia	Rabbit	Nishikage et al. (2004)
VEGF receptor (Flt-1)	Peritoneal fibrosis	Mouse	Motomura et al. (2004)
PDGF receptor	Glomerulonephritis	Rat	Nakamura et al. (2001)
Recombinant mAb	Vaccination, arthritis	Mouse, rat	Adachi et al. (2002); Kim et al. (2003); Perez et al. (2004); Tjelle et al. (2004); Yang et al. (2003)
Virus capside antigens	Vaccination	Mouse, rat, sheep, goat, cattle, rabbit, guinea pig	Babiuk et al. (2002); Bachy et al. (2001); Kadowaki et al. (2000); Nomura et al. (1996); Selby et al. (2000); Tollefsen et al. (2003); Widera et al. (2000); Wu et al. (2004); Zucchelli et al. (2000)

2001), lung epithelia (Dean *et al.*, 2003), or ovaries (Sato *et al.*, 2003), gene expression was lost within a few weeks.

B. Preclinical studies with therapeutic genes

Tables 5.2, 5.3, and 5.4 also summarize results of an extensive search for publications reporting the *in vivo* delivery of genes of therapeutic interest by means of DNA electrotransfer. Most of the experiments deal with gene transfer to the skeletal muscle in mice (Table 5.2). Experiments have been classified according to the gene used, and the main applications foreseen by their authors have been quoted. However, gene transfer of therapeutic genes to tumors (Table 5.3) and other tissues, e.g., spinal cord, skin, penile corpora cavernosa, cornea, and liver, have also been reported (Table 5.4). About 80% of the publications report experiments performed in mice; however, experiments have also been performed in rats, pigs, rabbits, hamster, guinea pigs, sheep, goats, dogs, and cattle. The experiments demonstrate the safety of the procedure and the possibility of repeating the treatment (Rizzuto *et al.*, 1999), as well as the possibility of coelectrotransferring up to three plasmids into the same skeletal muscle fibers (Martel-Renoir *et al.*, 2003).

The main application is immunotherapy, especially in tumors. Cancer treatment, correction of metabolic disorders, and correction of organ- or site-specific diseases are the three other frequent applications. Furthermore, correction of monogenetic diseases, treatment of cardiovascular diseases, and induction of analgesia are other applications found in the literature. It must be noted that each of these applications includes the use of a large variety of genes. The most frequently used genes are the ones involved in the immune system, e.g., genes encoding the interleukins (IL) 2, 10, 12, and so on. In many cases these genes have been transferred to the skeletal muscle, which has been used as a cell factory for the production of factors that will act systemically on distant targets.

For systemic proteins, therapeutic levels have been achieved after EGT into the muscle. The principal example is erythropoietin (EPO), where increases in the hematocrit level have been achieved in many cases (Ataka *et al.*, 2003; Dalle *et al.*, 2001; Kreiss *et al.*, 1999; Lamartina *et al.*, 2002; Maruyama *et al.*, 2001a; Nordstrom, 2003; Rizzuto *et al.*, 1999, 2000; Terada *et al.*, 2001). Another example is antiangiogenic factors, which, after gene transfer to the muscle, have demonstrated distant antitumor effects as well as antimetastatic effects in a murine model of B16F10 melanomas (Martel-Renoir *et al.*, 2003; Trochon-Joseph *et al.*, 2004). Moreover, noticeable elevated concentrations of the cytokines IL-2 and granulocyte–macrophage colony-stimulating factor have been measured in tumors transfected with the corresponding genes (Heller *et al.*, 2000b). An example of correction of monogenetic diseases is the EGT of

Table 5.3. *In Vivo* Delivery of Genes to Tumors by Means of DNA Electrotransfer

Gene	Tumor[a]	Parameters	Ref.
Green fluorescent protein, luciferase, β-galactosidase	Murine melanoma, hepatocellular carcinoma, mammary tumors	6–10 × 5–50 ms, 400–800 V/cm	Bettan et al. (2000b); Cemazar et al. (2002); Heller et al. (2000a); Kishida et al. (2001); Lohr et al. (2001); Rols et al. (1998); Wells et al. (2000)
			Heller and Coppola (2002)
Interleukin-2	Murine melanoma, murine hepatoma	6 × 100 μs, 1500 V/cm	Chi et al. (2003);
		6 × 100 μs, 1500 V/cm	Heller et al. (2000b);
			Lohr et al. (2001)
Interleukin-12	Bladder, colon, and renal cancer, hepatocellular carcinoma, murine melanoma	6–10 × 50 ms, 50–200 V/cm	Heller et al. (2000b); Kishida et al. (2001, 2003); Lee et al. (2004); Li et al. (2002b); Lohr et al. (2001); Lucas and Heller (2003); Lucas et al. (2002); Tamura et al. (2001, 2003); Yamashita et al. (2001); Yu et al. (2004)
		or	
		6 × 100 μs, 1500 V/cm	Lucas et al. (2002)

Interleukin-18	Murine melanoma, adenocarcinoma	6 × 100 ms, 100 V/cm	Kishida et al. (2001); Tamura et al. (2003)
Interferon-α	Murine melanoma	2–6 × 10 ms, 400 V/cm	Heller et al. (2002); Li et al. (2002a)
Granulocyte–macrophage colony-stimulating factor	Human esophageal tumors, murine melanoma	n.d.	Chi et al. (2003); Heller et al. (2000b); Matsubara et al. (2001a)
Herpes simplex virus – thymidine kinase	Adenocarcinoma, mammary carcinoma, bladder carcinoma	8 × 50 ms, 66 V/cm, 8 × 20 ms, 200 V/cm, 50 ms, 300 V/cm	Goto et al. (2000) Shibata et al. (2002b) Hsieh et al. (2003); Shibata et al. (2002a)
p53	Human prostata cancer, colorectal carcinoma, human esophageal cancer	8 × 99 μs, 1 kV/cm 8 × 5 ms, 600 V/cm	Cemazar et al. (2003); Matsubara et al. (2001b); Mikata et al. (2002)
Antisense methylated-DNA-binding protein-2	Non-small cell lung cancer	8 × 20 ms, 500 V/cm	Ivanov et al. (2003); Slack et al. (2002)
Endostatin	Murine melanoma	100 ms, 700 V/cm	Cichon et al. (2002)
Stat3	Murine melanoma	14 × 100 μs, 1500 V/cm	Niu et al. (1999)
Diphtheria toxin	Gliomas (in rats)	n.d.	Yoshizato et al. (2000)

[a]Unless otherwise stated, tumors were transplanted in immunocompetent or immunodeficient mice.
[b]n.d. = not described.

Table 5.4. Electro Gene Transfer in Various Tissues Except Muscle and Tumor

Tissue (animal)	Gene	Parameters	Duration of expression	Fold increase in Luc expression[a]	Ref.
Liver (mouse, rat)	GFP, Luc, β-gal, bcl-xs	8 × 20–50 ms, 250 V/cm, 1 Hz or 6 × 99 μs, 1000 V/cm, 1 Hz	3 weeks to 60 days	5-fold to 4 log	Baba et al. (2001); Heller et al. (1996); Liu and Huang (2002b); Suzuki et al. (1998)
Skin (rat)	GFP, Luc, IL-12, EPO	10 × 5 ms, 560 V/cm or 6 × 60 ms, 80-V bipolar pulses, 4 Hz	7 days	4-fold to 2 log	Babiuk et al. (2003); Dujardin et al. (2001); Heller et al. (2001); Maruyama et al. (2001b)
Bladder (rat)	GFP, Luc, β-gal	8 × 50 ms, 225 V/cm, 1 Hz	n.d.	1–2 log	Harimoto et al. (1998); Iwashita et al. (2004)
Kidney (rat)	Luc, β-gal, TGF-β,[d] PDGF[f]	6 × 50 ms, 100 V,[b] 1 Hz	n.d.	8-fold compared to liposome transfection	Isaka et al. (2004); Nakamura et al. (2001); Tsujie et al. (2004)
Testis (mouse)	GFP, Luc, β-gal	8 × 5–50 ms, 50 V[b,c]	At least 1 month		Muramatsu et al. (1997); Widlak et al. (2003); Yamazaki et al. (1998)
Penile corpus cavernosus (rat)	β-gal, NOS[e]	8 × 40 ms, 200 V/cm, 1 Hz	56 days		Magee et al. (2002)
Ovary (mouse)	β-gal	8 × 50 ms, 50 V[b,c]	2 weeks		Sato et al. (2003)
Lung (mouse)	Luc, β-gal	8 × 10 ms, 200 V/cm, 1 Hz	7 days	2–3 log	Dean et al. (2003)
Spleen (mouse)	Luc, SEAP	6 × 20 ms, 500 V/cm, 1–2 Hz	at least 30 days		Tupin et al. (2003)

Tissue (organism)	Reporter	Parameters	Duration	Increase	Reference
Tendon (mouse, rat, rabbit)	β-gal	10 ms, 200 V[b] or 100 μs, 1200 V,[b] 1 Hz	42 days		Jayankura et al. (2003)
Vessels – mesenteric artery (rat)	GFP, Luc, β-gal	8 × 10 ms, 200 V/cm[c]	5 days	2 log	Martin et al. (2000); Matsumoto et al. (2001)
Cartilage (rat)	GFP	8 × 20 ms, 250 V/cm,[c] 1 Hz	at least 2 months		Grossin et al. (2003)
Brain – periventricular region (rat)	GFP	1 × 50 ms, 500–1000 V/cm	2 weeks		Kondoh et al. (2000)
Spinal cord (rat)	GFP, POMC	5 × 50 ms, 200 V,[b] 1 Hz	2–3 weeks		Lee et al. (2003b); Lin et al. (2002a,b)
Synovium (rat)	Luc	6 × 100 ms, 215-V/cm bipolar pulses, 1 Hz	2 weeks	3 log	Ohashi et al. (2002)
Cornea (rat)	GFP, Luc, tPA	8 × 50 ms, 400 V/cm[c]	7–21 days	3 log	Blair-Parks et al. (2004); Oshima et al. (1998); Sakamoto et al. (1999)
Retinal ganglion cells (rat)	GFP	10 × 99 ms, 12 V/cm, 1 Hz	At least 21 days		Dezawa et al. (2002)
Retina (mouse, rat)	GFP	5 × 50 ms, 100 V,[b] 1 Hz	At least 50 days		Matsuda and Cepko (2004)
Nostrils, nasal mucosa (mouse)	Influenza virus HA	3 × 100 ms, 200 V/cm, 1 Hz	n.d.		Kadowaki et al. (2000)
Adipocytes	GFP	7 × 20 ms, 50 V,[b] 2 Hz	n.d.		Granneman et al. (2004)
Teeth (dog)	GFP	10 × 1 ms, 10 V,[b] 1 Hz	n.d.		Nakashima et al. (2002)
Embryonic chick heart	GFP	12 triplets of 10-ms pulses 10 s apart (5 min between triplets), 200 V/cm	n.d.		Harrison et al. (1998)
Embryonic mouse brain	GFP	5 × 50 ms, 50 V[b,c]	n.d.		Saito and Nakatsuji (2001)

(Continues)

Table 5.4. (Continued)

Tissue (animal)	Gene	Parameters	Duration of expression	Fold increase in Luc expression[a]	Ref.
Mononuclear cells (in vivo)	GFP	10 trains (at 1 Hz) of 1000 bipolar pulses of 400 μs, 150 V/cm	n.d.		Grønevik et al. (2004)
Oviduct (laying hens)	Chloramphenicol acetyltransferase, EPO	5×50 ms, 50 V[b,c]	n.d.		Ochiai et al. (1998)

[a]Fold increase with respect to injection alone is reported only for data concerning luciferase because only this reporter gene allows precise quantitative efficacy determination.

[b]V, not V/cm, is reported because tweezer electrodes were used in these experiments, and the authors did not report precisely the distance between the electrodes. To make experiments reproducible and to obtain less variable results, it is recommended to measure the distance between the electrodes or to use a fixed distance between the electrodes. Moreover, it is recalled that the achievement of a homogeneous electric field needs the use of parallel electrodes.

[c]The frequency of pulse delivery was not reported in these studies.

[d]Transforming growth factor-β.

[e]Nitric oxide synthase.

[f]Plateler-derived growth factor.

[g]n.d. = not described.

minidystrophin into myodystrophic muscles, followed by the improvement of motor function in the treated muscle (Murakami *et al.*, 2003; Vilquin *et al.*, 2001).

V. PERSPECTIVES FOR CLINICAL USE OF ELECTRO GENE TRANSFER

Already at this time, extensive knowledge on the use of electroporation in the treatment of cancer patients has been accumulated (Gothelf *et al.*, 2003; Sersa *et al.*, 2003). Certain cytostatic drugs have a poor penetrance of the cell membrane, and electroporation can be used to cause increased access to the cell cytosol in an area with tumor tissue. Two drugs have been used: bleomycin [as reviewed by Gothelf *et al.* (2003)] and cisplatinum [as reviewed by Sersa *et al.* (2003)]. The experience is that electroporation can be performed using local anesthesia, with very limited side effects. The treatment can be performed as an out-patient regimen with an electroporator approved for clinical use.

Therefore, the road for electrogenetherapy is already paved. With modification of the pulses to be used, the treatment protocols for electrochemotherapy can essentially be transferred to EGT in the clinical setting.

There are several interesting strategies to pursue.

- Production of missing proteins. EGT offers the possibility to transfect one tissue for use as a protein factory for erythropoietin or coagulation factors. The obvious candidate tissue for this would be the muscle.
- Vaccination. Short-term expression of an antigen can be obtained through EGT to skin.
- Cancer therapy. Gene transfer to tumors can be envisaged for (1) paracrine secretion of cytokines, (2) genes coding for costimulatory molecules, or (3) local secretion of toxic products.

VI. CONCLUSION

In conclusion, DNA electrotransfer or electrogenetherapy constitutes a real alternative to viral approaches for gene transfer *in vivo*. Its efficacy is proven and there is no doubt about its biological safety. Moreover, DNA preparation is easy and secure, the roles of the electric pulses are described, the control of transfer conditions is achievable, and appropriate equipment is available.

Acknowledgments

The authors acknowledge the financial support of the Danish Medical Research Council, CNRS, IGR, AFM (Association Française contre les Myopathies), and the EU Commission through the projects Cliniporator (QLK3-1999-00484) and Esope (QLK3-2002-02003) coordinated by L.M.M. The authors also acknowledge all their colleagues for fruitful discussions and collaborative work. F.A. is the recipient of an "aide aux etudes" from the AFM.

References

Adachi, O., Nakano, A., Sato, O., Kawamoto, S., Tahara, H., Toyoda, N., Yamato, E., Matsumori, A., Tabayashi, K., and Miyazaki, J. (2002). Gene transfer of Fc-fusion cytokine by *in vivo* electroporation: Application to gene therapy for viral myocarditis. *Gene Ther.* **9**, 577–583.

Aihara, H., and Miyazaki, J. (1998). Gene transfer into muscle by electroporation *in vivo*. *Nature Biotechnol.* **16**, 867–870.

Andre, F., and Mir, L. M. (2004). DNA electrotransfer: Its principles and an updated review of its therapeutic applications. *Gene Ther* **11**(1), S33–S42.

Ataka, K., Maruyama, H., Neichi, T., Miyazaki, J., and Gejko, F. (2003). Effects of erythropoietin-gene electrotransfer in rats with adenine-induced renal failure. *Am. J. Nephrol.* **23**, 315–323.

Aurisicchio, L., Ceccacci, A., La Monica, N., Palombo, F., and Traboni, C. (2001). Tamarin alpha-interferon is active in mouse liver upon intramuscular gene delivery. *J. Gene Med.* **3**, 394–402.

Baba, M., Iishi, H., and Tatsuta, M. (2001). Transfer of bcl-xs plasmid is effective in preventing and inhibiting rat hepatocellular carcinoma induced by N-nitrosomorpholine. *Gene Ther.* **8**, 1149–1156.

Babiuk, S., Baca-Estrada, M. E., Foldvari, M., Storms, M., Rabussay, D., Widera, G., and Babiuk, L. (2002). Electroporation improves the efficacy of DNA vaccines in large animals. *Vaccine* **20**, 3399–3408.

Babiuk, S., Baca-Estrada, M. E., Foldvari, M., Baizer, L., Stout, R., Storms, M., Rabussay, D., Widera, G., and Babiuk, L. (2003). Needle-free topical electroporation improves gene expression from plasmids administered in porcine skin. *Mol. Ther.* **8**, 992–998.

Bachy, M., Boudet, F., Bureau, M., Girerd-Chambaz, Y., Wils, P., Scherman, D., and Meric, C. (2001). Electric pulses increase the immunogenicity of an influenza DNA vaccine injected intramuscularly in the mouse. *Vaccine* **19**, 1688–1693.

Bettan, M., Emmanuel, F., Darteil, R., Caillaud, J. M., Soubrier, F., Delaère, P., Branelec, D., Mahfoudi, A., Duverger, N., and Scherman, D. (2000a). High-level protein secretion into blood circulation after electric pulse-mediated gene transfer into skeletal muscle. *Mol. Ther.* **2**, 204–210.

Bettan, M., Ivanov, M.-A., Mir, L. M., Boissiere, F., Delaere, P., and Scherman, D. (2000b). Efficient DNA electrotransfer into tumors. *Bioelectrochemistry* **52**, 83–90.

Blair-Parks, K., Weston, B. C., and Dean, D. A. (2004). High-level gene transfer to the cornea using electroporation. *J. Gene Med.* **4**, 92–100.

Bureau, M. F., Gehl, J., Deleuze, V., Mir, L. M., and Scherman, D. (2000). Importance of association between permeabilisation and electrophoretic forces for intramuscular DNA electrotransfer. *Biochim. Biophys. Acta* **1474**, 353–359.

Cemazar, M., Grosel, A., Glavac, D., Kotnik, V., Skobrne, M., Kranjc, S., Mir, L. M., Andre, F., Opolon, P., and Sersa, G. (2003). Effects of electrogenetherapy with p53wt combined with cisplatin on curvival of human tumor cell lines with different p53 status. *DNA Cell Biol.* **22**, 765–775.

Cemazar, M., Sersa, G., Wilson, J., Tozer, G., Hart, S., Grosel, A., and Dachs, G. (2002). Effective gene transfer to solid tumors using different non-viral gene delivery techniques: Electroporation, liposomes, and integrin-targeted vector. *Cancer Gene Ther.* **9**, 399–406.

Chi, C. H., Wang, Y. S., Lai, Y. S., and Chi, K. H. (2003). Anti-tumor effect of *in vivo* IL-2 and GM-CSF electrogene therapy in murine hepatoma model. *Anticancer Res.* **23**, 315–321.

Chuang, I. C., Jhao, C. M., Yang, C. H., Chang, H. C., Wang, C. W., Lu, C. Y., Chang, Y. J., Lin, S. H., Huang, P. L., and Yang, L. C. (2004). Intramuscular electroporation with the pro-opiomelanocortin gene in rat adjuvant arthritis. *Arthritis Res. Ther.* **6**, R7–R14.

Cichon, T., Jamrozy, L., Glogowska, J., Missol-Kolka, E., and Szala, S. (2002). Electrotransfer of gene encoding endostatin into normal and neoplastic mouse tissues: Inhibition of primary tumor growth and metastatic spread. *Cancer Gene Ther.* **9**, 771–777.

Croze, F., and Prud'homme, G. J. (2003). Gene therapy of streptozotocin-induced diabetes by intramuscular delivery of modified preproinsulin genes. *J. Gene Med.* **5**, 425–437.

Dalle, B., Henri, A., Rouyer-Fessard, P., Bettan, M., Scherman, D., Beuzard, Y., and Payen, E. (2001). Dimeric erythropoietin fusion protein with enhanced erythropoietic activity *in vitro* and *in vivo*. *Blood* **97**, 3776–3782.

Dean, D. A., Machado-Aranda, D., Blair-Parks, K., Yeldandi, A. V., and Young, J. L. (2003). Electroporation as a method for high-level non-viral gene transfer to the lung. *Gene Ther.* **10**, 1608–1615.

Dekeyser, R. A., Claes, B., De Rycke, R., Habets, M. E., Van Montagne, M. C., and Caplan, H. B. (1990). Transient gene expression in the intact and organized rice tissues. *Plant Cell* **2**, 591–602.

Deleuze, V., Lefort, J., Bureau, M. F., Scherman, D., and Vargaftig, B. B. (2004). LPS-induced bronchial hyperreactivity: Interference by mIL-10 differs according to site of delivery. *Am. J. Physiol. Lung Cell. Mol. Physiol.* **286**, L98–L105.

Deleuze, V., Scherman, D., and Bureau, M. F. (2002). Interleukin-10 expression after intramuscular DNA electrotransfer: Kinetic studies. *Biochem. Biophys. Res. Commun.* **299**, 29–34.

Dezawa, M., Takano, M., Negishi, H., Mo, X., Oshitari, T., and Sawada, H. (2002). Gene transfer into retinal ganglion cells by *in vivo* electroporation: A new approach. *Micron* **33**, 1–6.

Dujardin, N., Van der Smissen, P., and Preat, V. (2001). Topical gene transfer into rat skin using electroporation. *Pharm. Res.* **18**, 61–66.

Faurie, C., Golzio, M., Moller, P., Teissié, J., and Rols, M. P. (2004). Cell and animal imaging of electrically mediated gene transfer. *DNA Cell Biol.* **22**, 777–783.

Ferrer, A., Foster, H., Wells, K. E., Dickson, G., and Wells, D. J. (2004). Long-term expression of full-length human dystrophin in transgenic mdx mice expressing internally deleted human dystrophins. *Gene Ther.* **11**, 884–893.

Fewell, J. G., MacLaughlin, F., Mehta, V., Gondo, M., Nicol, F., Wilson, E., and Smith, L. C. (2001). Gene therapy for the treatment of hemophilia B using PINC-formulated plasmid delivered to muscle with electroporation. *Mol. Ther.* **3**, 574–583.

Furth, P. A., Shamay, A., Wall, R. J., and Hennighausen, L. (1992). Gene transfer into somatic tissues by jet injection. *Anal. Biochem.* **205**, 365–368.

Gabriel, B., and Teissie, J. (1997). Direct observation in the millisecond time range of fluorescent molecule asymmetrical interaction with the electropermeabilized cell membrane. *Biophys. J.* **73**, 2630–2637.

Gehl, J., Skovsgaard, T., and Mir, L. M. (2002). Vascular reactions to *in vivo* electroporation: characterization and consequences for drug and gene delivery. *Biochim. Biophys. Acta* **1569**, 51–58.

Gehl, J., and Mir, L. M. (1999). Determination of optimal parameters for *in vivo* gene transfer by electroporation, using a rapid *in vivo* test for cell permeabilization. *Biochem. Biophys. Res. Commun.* **261**, 377–380.

Gehl, J., Sorensen, T. H., Nielsen, K., Raskmark, P., Nielsen, S. L., Skovsgaard, T., and Mir, L. M. (1999). *In vivo* electroporation of skeletal muscle: Threshold, efficacy and relation to electric field distribution. *Biochim. Biophys. Acta* **1428,** 233–240.

Gilbert, R. A., Jaroszeski, M. J., and Heller, R. (1997). Novel electrode designs for electrochemotherapy. *Biochim. Biophys. Acta* **1334,** 9–14.

Godon, C., Caboche, M., and Daniel-Vedele, F. (1993). Transient plant gene expression: A simple and reproducible method based on flowing particle gun. *Biochimie* **75,** 591–595.

Gollins, H., McMahon, J., Wells, K. E., and Wells, D. J. (2003). High-efficiency plasmid gene transfer into dystrophic muscle. *Gene Ther.* **10,** 504–512.

Gothelf, A., Mir, L. M., and Gehl, J. (2003). Electrochemotherapy: Results of cancer treatment using enhanced delivery of bleomycin by electroporation. *Cancer Treat. Rev.* **29,** 371–387.

Goto, T., Nishi, T., Tamura, T., Dev, S. B., Takeshima, H., Kochi, M, Yoshizato,K., Kuratsu, J., Sakata, T., Hofmann, G. A., and Ushio, Y. (2000). Highly efficient electro-gene therapy of solid tumor by using an expression plasmid for the herpes simplex virus thymidine kinase gene. *Proc. Natl. Acad. Sci. USA* **97,** 354–359.

Granneman, J. G., Li, P., Lu, Y., and Tilak, J. (2004). Seeing the trees in the forest: Selective electroporation of adipocytes within adipose tissue. *Am. J. Physiol. Endocrinol. Metab.* **287**(3), E574–E582.

Grønevik, E., Tollefsen, S., Sikkeland, L., Haug, T., Tjelle, T. E., and Mathiesen, I. (2004). DNA transfection of mononuclear cells in muscle tissue. *J. Gene Med.* **5,** 909–917.

Grossin, L., Gaborit, N., Mir, L. M., Netter, P., and Gillet, P. (2003). Gene therapy in cartilage using electroporation. *Joint Bone Spine* **70,** 480–482.

Harimoto, K., Sugimura, K., Lee, C. R., Kuratsukuri, K., and Kishimoto, T. (1998). *In vivo* gene transfer methods in the bladder without viral vectors. *Br. J. Urol.* **81,** 870–874.

Harrison, R. L., Byrne, B. J., and Tung, L. (1998). Electroporation-mediated gene transfer in cardiac tissue. *FEBS Lett.* **435,** 1–5.

Hase, M., Tanaka, M., Yokota, M., and Yamada, Y. (2002). Reduction in the extent of atherosclerosis in apolipoprotein E-deficient mice induced by electroporation-mediated transfer of the human plasma platelet-activating factor acetylhydrolase gene into skeletal muscle. *Prostagland. Other Lipid Mediat.* **70,** 107–118.

Heller, L., and Coppola, D. (2002). Electrically mediated delivery of vector plasmid DNA elicits an antitumour effect. *Gene Ther.* **9,** 1321–1325.

Heller, L., Jaroszeski, M. J., Coppola, D., Pottinger, C., Gilbert, R., and Heller, R. (2000a). Electrically mediated plasmid DNA delivery to hepatocellular carcinomas. *in vivo. Gene Ther.* **7,** 826–829.

Heller, L., Pottinger, C., Jaroszeski, M. J., Gilbert, R., and Heller, R. (2000b). *In vivo* electroporation of plasmids encoding GM-CSF or interleukin-2 into existing B16 melanomas combined with electrochemotherapy induces long-term antitumour immunity. *Melanoma Res.* **10,** 577–583.

Heller, L. C., Ingram, S. F., Lucas, M. L., Gilbert, R. A., and Heller, R. (2002). Effect of electrically mediated intratumor and intramuscular delivery of a plasmid encoding IFN alpha on visible B16 mouse melanomas. *Technol. Cancer Res. Treat.* **1,** 205–209.

Heller, R., Jaroszeski, M. J., Atkin, A., Moradpour, D., Gilbert, R. A., Wands, J., and Nicolau, C. (1996). *In vivo* gene electroinjection and expression in rat liver. *FEBS Lett.* **389,** 225–228.

Heller, R., Schultz, J., Lucas, M. L., Jaroszeski, M. J., Heller, L. C., Gilbert, R. A., Moelling, K., and Nicolau, C. (2001). Intradermal delivery of interleukin-12 plasmid DNA by *in vivo* electroporation. *DNA Cell Biol.* **20,** 21–26.

Hoover, F., and Kalhovde, J. M. (2000). A double-injection DNA electroporation protocol to enhance *in vivo* gene delivery in skeletal muscle. *Anal. Biochem.* **285,** 175–178.

Horiki, M., Yamato, E., Noso, S., Ikegami, H., Ogihara, T., and Miyazaki, J. (2003). High-level expression of interleukin-4 following electroporation-mediated gene transfer accelerates type 1 diabetes in NOD mice. *J. Autoimmun.* **20,** 111–117.

Hsieh, Y. H., Wu, C. J., Chow, K. P., Tsai, C. L., and Chang, Y. S. (2003). Electroporation-mediated and EBV LMP1-regulated gene therapy in a syngenic mouse tumor model. *Cancer Gene Ther.* **10,** 626–636.

Isaka, Y., Nakamura, H., Mizui, M., Takabatake, Y., Horio, M., Kawachi, Shimizu, F., Imai, E., and Hori, M. (2004). DNAzyme for TGF-beta suppressed extracellular matrix accumulation in experimental glomerulonephritis. *Kidney Int.* **66,** 586–590.

Ivanov, M. A., Lamrihi, B., Szyf, M., Scherman, D., and Bigey, P. (2003). Enhanced antitumor activity of a combination of MBD2-antisense electrotransfer gene therapy and bleomycin electrochemotherapy. *J. Gene Med.* **5,** 893–899.

Iwashita, H., Yoshida, M., Nishi, T., Otani, M., and Ueda, S. (2004). *In vivo* transfer of a neuronal nitric oxide synthase expression vector into the rat bladder by electroporation. *BJU Int.* **93,** 1098–1103.

Jayankura, M., Boggione, C., Frisén, C., Boyer, O., Fouret, P., Saillant, G., and Klatzmann, D. (2003). *In situ* gene transfer into animal tendons by injection of naked DNA and electrotransfer. *J. Gene Med.* **5,** 618–624.

Jiang, Y., Wang, M., Celiker, M. Y., Liu, Y. E., Sang, Q. X., Goldberg, I. D., and Shi, Y. E. (2001). Stimulation of mammary tumorigenesis by systemic tissue inhibitor of matrix metalloproteinase 4 gene delivery. *Cancer Res.* **61,** 2365–2370.

Kadowaki, S., Chen, Z., Asanuma, H., Aizawa, C., Kurata, T., and Tamura, S. (2000). Protection against influenza virus infection in mice immunized by administration of hemagglutinin-expressing DNAs with electroporation. *Vaccine* **18,** 2779–2788.

Kim, J. M., Ho, S. H., Hahn, W., Jeong, J. G., Park, E. J., Lee, H. J., Yu, S. S., Lee, C. S., Lee, Y. W., and Kim, S. (2003). Electro-gene therapy of collagen-induced arthritis by using an expression plasmid for the soluble p75 tumor necrosis factor receptor-Fc fusion protein. *Gene Ther.* **10,** 1216–1224.

Kishida, T., Asada, H., Itokawa, Y., Yasutomi, K., Shin-Ya, M., Gojo, S., Cui, F. D., Ueda, Y., Yamagishi, H., Imanishi, J., and Mazda, O. (2003). Electrochemo-gene therapy of cancer: Intratumoral delivery of interleukin-12 gene and bleomycin synergistically induced therapeutic immunity and suppressed subcutaneous and metastatic melanomas in mice. *Mol. Ther.* **8,** 738–745.

Kishida, T., Asada, H., Satoh, E., Tanaka, S., Shinya, M., Hirai, H., Iwai, M., Tahara, H., Imanishi, J., and Mazda, O. (2001). *In vivo* electroporation-mediated transfer of interleukin-12 and interleukin-18 genes induces significant antitumor effects against melanoma in mice. *Gene Ther.* **8,** 1234–1240.

Kishimoto, K. N., Watanabe, Y., Nakamura, H., and Kokubun, S. (2002). Ectopic bone formation by electroporatic transfer of bone morphogenetic protein-4 gene. *Bone* **31,** 340–347.

Komamura, K., Tatsumi, R., Miyazaki, J., Matsumoto, K., Yamato, E., Nakamura, T., Shimizu, Y., Nakatani, T., Kitamura, S., Tomoike, H., Kitakaze, M., Kangawa, K., and Miyatake, K. (2004). Treatment of dilated cardiomyopathy with electroporation of hepatocyte growth factor gene into skeletal muscle. *Hypertension* **44**(3), 365–371.

Kondoh, T., Motooka, Y., Bhattacharjee, A. K., Kokunai, T., Saito, N., and Tamaki, N. (2000). *In vivo* gene transfer into the periventricular region by electroporation. *Neurol. Med Chir.* **40,** 618–622.

Kreiss, P., Bettan, M., Crouzet, J., and Scherman, D. (1999). Erythropoietin secretion and physiological effect in mouse after intramuscular plasmid DNA electrotransfer. *J. Gene Med.* **1,** 245–250.

Lamartina, S., Roscilli, G., Rinaudo, C. D., Sporeno, E., Silvi, L., Hillen, W., Bujard, H., Cortese, R., Ciliberto, G., and Toniatti, C. (2002). Stringent control of gene expression *in vivo* by using novel doxycycline-dependent trans-activators. *Hum. Gene Ther.* **13,** 199–210.

Lee, C. F., Chang, S. Y., Hsieh, D. S., and Yu, D. S. (2004). Immunotherapy for bladder cancer using recombinant bacillus Calmette-Guerin DNA vaccines and interleukin-12 DNA vaccine. *J. Urol.* **171,** 1343–1347.

Lee, S. C., Wu, C. J., Wu, P. Y., Huang, Y. L., Wu, C. W., and Tao, M. H. (2003a). Inhibition of established subcutaneous and metastatic murine tumors by intramuscular electroporation of the interleukin-12 gene. _J. Biomed. Sci._ **10,** 73–86.

Lee, T. H., Yang, L. C., Chou, A. K., Wu, P. C., Lin, C. R., Wang, C. H., Chen, J. T., and Tang, C. S. (2003b). _In vivo_ electroporation of proopiomelanocortin induces analgesia in a formalin-injection pain model in rats. _Pain_ **104,** 159–167.

Lesbordes, J. C., Bordet, T., Haase, G., Castelnau-Ptakhine, L., Rouhani, S., Gilgenkrantz, H., and Kahn, A. (2002). _In vivo_ electrotransfer of the cardiotrophin-1 gene into skeletal muscle slows down progression of motor neuron degeneration in pmn mice. _Hum. Mol. Genet._ **11,** 1615–1625.

Li, S., Xia, X., Zhang, X., and Suen, J. (2002a). Regression of tumors by IFN-alpha electroporation gene therapy and analysis of the responsible genes by cDNA array. _Gene Ther._ **9,** 390–397.

Li, S., Zhang, X., and Xia, X. (2002b). Regression of tumor growth and induction of long-term antitumor memory by interleukin 12 electro-gene therapy. _J. Natl. Cancer Inst._ **94,** 762–768.

Li, S., Zhang, X., Xia, X., Zhou, L., Breau, R., Suen, J., and Hanna, E. (2001). Intramuscular electroporation delivery of IFN-alpha gene therapy for inhibition of tumor growth located at a distant site. _Gene Ther._ **8,** 400–407.

Lin, C. R., Tai, M. H., Cheng, J. T., Zhou, A. K., Yang, J. J., Tan, P. H., Marsala, M., and Yang., L. C. (2002a). Electroporation for direct spinal gene transfer in rats. _Neurosci. Lett._ **317,** 1–4.

Lin, C. R., Yang, L. C., Lee, T. H., Lee, C. T., Huang, H. T., Sun, W. Z., and Cheng, J. T. (2002b). Electroporation-mediated pain-killer gene therapy for mononeuropathic rats. _Gene Ther._ **9,** 1247–1253.

Liu, F., and Huang, L. (2002a). A syringe electrode device for simultaneous injection of DNA and electrotransfer. _Mol. Ther._ **5,** 323–328.

Liu, F., and Huang, L. (2002b). Electric gene transfer to the liver following systemic administration of plasmid DNA. _Gene Ther._ **9,** 1116–1119.

Liu, F., Song, Y., and Liu, D. (1999). Hydrodynamics-based transfection in animals by systemic administration of plasmid DNA. _Gene Ther._ **6,** 1258–1266.

Lohr, F., Lo, D. Y., Zaharoff, D. A., Hu, K., Zhang, X., Li, Y., Zhao, Y., Dewhirst, M. W., Yuan, F., and Li, C. Y. (2001). Effective tumor therapy with plasmid-encoded cytokines combined with _in vivo_ electroporation. _Cancer Res._ **61,** 3281–3284.

Lucas, M. L., and Heller, R. (2001). Immunomodulation by electrically enhanced delivery of plasmid DNA encoding IL-12 to murine skeletal muscle. _Mol. Ther._ **3,** 47–53.

Lucas, M. L., and Heller, R. (2003). IL-12 gene therapy using an electrically mediated non-viral approach reduces metastatic growth of melanoma. _DNA Cell Biol._ **22,** 755–763.

Lucas, M. L., Heller, L., Coppola, D., and Heller, R. (2002). IL-12 plasmid delivery by _in vivo_ electroporation for the successful treatment of established subcutaneous B16.F10 melanoma. _Mol. Ther._ **5,** 668–675.

Magee, T. R., Ferrini, M., Garban, H.J, Vernet, D., Mitani, K., Rajfer, J., and Gonzalez-Cadavid, N. F. (2002). Gene therapy of erectile dysfunction in the rat with penile neuronal nitric oxide synthase. _Biol. Rep._ **67,** 1032–1041.

Mallat, Z., Corbaz, A., Scoazec, A., Graber, P., Alouani, S., Esposito, B., Humbert, Y., Chvatchko, Y., and Tedgui, A. (2001). Interleukin-18/interleukin-18 binding protein signaling modulates atherosclerotic lesion development and stability. _Circ. Res._ **89,** 41–45.

Mallat, Z., Silvestre, J. S., Le Ricousse-Roussanne, S., Lecomte-Raclet, L., Corbaz, A., Clergue, M., Duriez, M., Barateau, V., Akira, S., Tedgui, A., Tobelem, G., Chvatchko, Y., and Levy, B. I. (2002). Interleukin-18/interleukin-18 binding protein signaling modulates ischemia-induced neovascularization in mice hindlimb. _Circ. Res._ **91,** 441–448.

Martel-Renoir, D., Trochon-Joseph, V., Galaup, A., Bouquet, C., Griscelli, F., Opolon, P., Opolon, D., Connault, E., Mir, L. M., and Perricaudet, M. (2003). Coelectrotransfer to skeletal muscle of three plasmids coding for antiangiogenic factors and regulatory factors of the tetracycline-inducible system: Tightly regulated expression, inhibition of transplanted tumor growth, and antimetastatic effect. *Mol. Ther.* **8,** 425–433.

Martin, J. B., Young, J. L., Benoit, J. N., and Dean, D. A. (2000). Gene transfer to intact mesenteric arteries by electroporation. *J. Vasc. Res.* **37,** 372–380.

Martinenghi, S., Cusella De Angelis, G., Biressi, S., Amadio, S., Bifari, F., Roncarolo, M. G., Bordignon, C., and Falqui, L. (2002). Human insulin production and amelioration of diabetes in mice by electrotransfer-enhanced plasmid DNA gene transfer to the skeletal muscle. *Gene Ther.* **9,** 1429–1437.

Maruyama, H., Ataka, K., Gejyo, F., Higuchi, N., Ito, Y., Hirahara, H., Imazeki, I., Hirata, M., Ichikawa, F., Neichi, T., Kikuchi, H., Sugawa, M., and Miyazaki, J. (2001a). Long-term production of erythropoietin after electroporation-mediated transfer of plasmid DNA into the muscles of normal and uremic rats. *Gene Ther.* **8,** 461–468.

Maruyama, H., Ataka, K., Higuchi, N., Sakamoto, F., Gejyo, F., and Miyazaki, J. (2001b). Skin-targeted gene transfer using *in vivo* electroporation. *Gene Ther.* **8,** 1808–1812.

Maruyama, H., Sugawa, M., Moriguchi, Y., Imazeki, I., Ishikawa, Y., Ataka, K., Hasegawa, S., Ito, Y., Higuchi, N., Kazama, J. J., Gejyo, F., and Miyazaki, J. I. (2000). Continuous erythropoietin delivery by muscle-targeted gene transfer using *in vivo* electroporation. *Hum. Gene Ther.* **11,** 429–437.

Mathiesen, I. (1999). Electropermeabilisation of skeletal muscle enhances gene transfer *in vivo*. *Gene Ther.* **6,** 508–514.

Matsubara, H., Gunji, Y., Maeda, T., Tasaki, K., Koide, Y., Asano, T., Ochiai, T., Sakiyama, S., and Tagawa, M. (2001a). Electroporation-mediated transfer of cytokine genes into human esophageal tumors produces anti-tumor effects in mice. *Anticancer Res.* **21,** 2501–2503.

Matsubara, H., Maeda, T., Gunji, Y., Koide, Y., Asano, T., Ochiai, T., Sakiyama, S., and Tagawa, M. (2001b). Combinatory anti-tumor effects of electroporation-mediated chemotherapy and wild-type p53 gene transfer to human esophageal cancer cells. *Int. J. Oncol.* **18,** 825–829.

Matsuda, T., and Cepko, C. L. (2004). Electroporation and RNA interference in the rodent retina *in vivo* and *in vitro*. *Proc. Natl. Acad. Sci. USA* **101,** 16–22.

Matsumoto, T., Komori, K., Shoji, T., Kuma, S., Kume, M., Yamaoka, T., Mori, E., Furuyama, T., Yonemitsu, Y., and Sugimachi, K. (2001). Successful and optimized *in vivo* gene transfer to rabbit carotid artery mediated by electronic pulse. *Gene Ther.* **8,** 1174–1179.

Mennuni, C., Calvaruso, F., Zampaglione, I., Rizzuto, G., Rinaudo, C. D., Dammassa, E., Ciliberto, G., Fattori, E., and La Monica, N. (2004). Hyaluronidase increases electrogene transfer efficiency in skeletal muscle. Hum. *Gene Ther.* **13,** 355–365.

Mikata, K., Uemura, H., Ohuchi, H., Ohta, S., Nagashima, Y., and Kubota, Y. (2002). Inhibition of growth of human prostate cancer xenograft by transfection of p53 gene: Gene transfer by electroporation. *Mol. Cancer Ther.* **1,** 247–252.

Miklavcic, D., Semrov, D., Mekid, H., and Mir, L. M. (2000). A validated model of *in vivo* electric field distribution in tissues for electrochemotherapy and for DNA electrotransfer for gene therapy. *Biochim. Biophys. Acta* **1523,** 73–83.

Mir, L. M., Bureau, M. F., Gehl, J., Rangara, R., Rouy, D., Caillaud, J. M., Delaere, P., Branellec, D., Schwartz, B., and Scherman, D. (1999). High-efficiency gene transfer into skeletal muscle mediated by electric pulses. *Proc. Natl. Acad. Sci. USA* **96,** 4262–4267.

Mir, L. M., Bureau, M. F., Rangara, R., Schwartz, B., and Scherman, D. (1998). Long term, high level *in vivo* gene expression after electric pulse-mediated gene transfer into skeletal muscle. *C. R. Acad. Sci. Paris* **321,** 893–899.

Motomura, Y., Kanbayashi, H., Khan, W. I., Deng, Y., Blennerhassett, P. A. M. P. J., Gauldie, J., Egashira, K., and Collins, S. M. (2005). The gene transfer of soluble VEGF type 1 receptor (Flt-1) attentuates peritoneal fibrosis formation in mice but not soluble TGF-beta type II receptor gene transfer. *Am. J. Gastrointest. Liver Physiol.* **288**(1), G143–G150.

Murakami, T., Nishi, T., Kimura, E., Goto, T., Maeda, Y., Ushio, Y., Uchino, M., and Sunada, Y. (2003). Full-length dystrophin cDNA transfer into skeletal muscle of adult mdx mice by electroporation. *Muscle Nerve* **27**, 237–241.

Muramatsu, T., Shibata, O., Ryoki, S., Ohmori, Y., and Okumura, J. (1997). Foreign gene expression in the mouse testis by localized *in vivo* gene transfer. *Biochem. Biophys. Res. Commun.* **233**, 45–49.

Nakamura, H., Isaka, Y., Tsujie, M., Akagi, Y., Sudo, T., Ohno, N., Imai, E., and Hori, M. (2001). Electroporation-mediated PDGF receptor-IgG chimera gene transfer ameliorates experimental glomerulonephritis. *Kidney Int.* **59**, 2134–2145.

Nakano, A., Matsumori, A., Kawamoto, S., Tahara, H., Yamato, E., Sasayama, S., and Miyazaki, J. I. (2001). Cytokine gene therapy for myocarditis by *in vivo* electroporation. *Hum. Gene Ther.* **12**, 1289–1297.

Nakashima, M., Mizunuma, K., Murakami, T., and Akamine, A. (2002). Induction of dental pulp stem cell differentiation into odontoblasts by electroporation-mediated gene delivery of growth/differentiation factor 11 (Gdf11). *Gene Ther.* **9**, 814–818.

Neumann, E., Schaefer-Ridder, M., Wang, Y., and Hofschneider, H. (1982). Gene transfer into mouse lyoma cells by electroporation in high electric fields. *EMBO J.* **1**, 841–845.

Nishikage, S., Koyama, H., Miyata, T., Ishii, S., Hamada, H., and Shigematsu (2004). *In vivo* electroporation enhances plasmid-based gene transfer of basic fibroblast growth factor for the treatment of ischemia. *J. Surg. Res.* **120**, 37–46.

Niu, G., Heller, R., Catlett-Falcone, R., Coppola, D., Jaroszeski, M., Dalton, W., Jove, R., and Yu, H. (1999). Gene therapy with dominant-negative Stat3 suppresses growth of the murine melanoma B16 tumor *in vivo*. *Cancer Res.* **59**, 5059–5063.

Nomura, M., Nakata, Y., Inoue, T., Uzawa, A., Itamura, S., Nerome, K., Akashi, M., and Suzuki, G. (1996). *In vivo* induction of cytotoxic T lymphocytes specific for a single epitope introduced into an unrelated molecule. *J. Immunol. Methods* **193**, 41–49.

Nordstrom, J. N. (2003). The antiprogestin-dependent GeneSwitch system for regulated gene therapy. *Steroids* **68**, 1085–1094.

Ochiai, H., Park, H. M., Nakamura, A., Sasaki, R., Okumura, J. I., and Muramatsu, T. (1998). Synthesis of human erythropoietin *in vivo* in the oviduct of laying hens by localized *in vivo* gene transfer using electroporation. *Poult. Sci.* **77**, 299–302.

Ohashi, S., Kubo, T., Kishida, T., Ikeda, T., Takahashi, K., Arai, Y., Terauchi, R., Asada, H., Imanishi, J., and Mazda, O. (2002). Successful genetic transduction *in vivo* into synovium by means of electroporation. *Biochem. Biophys. Res. Commun.* **293**, 1530–1535.

Oshima, Y., Sakamoto, T., Yamanaka, I., Nishi, T., Ishibashi, T., and Inomata, H. (1998). Targeted gene transfer to corneal endothelium *in vivo* by electric pulse. *Gene Ther.* **5**, 1347–1354.

Ostapenko, O. V., Boitsov, A. S., Baranov, A. N., Lesina, E. A., Mikhailov, V. M., and Baranov, V. S. (2004). Dependence of the efficacy of transfecting muscle fibers *in vivo* on electroporation conditions. *Genetika* **40**, 41–48.

Payen, E., Bettan, M., Rouyer-Fessard, P., Beuzard, Y., and Scherman, D. (2001). Improvement of mouse [beta]-thalassemia by electrotransfer of erythropoietin cDNA. *Exp. Hematol.* **29**, 295–300.

Peretz, Y., Zhou, Z. F., Halwani, F., and Prud'homme, G. J. (2002). *In vivo* generation of dendritic cells by intramuscular codelivery of FLT3 ligand and GM-CSF plasmids. *Mol. Ther.* **6**, 407–414.

Perez, N., Bigey, P., Scherman, D., Danos, O., Piechaczyk, M., and Pelegrin, M. (2004). Regulatable systemic production of monoclonal antibodies by *in vivo* muscle electroporation. *Genet. Vaccines Ther.* **2**, 2.

Perez, N., Plence, P., Millet, V., Greuet, D., Minot, C., Noel, D., Danos, O., Jorgensen, C., and Apparailly, F. (2002). Tetracycline transcriptional silencer tightly controls transgene expression after *in vivo* intramuscular electrotransfer: Application to interleukin 10 therapy in experimental arthritis. *Hum. Gene Ther.* **13**, 2161–2172.

Potter, H., Weir, L., and Leder, P. (1984). Enhancer-dependent expression of human kappa immunoglobulin genes introduced into mouse pre-B lymphocytes by electroporation. *Proc. Natl. Acad. Sci. USA* **81**, 7161–7165.

Pradat, P. F., Finiels, F., Kennel, P., Naimi, S., Orsini, C., Delaere, P., Revah, F., and Mallet, J. (2001). Partial prevention of cisplatin-induced neuropathy by electroporation-mediated non-viral gene transfer. *Hum. Gene Ther.* **12**, 367–375.

Prud'homme, G. J., Chang, Y., and Li, X. (2002). Immunoinhibitory DNA vaccine protects against autoimmune diabetes through cDNA encoding a selective CTLA-4 (CD152) ligand. *Hum. Gene Ther.* **13**, 395–406.

Rabinovsky, E. D., and Draghia-Akli, R. (2004). Insulin-like growth factor I plasmid therapy promotes *in vivo* angiogenesis. *Mol. Ther.* **9**, 46–55.

Ramirez, L. H., Orlowski, S., An, D., Bindoula, G., Dzodic, R., Ardouin, P., Bognel, C., Belehradek, J., Jr., Munck, J. N., and Mir, L. M. (1998). Electrochemotherapy on liver tumours in rabbits. *Br. J. Cancer* **77**, 2104–2111.

Riera, M., Chillon, M., Aran, J. M., Cruzado, J. M., Torras, J., Grinyo, J. M., and Fillat, C. (2004). Intramuscular SP1017-formulated DNA electrotransfer enhances transgene expression and distributes hHGF to different rat tissues. *J. Gene Med.* **6**, 111–118.

Rizzuto, G., Cappelletti, M., Maione, D., Savino, R., Lazzaro, D., Costa, P., Mathiesen, I., Cortese, R., Ciliberto, G., Laufer, R., Monica, N., and Fattori, E. (1999). Efficient and regulated erythropoietin production by naked DNA injection and muscle electroporation. *Proc. Natl. Acad. Sci. USA* **96**, 6417–6422.

Rizzuto, G., Cappelletti, M., Mennuni, C., Wiznerowicz, M., DeMartis, A., Maione, D., Ciliberto, G., La Monica, N., and Fattori, E. (2000). Gene electrotransfer results in a high-level transduction of rat skeletal muscle and corrects anemia of renal failure. *Hum. Gene Ther.* **11**, 1891–1900.

Robinson, M., and Sharon, A. (1999). Transformation of the bioherbicide Colletotrichum gloeosporioides f. sp Aschynomene by electroporation of germinated conidia. *Curr. Genet.* **36**, 98–104.

Rols, M. P., Delteil, C., Golzio, M., Dumond, P., Cros, S., and Teissie, J. (1998). *In vivo* electrically mediated protein and gene transfer in murine melanoma. *Nature Biotechnol.* **16**, 168–171.

Saidenberg-Kermanach, N., Bessis, N., Deleuze, V., Bloquel, C., Bureau, M., Scherman, D., and Boissier, M. C. (2003). Efficacy of interleukin-10 gene electrotransfer into skeletal muscle in mice with collagen-induced arthritis. *J. Gene Med.* **5**, 164–171.

Saito, T., and Nakatsuji, N. (2001). Efficient gene transfer into the embryonic mouse brain using *in vivo* electroporation. *Dev. Biol.* **240**, 237–246.

Sakamoto, T., Oshima, Y., Nakagawa, K., Ishibashi, T., Inomata, H., and Sueishi, K. (1999). Target gene transfer of tissue plasminogen activator to cornea by electric pulse inhibits intracameral fibrin formation and corneal cloudiness. *Hum. Gene Ther.* **10**, 2551–2557.

Samakoglu, S., Fattori, E., Lamartina, S., Toniatti, C., Stockholm, D., Heard, J. M., and Bohl, D. (2001). betaMinor-globin messenger RNA accumulation in reticulocytes governs improved erythropoiesis in beta thalassemic mice after erythropoietin complementary DNA electrotransfer in muscles. *Blood* **97**, 2213–2220.

Satkauskas, S., Bureau, M. F., Puc, M., Mahfoudi, A., Scherman, D., Miklavcic, D., and Mir, L. M. (2002). Mechanisms of *in vivo* DNA electrotransfer: Respective contributions of cell electropermeabilisation and DNA electrophoresis. *Mol. Ther.* **5**, 133–140.

Sato, M., Tanigawa, M., Kikuchi, N., Nakamura, S., and Kimura, M. (2003). Efficient gene delivery into murine ovarian cells by intraovarian injection of plasmid DNA and subsequent *in vivo* electroporation. *Genesis* **35**, 169–174.

Selby, M., Goldbeck, C., Pertile, T., Walsh, R., and Ulmer, J. (2000). Enhancement of DNA vaccine potency by electroporation *in vivo*. *J. Biotechnol.* **83**, 147–152.

Sersa, G., Cemazar, M., and Rudolf, Z. (2003). Electrochemotherapy: Advantages and drawbacks in treatment of cancer patients. *Cancer Ther.* **1,** 133–142.

Shibata, M., Horiguchi, T., Morimoto, J., and Otsuki, Y. (2002a). Massive apoptotic cell death in chemically induced rat urinary bladder carcinomas following *in situ* HSVtk electrogene transfer. *J. Gene Med.* **5,** 219–231.

Shibata, M., Morimoto, J., and Otsuki, Y. (2002b). Suppression of murine mammary carcinoma growth and metastasis by HSVtk/GCV gene therapy using *in vivo* electroporation. *Cancer Gene Ther.* **9,** 16–27.

Silvestre, J. S., Mallat, Z., Duriez, M., Tamarat, R., Bureau, M. F., Scherman, D., Duverger, N., Branellec, D., Tedgui, A., and Levy, B. I. (2000). Antiangiogenic effect of interleukin-10 in ischemia-induced angiogenesis in mice hindlimb. *Circ. Res.* **87,** 448–452.

Silvestre, J. S., Tamarat, R., Ebrahimian, T. G., Le-Roux, A., Clergue, M., Emmanuel, F., Duriez, M., Schwartz, B., Branellec, D., and Levy, B. I. (2003). Vascular endothelial growth factor-B promotes *in vivo* angiogenesis. *Circ. Res.* **93,** 114–123.

Slack, A., Bovenzi, V., Bigey, P., Ivanov, M. A., Ramchandani, S., Bhattacharya, S., Tenever, B., Lamrihi, B., Scherman, D., and Szyf, M. (2002). Antisense MBD2 gene therapy inhibits tumorigenesis. *J. Gene Med.* **4,** 381–389.

Suzuki, T., Shin, B., Fujikura, K., Matsuaki, T., and Takata, K. (1998). Direct gene transfer into rat liver cells by *in vivo* electroporation. *FEBS Lett.* **425,** 436–440.

Takahashi, T., Ishida, K., Itoh, K., Konishi, Y., Yagyu, K., Tominaga, A., Miyazaki, J. I., and Yamamoto., H. (2003). IGF-I gene transfer by electroporation promotes regeneration in a muscle injury model. *Genesis* **10,** 612–620.

Takeshita, S., Isshiki, T., and Sato, T. (1996). Increased expression of direct gene transfer into skeletal muscles observed after acute ischemia injury in rats. *Lab. Invest.* **74,** 1061–1065.

Tamura, T., Nishi, T., Goto, T., Takeshima, H., Dev, S. B., Ushio, Y., and Sakata, T. (2001). Intratumoral delivery of interleukin 12 expression plasmids with *in vivo* electroporation is effective for colon and renal cancer. *Hum. Gene Ther.* **12,** 1265–1276.

Tamura, T., Nishi, T., Goto, T., Takeshima, H., Ushio, Y., and Sakata, T. (2003). Combination of IL-12 and IL-18 of electro-gene therapy synergistically inhibits tumor growth. *Anticancer Res.* **23,** 1173–1179.

Tanaka, M., Yamada, M., Ono, T., Noguchi, Y., Uenaka, A., Ota, S., Hata, H., Harada, M., Tanimoto, M., and Nakayama, E. (2004). Inhibition of RL male 1 tumor growth in BALB/c mice by introduction of the RLakt gene coding for antigen recognized by cytotoxic T-lymphocytes and the GM-CSF gene by *in vivo* electroporation. *Cancer Sci.* **95,** 154–159.

Tanaka, T., Ichimaru, N., Takahara, S., Yazawa, K., Hatori, M., Suzuki, K., Isaka, Y., Moriyama, T., Imai, E., Azuma, H., Nakamura, T., Okuyama, A., and Yamanaka, H. (2002). *In vivo* gene transfer of hepatocyte growth factor to skeletal muscle prevents changes in rat kidneys after 5/6 nephrectomy. *Am. J. Transplant.* **2,** 828–836.

Terada, Y., Tanaka, H., Okado, T., Inoshita, S., Kuwahara, M., Akiba, T., Sasaki, S., and Marumo, F. (2001). Efficient and ligand-dependent regulated erythropoietin production by naked dna injection and *in vivo* electroporation. *Am. J. Kidney Dis.* **38,** S50–S53.

Tieleman, D. P., Leontiadou, H., Mark, A. E., and Marrink, S. J. (2003). Simulation of pore formation in lipid bilayers by mechanical stress and electric fields. *J. Am. Chem. Soc.* **125,** 6382–6383.

Tjelle, T. E., Corthay, A., Lunde, E., Sandlie, I., Michaelsen, T. E., Mathiesen, I., and Bogen, B. (2004). Monoclonal antibodies produced by muscle after plasmid injection and electroporation. *Mol. Ther.* **9,** 328–336.

Tollefsen, S., Vordermeier, M., Olsen, I., Storset, A. K., Reitan, L. J., Clifford, D., Lowrie, D. B., Wiker, H. G., Huygen, K., Hewinson, G., Mathiesen, I., and Tjelle, T. E. (2003). DNA injection in combination with electroporation: A novel method for vaccination of farmed ruminants. *Scand. J. Immunol.* **57,** 229–238.

Tomanin, R., Friso, A., Alba, S., Piller Puicher, E., Mennuni, C., La Monica, N., Hortelano, G., Zacchello, F., and Scarpa, M. (2002). Non-viral transfer approaches for the gene therapy of mucopolysaccharidosis type II (Hunter syndrome). *Acta Paediatr. Suppl.* **91,** 100–104.

Trochon-Joseph, V., Martel-Renoir, D., Mir, L. M., Thomaidis, A., Opolon, P., Connault, E., Li, H., Grenet, C., Fauvel-Lafeve, F., Soria, J., Legrand, C., Soria, C., Perricaudet, M., and Lu, H. (2004). Evidence of antiangiogenic and antimetastatic activities of the recombinant disintegrin domain of metargidin. *Cancer Res.* **64,** 2062–2069.

Tsujie, M., Isaka, Y., Nakamura, H., Imai, E., and Hori, M. (2004). Electroporation-mediated gene transfer that targets glomeruli. *J. Am. Soc. Nephrol.* **12,** 949–954.

Tsurumi, Y., Takeshita, S., Chen, D., Kearney, M., Rossow, S. T., Passeri, J., Horowitz, J. R., Symes, J. F., and Isner, J. M. (1996). Direct intramuscular gene transfer of naked DNA encoding vascular endothelial growth factor augments collateral development and tissue perfusion. *Circulation* **94,** 3281–3290.

Tupin, E., Poirier, B., Bureau, M. F., Khallou-Laschet, J., Vranckx, R., Caligiuri, G., Gaston, A. T., Van Huyen, J., Scherman, D., Bariéty, J., Michel, J. B., and Nicoletti, A. (2003). Non-viral gene transfer of murine spleen cells achieved by *in vivo* electroporation. *Gene Ther.* **10,** 569–579.

Umeda, Y., Marui, T., Matsuno, Y., Shirahashi, K., Iwata, H., Takagi, H., Matsumoto, K., Nakamura, T., Kosugi, A., Mori, Y., and Takemura, H. (2004). Skeletal muscle targeting *in vivo* electroporation-mediated HGF gene therapy of bleomycin-induced pulmonary fibrosis in mice. *Lab. Invest.* **84,** 836–844.

Vicat, J. M., Boisseau, S., Jourdes, P., Lainé, M., Wion, D., Bouali-Benazzouz, R., Benabid, A. L., and Berger, F. (2000). Muscle transfection by electroporation with high-voltage and short-pulse currents provides high-level and long-lasting gene expression. *Hum. Gene Ther.* **11,** 909–916.

Vilquin, J. T., Kennel, P. F., Paturneau-Jouas, M., Chapdelaine, P., Boissel, N., Delaère, P., Tremblay, J. P., Scherman, D., Fiszman, M. Y., and Schwartz, K. (2001). Electrotransfer of naked DNA in the skeletal muscles of animal models of muscular dystrophies. *Gene Ther.* **8,** 1097–1107.

Walther, W., Stein, U., Fichtner, I., Voss, C., Schmidt, T., Schleef, M., Nellessen, T., and Schlag, P. M. (2002). Intratumoral low-volume jet-injection for efficient non-viral gene transfer. *Mol. Biotechnol.* **21,** 105–115.

Wang, G., Williamson, R., Mueller, G., Thomas, P., Davidson, B. L., and McCray, P. B., Jr. (1998). Ultrasound-guided gene transfer to hepatocytes in utero. *Fetal Diagn. Ther.* **13,** 197–205.

Wang, L. Y., Sun, W., Chen, M. Z., and Wang, X. (2003). Intramuscular injection of naked plasmid DNA encoding human preproinsulin gene in streptozotocin-diabetes mice results in a significant reduction of blood glucose level. *Sheng Li Xue Bao* **55,** 641–647.

Watanabe, K., Nakazawa, M., Fuse, K., Hanawa, H., Kodama, M., Aizawa, Y., Ohnuki, T., Gejyo, F., Maruyama, H., and Miyazaki, J. (2001). Protection against autoimmune myocarditis by gene transfer of interleukin-10 by electroporation. *Circulation* **104,** 1098–1100.

Wells, J. M., Li, L. H., Sen, A., Jahreis, G. P., and Hui, S. W. (2000). Electroporation-enhanced gene delivery in mammary tumors. *Gene Ther.* **7,** 541–547.

Widera, G., Austin, M., Rabussay, D., Goldbeck, C., Barnett, S. W., Chen, M., Leung, L., Otten, G. R., Thudium, K., Selby, M. J., and Ulmer, J. B. (2000). Increased DNA vaccine delivery and immunogenicity by electroporation *in vivo*. *J. Immunol.* **164,** 4635–4640.

Widlak, W., Scieglinska, D., Vydra, N., Malusecka, E., and Krawczyk, Z. (2003). *In vivo* electroporation of the testis versus transgenic mice model in functional studies of spermatocyte-specific hst70 gene promoter: A comparative study. *Mol. Rep. Dev.* **65,** 382–388.

Wolff, J. A., Malone, R. W., Williams, P., Chong, W., Acsadi, G., Jani, A., and Felgner, P. L. (1990). Direct gene transfer into mouse muscle *in vivo*. *Science* **247,** 1465–1468.

Wu, C. J., Lee, S. C., Huang, H. W., and Tao, M. H. (2004). *In vivo* electroporation of skeletal muscles increases the efficacy of Japanese encephalitis virus DNA vaccine. *Vaccine* **22,** 1457–1464.

Xue, F., Takahara, T., Yata, Y., Minemura, M., Morioka, C. Y., Takahara, S., Yamato, E., Dono, K., and Watanabe, A. (2002). Attenuated acute liver injury in mice by naked hepatocyte growth factor gene transfer into skeletal muscle with electroporation. *Gut* **50,** 558–562.

Yamashita, Y. I., Shimada, M., Hasegawa, H., Minagawa, R., Rikimaru, T., Hamatsu, T., Tanaka, S., Shirabe, K., Miyazaki, J. I., and Sugimachi, K. (2001). Electroporation-mediated interleukin-12 gene therapy for hepatocellular carcinoma in the mice model. *Cancer Res.* **61,** 1005–1012.

Yamazaki, Y., Fujimoto, H., Ando, H., Ohyama, T., Hirota, Y., and Noce, T. (1998). *In vivo* gene transfer to mouse spermatogenic cells by deoxyribonucleic acid injection into seminiferous tubules and subsequent electroporation. *Biol. Rep.* **59,** 1439–1444.

Yang, L., Cheong, N., Wang, D. Y., Lee, B. W., Kuo, I. C., Huang, C. H., and Chua, K. Y. (2003). Generation of monoclonal antibodies against Blo t 3 using DNA immunization with *in vivo* electroporation. *Clin. Exp. Allergy* **33,** 663–668.

Yasui, A., Oda, K., Usunomiya, H., Kakudo, K., Suzuki, T., Yoshida, T., Park, H. M., Fukazawa, K., and Muramatsu, T. (2001). Elevated gastrin secretion by *in vivo* gene electroporation in skeletal muscle. *Int. J. Mol. Med.* **8,** 489–494.

Yin, D., and Tang, J. G. (2001). Gene therapy for streptozotocin-induced diabetic mice by electroporational transfer of naked human insulin precursor DNA into skeletal muscle *in vivo*. *FEBS Lett.* **495,** 16–20.

Yoshizato, K., Nishi, T., Goto, T., Dev, S. B., Takeshima, H., Kino, T., Tada, K., Kimura, T., Shiraishi, S., Kochi, M., Kuratsu, J. I., Hofmann, G. A., and Ushio, Y. (2000). Gene delivery with optimized electroporation parameters shows potential for treatment of gliomas. *Int. J. Oncol.* **16,** 899–905.

Yu, D. S., Lee, C. F., Hsieh, D. S., and Chang, S. Y. (2004). Antitumor effects of recombinant BCG and interleukin-12 DNA vaccines on xenografted murine bladder cancer. *Urology* **63,** 596–601.

Zaharoff, D. A., Barr, R. C., Li, C. Y., and Yaun, F. (2002). Electromobility of plasmid DNA in tumor tissues during electric field-mediated gene delivery. *Gene Ther.* **9,** 1286–1290.

Zang, W. P., Wei, X. D., Fu, S., Wang, S. W., Tang, J., and Wang, D. B. (2001a). Transfer and expression of recombinant human thrombopoietin gene in cos-7 cells and mice *in vivo*. *Zhongguo Shi Yan Xue Ye Xue Za Zhi* **9,** 14–17.

Zang, W. P., Wei, X. D., Wang, S. W., and Wang, D. B. (2001b). Thrombopoietic effect of recombinant human thrombopoietin gene transferred to mice mediated by electric pulse on normal and experimental thrombocytopenia mice. *Zhonghua Xue Ye Xue Za Zhi* **22,** 128–131.

Zerbib, D., Amalric, F., and Teissie, J. (1985). Electric field mediated transformation: Isolation and characterization of a TK+ subclone. *Biochem. Biophys. Res. Commun.* **129,** 611–618.

Zhang, G., Gao, X., Song, Y. K., Vollmer, R., Stolz, D. B., Gasiorowski, J. Z., Dean, D. A., and Liu, D. (2004). Hydroporation as the mechanism of hydrodynamic delivery. *Gene Ther.* **11,** 675–682.

Zhang, G. H., Tan, X. F., Shen, D., Zhao, S. Y., Shi, Y. L., Jin, C. K., Sun, W. G., Guo, Y. H., Chen, K. H., and Tang, J. (2003). Gene expression and antitumor effect following im electroporation delivery of human interferon alpha 2 gene. *Acta Pharmacol. Sin.* **24,** 891–896.

Zucchelli, S., Capone, S., Fattori, E., Folgori, A., Di Marco, A., Casimiro, D., Simon, A. J., Laufer, R., La Monica, N., Cortese, R., and Nicosia, A. (2000). Enhancing B- and T-cell immune response to a hepatitis C virus E2 DNA vaccine by intramuscular electrical gene transfer. *J. Virol.* **74,** 11598–11607.

Section 2

GENE REGULATION

6

In Vivo Application of RNA Interference: From Functional Genomics to Therapeutics

Patrick Y. Lu, Frank Xie, and Martin C. Woodle
Intradigm Corporation, Rockville, Maryland 20852

ABSTRACT

RNAi has rapidly become a powerful tool for drug target discovery and validation in cell culture, and now has largely displaced efforts with antisense and ribozymes. Consequently, interest is rapidly growing for extension of its application to *in vivo* systems, such as animal disease models and human therapeutics. Studies on RNAi have resulted in two basic methods for its use for gene selective inhibition: 1) cytoplasmic delivery of short dsRNA oligonucleotides (siRNA), which mimics an active intermediate of an endogenous RNAi mechanism and 2) nuclear delivery of gene expression cassettes that express a short hairpin RNA (shRNA), which mimics the micro interfering RNA (miRNA) active

Advances in Genetics, Vol. 54
0065-2660/05 $35.00
DOI: 10.1016/S0065-2660(05)54006-9

intermediate of a different endogenous RNAi mechanism. Non-viral gene delivery systems are a diverse collection of technologies that are applicable to both of these forms of RNAi. Importantly, unlike antisense and ribozyme systems, a remarkable trait of siRNA is a lack of dependence on chemical modifications blocking enzymatic degradation, although chemical protection methods developed for the earlier systems are being incorporated into siRNA and are generally compatible with non-viral delivery systems. The use of siRNA is emerging more rapidly than for shRNA, in part due to the increased effort required to construct shRNA expression systems before selection of active sequences and verification of biological activity are obtained. In contrast, screens of many siRNA sequences can be accomplished rapidly using synthetic oligos. It is not surprising that the use of siRNA *in vivo* is also emerging first. Initial *in vivo* studies have been reported for both viral and non-viral delivery but viral delivery is limited to shRNA. This review describes the emerging *in vivo* application of non-viral delivery systems for RNAi for functional genomics, which will provide a foundation for further development of RNAi therapeutics. Of interest is the rapid adaptation of ligand-targeted plasmid-based nano-particles for RNAi agents. These systems are growing in capabilities and beginning to pose a serious rival to viral vector based gene delivery. The activity of siRNA in the cytoplasm may lower the hurdle and thereby accelerate the successful development of therapeutics based on targeted non-viral delivery systems. © 2005, Elsevier Inc.

I. INTRODUCTION

A. RNA interference (RNAi): strong and selective gene inhibition

RNA interference has rapidly displaced antisense and ribozymes as the preferred means for sequence-specific gene inhibition in cell culture studies (Bantounas *et al.*, 2004; Lu *et al.*, 2003; McManus and Sharp, 2002). Along with rapid adoption as a tool for functional genomics, expanding studies on RNAi itself have greatly enhanced our understanding of this endogenous gene inhibition process since discovery that posttranscriptional gene silencing (PTGS) in plants is also active in animal cells. Despite rapidly expanding studies, RNAi biology is still far from being well understood. A major challenge is the growing appreciation that the RNAi system is involved in several endogenous activities with differing roles.

One of the earliest recognized roles of RNAi is as an antiviral response triggered by the double-stranded (ds) RNA genome of the double-stranded RNA virus. For such a response to be effective, strong and selective gene inhibition is important. This gene inhibition function operates through an active intermediate that is short fragments of the dsRNA genome, now called short interfering RNA (siRNA), as shown in Fig. 6.1. The RNAi antiviral response generally is accompanied by interferon and other events induced by recognition of the dsRNA viral genome. An important finding has been that introduction of siRNA with sequences matching endogenous genes, instead of invading virus genomes, leads to activation of the RNAi system and inhibition of that endogenous gene. Importantly, studies have shown that introduction of these artificial siRNA usually avoids the interferon and other concomitant biological responses (McManus and Sharp, 2002). Also, the RNAi machinery has been found to perform its function in the cytoplasmic compartment, reducing the hurdles for intracellular siRNA delivery compared to gene therapy requirements for delivery to the nucleus.

Another major RNAi role, but one that remains poorly understood, is the regulation of cellular activity by modulating endogenous gene expression. This function operates, at least in part, through an active intermediate that is a short expressed RNA with an imperfect palindrome sequence, now called

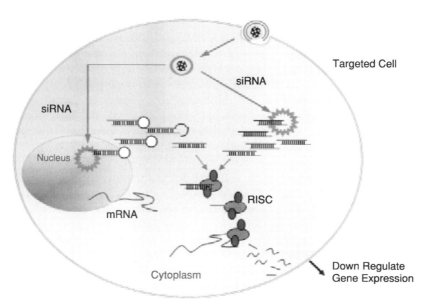

Figure 6.1. The two major pathways of RNAi can be invoked by delivering synthetic siRNA duplexes to the cytoplasm or delivering expression cassettes to the nucleus that produce "short hairpin" shRNA exported to the cytoplasm, either of which are taken up by cytoplasmic RISC machinery to down-regulate expression of the targeted gene. (See Color Insert.)

microinterfering RNA (miRNA). The palindrome sequence forms an imperfect dsRNA segment with a loop of single-stranded (ss) RNA at one end, or a hairpin type of structure, mimicked by expressed short hairpin RNA (shRNA). This RNAi role regulating endogenous gene expression may help explain why the introduction of siRNA matching endogenous genes shows strong and selective inhibition of matching endogenous genes.

Thus while the specific role of RNAi can vary considerably, the fundamental activity of RNAi is blocking expression of specific genes. Consequently, it is not surprising that different RNAi processes have very similar mechanisms of action and share machinery. The shared machinery may aid efforts to harness RNAi for our own purposes by allowing greater utility of the techniques developed. However, with versatility comes complexity and increased risk of inducing unwanted or unexpected RNAi activities. Ultimately, though, what makes the RNAi system so attractive is its fundamental capabilities to achieve very strong inhibition of specific genes, and with striking selectivity. This is clearly essential for biological systems to regulate themselves. It is also the goal of targeted therapeutic strategies, driven by realization that the most successful drugs act through inhibition of specific target proteins. A number of studies have evaluated the extent of gene inhibition specificity achieved with siRNA, in most cases indicating selectivity but not an absolute selectivity (Chi *et al.*, 2003; Jackson *et al.*, 2003; Kariko *et al.*, 2004; Semizarov *et al.*, 2003; Sledz *et al.*, 2003). In fact, it can be difficult to distinguish between an off-target effect on the wrong gene and downstream biological consequences from RNAi silenced genes. An important aspect that has yet to be addressed is the relative selectivity by siRNA versus other gene sequence-based methods for inhibiting gene expression, i.e., antisense. Nonetheless, while off-target effects have been observed, the most common outcome is a finding of substantial siRNA gene selectivity. Thus the use of appropriate controls is important.

So far, the common feature found in RNAi processes is reliance on a short RNA oligonucleotide whose sequence determines the gene to be inhibited. This feature has fueled a rapid adoption of RNAi for biology research through the design of siRNA with a sequence matching genes of interest. The siRNA oligonucleotides were originally identified as short dsRNA fragments generated by RNAi machinery called Dicer that processes long dsRNA, such as from a viral infection, into siRNA fragments. These siRNA were found to be an active intermediate used a second piece of the machinery, called RISC, to select mRNA for degradation, as described elsewhere (Bantounas *et al.*, 2004; Lu *et al.*, 2003; McManus and Sharp, 2002). The mechanism for miRNA appears to be similar but not identical to that of siRNA. One similarity is that the miRNA also appear to invoke some of the same RISC components and degrade the homologous mRNA before ribosomal translation can occur. The endogenous activity of RNAi to regulate endogenous gene expression, which clearly depends on both good specificity and strong inhibition, may explain its robust potency and selectivity. But while the nature of RNAi remains an area of active investigation, the use of short dsRNA sequences is proving to be an extremely robust and effective method to inhibit specific genes of interest, simply, and depending largely on the identification of unique sequences for that gene or splice variant (Fig. 6.1).

B. RNAi applications

In a short period of time, the use of siRNA to down-regulate expression of a specific gene has become the method of choice for cell culture or *in vitro* studies. Delivery of the siRNA duplex specifically targeting certain genes of interest in the cell tissue typically has been performed with siRNA targeting an individual gene, but multiples of siRNA duplexes inhibiting groups of genes are also possible. Beneficial or detrimental effects of siRNA inhibition of specific genes and inducing phenotypic changes in cells are analyzed using various means, including biochemical, pharmacological, and histological assays.

Currently, the use of siRNA to characterize gene function, and in particular exploration for potential therapeutic drug targets, is spreading over every aspect of biological research (Fig. 6.2). This phenomenon results from two basic realities: (1) siRNA is proving to be a very potent, robust, and easy to use inhibitor and (2) down-regulation of individual genes is a powerful tool for understanding the biological function of genes and biochemical pathways in the

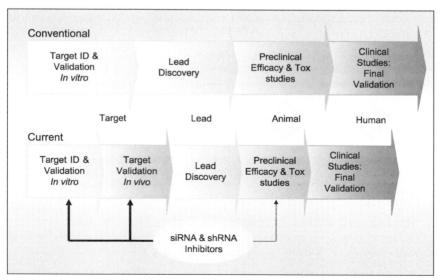

Figure 6.2. RNAi in Drug Discovery and Therapeutic Development. (See Color Insert.)

development and control of pathology. The latter has led to considerable adoption of siRNA for functional genomics looking for inhibition of genes that can generate therapeutic benefits reversing pathological effects, often caused by overexpression of those same genes. However, the approach clearly depends on the effectiveness of siRNA delivery to modulate, very selectively, the expression of specific genes and, as a result, to induce physiological or pharmacological effects. A key requirement is identification of an effective site in the target mRNA sequence for siRNA accessibility. Also, for most applications, effective transfection of the siRNA agent into the cytoplasmic compartment is critical (Bantounas *et al.*, 2004; McManus and Sharp, 2002).

Not surprisingly, cancer research is a dynamic and exciting area for the application of siRNA inhibitors (Lu *et al.*, 2003). However, research in many different therapeutic areas is invoking application of siRNA for use as a research tool for the validation of gene functions (Xie *et al.*, 2004). Regardless of therapeutic area, the success of *in vitro* transfection of siRNA into cells for target research has led to a strong interest in extending those studies to *in vivo* systems. However, *in vivo* delivery of siRNA into specific tissues of animal disease is much more complicated. Although increasing numbers of studies on target identification and validation using siRNA *in vitro* have been reported, limited reports of *in vivo* studies have indicated a lack of effective delivery methods for siRNA agents. The key to *in vivo* application is a delivery system that transports the siRNA into the target tissue and into the cell cytoplasm, or shRNA expression cassette to the nucleus much like the dependence of gene therapy on appropriate delivery methods. Given the extensive efforts in nucleic acid delivery for gene therapy, many delivery tools can be considered for siRNA or shRNA delivery, including biological-based vectors and synthetic chemical- and physical-based systems.

A consequence of the fast-growing literature on using siRNA as a research tool for functional genomics is an emerging interest in siRNA as a therapeutic. Therapeutic applications, even more than functional genomics, clearly depend on optimized local and systemic delivery of siRNA *in vivo*. Therefore, means for delivering siRNA into targeted tissues so as to maintain its activity within targeted cells and on the targeted gene sequence are the key aspects considered here in order to fulfill the goals of both functional genomic research and therapeutic development.

II. DELIVERING SHORT INTERFERING RNA (siRNA) *IN VIVO*

The effectiveness of siRNA *in vitro* has been demonstrated by inhibiting expression of many different genes and in many different cell types. In virtually all cases, the results depend on means for intracellular delivery, although transfection of siRNA into cultured cells is relatively easy compared with plasmids and other gene expression systems (Bantounas *et al.*, 2004; McManus and Sharp, 2002). A wide variety of nucleic acid delivery systems have been developed, including viral vectors and "non-viral" approaches, achieving efficient and significant modulation of gene expression for many types of cells. Regardless of whether the systems are biologics such as shRNA expression vectors (Davidson and Paulson, 2004; Pardridge, 2004) or chemical agents delivering synthetic siRNA, the RNAi tools can be used to decrease gene expression and, if the gene is important, result in phenotypic changes. The extension of *in vitro* success to *in vivo* systems is emerging, but further improvement remains a critical need for application of siRNA to drug target research, and potential clinical application.

A. "Non-viral" RNAi agent classes

Several methods have been developed to utilize the endogenous RNAi machinery to knock down specific genes of interest. So far, all the available methods rely on introducing, by one means or another, a short double-stranded RNA oligonucleotide matching a part of the gene of interest. The original method relied on introduction into cells of a synthetic dsRNA oligonucleotide that resembles the natural siRNA oligonucleotides. The artificial siRNA is taken up by the RISC machinery and used to block the expression of endogenous mRNA containing a matching sequence. This approach remains the most prominent method for high-throughput RNAi research, as short oligonucleotide synthesis can be fast and performed in high throughput. It is limited primarily by the cost of synthetic oligonucleotides and the need for repeated chemical synthesis.

A growing effort has developed around methods to deliver an expression cassette to produce the RNAi intermediate. An initial challenge for this approach was the difficulty of getting expression of two short complementary RNA oligonucleotides within the cell to result in their hybridization to yield the dsRNA oligonucleotide. One solution to this problem is expressing a single RNA with a palindromic sequence so that it can hybridize with itself and form a stem–loop "hairpin" structure much like miRNA, now called short hairpin RNA (shRNA). Theoretically, such expression cassettes can be introduced by any of the standard gene delivery systems, including viral and non-viral methods. A growing number of efforts are being reported for the construction of such systems and evaluation for *in vivo* application (Davidson and Paulson, 2004; Devroe and Silver, 2002; Kobayashi *et al.*, 2004; Pardridge, 2004; Tiscornia *et al.*, 2003; Xia *et al.*, 2002) However, the time and effort needed to construct such expression cassettes and incorporate them into delivery systems add considerably to the time and cost to set up a study for a gene of interest. This is particularly true when it is realized that several sequences must be tested to find an effective sequence. Unfortunately, identification of active sequences using synthetic siRNA oligonucleotides is not always predictive of active shRNA sequences, reducing the utility of preliminary siRNA screens to address this problem. Once an effective expression cassette is identified, the materials usually can be replenished through standard molecular biology techniques.

B. "Non-viral" delivery systems

A wide range of nucleic acid delivery systems have been devised without adaptation of virus and thus fall into the broad category called "non-viral" vectors. Lumped together in this poorly defined class are many different systems covering vastly different forms of nucleic acid. Likewise, this class covers a wide variety of methods to formulate and deliver the nucleic acid to enable intracellular delivery, including cationic complexes with lipids, polymers, or both, PLGA microsphere depot formulations, hydrophilic protective polymers, physical force-based delivery such as bombardment with nucleic acid-coated gold particles or electroporation, and many other delivery methods. Note that a common feature conveyed by the non-viral category is their operability with either natural or synthetic forms of the nucleic acid, e.g., from plasmids to phosphorothioate antisense oligonucleotides. In addition, the success of each for delivery *in vivo* depends on many factors, such as route and tissue uptake. The greatest success has been attained for local delivery to tissues, e.g., PLGA microspheres and electroporation for muscle delivery. Only a few successes have been attained for systemic delivery such as high-pressure tail vein administration or "hydrodynamic" delivery for liver hepatocytes and the DOTAP lipoplex for lung tissues. Some of these delivery methods have limited potential application in humans, e.g., hydrodynamic delivery, but their capabilities can be enabling for functional genomics research investigations *in vivo*, where the criteria for success are very different (Lu *et al.*, 2003).

One of the important non-viral delivery methods is based on the use of cationic transfection agents to form complexes with oligonucleotides, and these methods have been shown to be active with siRNA *in vivo*. The most commonly used transfection reagents have been cationic lipids, producing lipoplex complexes (Spagnou *et al.*, 2004). Sorensen *et al.* (2003) examined delivery with cationic liposome-based administration in mice of siRNAs combined with a matching plasmid encoding the green fluorescent protein (GFP) and found inhibition *in vivo* of the GFP plasmid gene expression. Furthermore, the studies evaluated the intraperitoneal injection of anti-TNF-α siRNA inhibited lipopolysaccharide-induced tumor necrosis factor-α (TNF-α) gene expression, whereas secretion of interleukin (IL)1-α was not inhibited. As found in gene therapy efforts, development of non-viral delivery methods that achieve tumor-targeted siRNA delivery and activity from an intravenous administration has proven to be difficult. One study reported use of a cationic chemical derivative of cardiolipin to form lipoplexes with siRNA targeting the cRAF (also known as RAF-1) oncogene that inhibited tumor growth in a sequence-specific manner but the study neglected to evaluate whether the target gene was specifically inhibited (Chien *et al.*, 2004). This confirmation of target inhibition can help reveal effects that cannot be attributed to an RNAi mechanism as found in earlier studies using an aqueous administration of siRNA (Filleur *et al.*, 2003).

The cationic lipoplex and polyplex systems have formed the foundation for numerous efforts to develop ligand-targeted gene delivery systems. Studies have begun to emerge for adaptation of these systems for siRNA (Kakizawa *et al.*, 2004; Khan *et al.*, 2004; Shiffeleres *et al.*, 2004b). Results obtained using ligand targeting of plasmids to tumor neovasculature suggest that systemic tumor therapeutics using gene delivery is a possibility (Hood *et al.*, 2002; Ogris *et al.*, 2003; Walker, 2005; Woodle *et al.*, 2001), but this approach focuses on delivery to endothelial cells of the tumor blood vessels. The efficiency of plasmid expression is also limited, as trafficking to the nucleus is thought to be inadequate. For siRNA, its activity in the cytoplasm reduces the challenge. Results have been published using systemic administration of a targeted nanoplex using an Arg-Gly-Asp (RGD) motif peptide ligand that combines selective localization at neovasculature with delivery of siRNA inhibiting the vascular endothelial growth factor (VEGF) pathway gene expression driving angiogenesis whether in tumors or eye disease (Kim *et al.*, 2004; Schiffelers *et al.*, 2004b). Importantly, in both studies, results showed siRNA sequence-specific target gene inhibition in the pathological tissue. Such systems introducing dual targeting offer the promise of increasing selectivity in targeted therapeutic development. These advances in cationic complexes forming layered nanoparticles appear promising for the therapeutic application of siRNA for metastatic cancer and many other angiogenesis-related diseases.

Non-viral vectors have many advantages, especially for therapeutics, over viral vectors. For example, immunogenicity of viral vectors has precluded multiple administrations and resulted in severe toxicity limitations. Chromosomal integration, and the resulting safety concerns, is also avoided by non-viral vectors. While potential safety issues are not a limitation for functional genomics, immunogenicity and other side effects can obscure the results and need to be avoided. Clearly, effective siRNA delivery is crucial for *in vivo* functional genomics, i.e., inhibition of gene expression in a significant number of cells as a means to emulate the biological effect of a drug targeting the same protein can be emulated. This means that the siRNA needs to be delivered to approximately the same percentage of the same cells and the cells in the same tissue as will be affected by the candidate drug. As mentioned for viral vectors, an equally important requirement for non-viral vectors is delivery without significant background activity from the delivery method itself. Non-viral delivery methods tend to have much lower levels of biological effects, other than that from the gene, relative to viral vectors with their many protein components.

C. *In vivo* siRNA delivery

Delivering RNAi agents *in vivo* basically falls into two approaches: local and systemic administration. Each of these approaches can be uniquely suitable for a particular tissue type or biological study system. Or, they can be used independently to reach the same targeted organ as a means to verify that the observed gene inhibition outcome is not due to the delivery method. For example, skin and muscle can be better accessed using local delivery, whereas lung and tumor can be reached efficiently by both local and systemic deliveries. The choice between local and systemic delivery largely depends on what tissues and cell types are targeted and the expected outcome for siRNA-mediated gene knockdown in terms of biological readout. In addition, the choice usually is only a part of the entire consideration of study design involving vector carriers, administration routes, and approaches for siRNA delivery *in vivo*. Increasing data show that siRNA is a very potent sequence-specific inhibitor in many tissue types.

1. Local administration

a. Ocular

An increasing number of clinical protocols are appearing for treating eye diseases with nucleic acid drugs such as antisense or RNA aptamers. These clinical delivery approaches appear suitable for siRNA administration but depend on local administration. Using a model of retinal neovascularization induced by laser damage, Reich and colleagues (2003) delivered nonformulated siRNA specific to murine VEGF to the subretinal space and observed significant reduction of eye angiogenesis. Importantly, unlike antisense or RNA aptamers, this work indicated that chemical protection of the siRNA was not essential, at least in the intravitreal compartment of the eye. Using a different murine model, one for herpetic stromal ketatitis, which develops from herpes simplex virus infection, the laboratory of Barry Rouse also reported the use of polyplexes for local administration of siRNA to inhibit the VEGF pathway (Kim *et al.*, 2004). This work compared local with systemic targeted delivery, as discussed later. It now seems clear that local ocular delivery of siRNA, through different routes with different formulations, can be used to inhibit genes sufficiently for gene function studies. The work also provides strong support for the therapeutic application of siRNA for ocular disease, especially ocular neovascularization.

b. Brain

The brain tissues are the foundation of the central nervous system (CNS), obviously a very important biological system and one representing considerable interest for both functional genomics and therapeutics. This tissue is one in which capabilities to inhibit genes selectively *in vivo*, for either application, are especially critical, as nerve function depends on the actual nerve network within the brain (Buckingham *et al.*, 2004). Initial efforts for brain delivery focused on expressed RNAi constructs (Davidson and Paulson, 2004), but Dorn *et al.* (2004) evaluated the infusion of aqueous solutions of chemically protected siRNA oligonucleotides directly into the brain and found that selective gene inhibition could be obtained. However, the special nature of the latter method, including specialized chemical modifications of the siRNA and surgically implanted infusion pumps delivering high doses, limits its usefulness for many laboratories interested in this area. Other studies on the use of cationic formulations for the delivery of siRNA to brain found that delivery was more effective with the tested lipoplex than with the tested polyplex (Hassani *et al.*, 2004). The success observed in these diverse studies shows the promise of siRNA delivery to brain tissues for functional genomics and, in the long term, good potential for therapeutic applications.

c. Tumoral

For functional validation of the tumorigenic genes, intratumoral delivery of siRNA is a very attractive approach. The rapid extension of siRNA functional genomics studies *in vitro*, now very well accepted, to studies in preclinical human xenograft tumor models (Ding *et al.*, 2004), also well accepted as a key stepping stone to clinical investigation, is limited only by a lack of good local tumor delivery. Unfortunately, like earlier gene therapy studies, local delivery of siRNA into tumors is not efficient with either aqueous solutions (so-called "naked" nucleic acid) or standard transfection reagents, including the best cationic lipids and polymers. In fact, local administration of aqueous siRNA into tumors was found less effective than a distal systemic administration on an endogenous reporter gene (Filleur *et al.*, 2003). In that study, effects observed from distal administration were not attributable to an RNAi activity and the authors concluded that tumor delivery was inadequate. Some work has begun to emerge for the local tumor delivery of siRNA, showing sufficient gene inhibition for target validation (Lu *et al.*, 2002; Minakuchi *et al.*, 2004).

d. Pulmonary

Non-viral delivery of nucleic acid into the airway has been a very active area, in part due to intense efforts for the development of therapy to treat cystic fibrosis patients. Because of an overwhelming immune response to adenoviral vectors, efforts for pulmonary nucleic acid delivery turned to non-viral carriers to draw upon their generally reduced toxicity and immune response. Unfortunately, non-viral methods have not yielded positive results in clinical investigations, but may be more effective with siRNA given its lack of dependence on nuclear delivery.

Evidence has appeared that siRNA can effectively knockdown endogenous genes (Zhang et al., 2003) and viral proteins of a group of RNA viruses, e.g., influenza (Ge et al., 2003; Tompkins et al., 2004) and SARS coronavirus (Li et al., 2005), resulting in significant effects of antiviral infection in various mammalian cell systems. The *in vitro* proven siRNA for influenza were evaluated for *in vivo* pulmonary delivery using standard non-viral vectors and found to inhibit influenza virus pulmonary infection (Ge et al., 2004; Tompkins et al., 2004). However, these delivery methods are not strong candidates for clinical application and better methods are needed. In a separate study with a clinically feasible intranasal delivery system and using SARS virus siRNA, results using a primate disease model found good evidence of both safety and efficacy (Li et al., 2005). Thus it appears that the many non-viral gene delivery systems are applicable for pulmonary delivery of siRNA.

e. Skeletal–muscular

The tissues of the skeletal–muscle system, the joints and skeletal muscles, are relatively accessible for local administration methods. Thus the key to using siRNA sequences directed to any gene target implicated in rheumatic and musculoskeletal diseases is effective local delivery enabling intracellular delivery of siRNA agents at the diseased site and, more specifically, the specific cellular components of the disease pathology (Rutz and Scheffold, 2004). Direct injection of siRNA molecules has been reported by hydrodynamic pressure in isolated limbs (Hagstrom et al., 2004) or formulated with cationic lipids or polymers can be considered for local delivery, but tend to induce inflammation, which is contraindicated for arthritic disease studies. Also, for plasmid delivery to skeletal muscle, eletroporation has proven the most robust method and recently applied to siRNA delivery, (Golzio et al., 2004). The extension of this method to inflamed joints in an arthritis model has shown promise as a method for functional genomic studies (Schiffelers et al., 2005), but clinical application of this approach has yet to be described. Ultimately, the biology of the inflammation and immune response, which is central to the rheumatic arthritic diseases, may benefit from cell targeting. To that end, the use of ligand-targeted nanoparticles, even with local intraarticular injection, is attractive as a potential means to achieve gene inhibition without exacerbating the pathological phenotype by the delivery system itself.

2. Systemic administration

a. Liver targeting

Some of the first published results showing activity of siRNA in mammals accomplished that feat by delivery into mouse liver. This was achieved using hydrodynamic delivery, a rapid injection of a large volume of aqueous solution into the mouse tail vein creating high pressure in the vascular circulation that leads to extensive delivery into hepatocytes (Lewis *et al.*, 2002; McCaffrey *et al.*, 2002; Song *et al.*, 2003; Zender *et al.*, 2003). Studies using hydrodynamic administration also found poor activity of aqueous siRNA when given by traditional intravenous administration, even when chemically stabilized siRNA were used (Layzer *et al.*, 2004), unlike earlier findings with antisense oligonucleotides. These studies showed that siRNA not only can inhibit exogenous plasmid expression, but also endogenous hepatocyte gene expression. The success implies that delivery of siRNA to the liver results in uptake by a majority of hepatocytes, as found for plasmid delivery. Interestingly, recent studies found that aqueous administration of siRNA didn't induce a non-specific interferon response (Heidel *et al.*, 2004). The hydrodynamic administration procedure allows the use of siRNA in mice for gene function and drug target validation studies (Sen *et al.*, 2003; Zender *et al.*, 2003), but is limited largely to research on liver function and metabolism or liver infectious diseases such as hepatitis. Although this particular method is not clinically feasible for human patients, it is an effective approach for gene function validation. Song *et al.* (2003) injected Fas siRNA and achieved down-regulation of both Fas mRNA and Fas protein in mouse hepatocytes, and the effects persisted without diminution for 10 days. Zender and colleagues (2003) delivered 21 nucleotide siRNAs against caspase 8, resulting in inhibition of caspase 8 gene expression in the liver, thereby preventing Fas (CD95)-mediated apoptosis. Giladi and colleagues (2004) used hydrodynamic delivery of siRNA to inhibit levels of hepatitis B viral transcripts, viral antigens, and viral DNA in liver and sera. Adenovirus has also been applied for liver delivery of expressed RNAi methods but mainly offers an orthogonal delivery approach to confirm that findings are due to gene inhibition and not specific to the delivery method. Although both hydrodynamic delivery and adenoviral vector are not clinically acceptable methods, these two systems have a relatively high hepatocyte efficiency, which is a very powerful tool for hepatocyte functional genomic studies.

Targeted systems for siRNA delivery into liver have also been a very attractive approach and are under development (Ren *et al.*, 2001). Clearly, such liver-targeted delivery systems are more clinically feasible for the development of siRNA-based therapeutics for the treatment of various liver-related metabolic and hepatitis viral diseases. One alternative method for liver-targeted delivery of siRNA is to use chemical modification of the oligonucleotide, as suggested by work with cholesterol conjugates (Soutschek et al., 2004). However, this work shows that at least three challenges must be addressed: adequate protection of the siRNA oligonucleotide from biological degradation en route to the liver, protection of the siRNA oligonucleotide from rapid glomerfiltration by the kidney into the urine (actually this is more important than stabilization), and selective uptake by the target hepatocytes. The use of cholesterol conjugates was found to require very high doses, suggesting that the distribution was widespread rather than liver targeted. Further efforts to attain liver targeted siRNA are warranted.

b. Tumor and neovasculature targeting

Although primary tumors often can be reached by local administration, malignant tumors grow fast and spread throughout the body via blood or the lymphatic system. Their unpredictable and uncontrolled growth makes malignant cancer the most dangerous and fatal form of cancer. Metastatic tumors established at distant locations usually are not encapsulated and thus are more amenable for systemic delivery. Therefore, local delivery of siRNA or any therapeutic agent is limited to a few tumor types, such as head-and-neck cancer and melanoma. For functional genomics, local administration methods (as discussed earlier) can meet the requirements for most studies by acting on primary tumors, even in xenograft models, which form the basis of most cancer biology research. However, for therapeutics, systemic delivery of siRNA is needed.

Systemic delivery imposes several requirements and greater hurdles than local delivery. For siRNA, it must provide increased oligonucleotide stability in the blood and in the local environment before it enters the target cells. In addition, it often requires the siRNA to pass through multiple tissue barriers to reach the target cell and may benefit from protection from loss of activity once within the target cells. Sorenson et al. (2003) described the use of lipoplex gene delivery systems for the systemic delivery of siRNA, attaining a strong lung delivery as found when plasmid DNA is used. Other work with siRNA lipoplexes formed with a cardiolipid analogue showed phenotypic effects on tumor growth but lack confirmation of an RNAi activity at the tumor and suggest that much of the distribution is to the lung as expected from a lipoplex. An emergence of targeted nanoparticle systems, initially for plasmids or DNA oligonucleotides for cancer, is much more promising (Hood et al., 2002; Pun et al., 2004; Woodle et al., 2001). The nature of these systems to protect and deliver isolated DNA suggests that they can be adapted for siRNA oligonucleotides. Results support this conclusion using an RGD peptide ligand-directed nanoparticle for antiangiogenic treatment for cancer (Dubey et al., 2004) by systemic siRNA delivery (Schiffelers *et al.*, 2004b). These results provide growing support for the development of systemic siRNA treatments for cancer, with a growing proof of concept for antiangiogenic modalities as reviewed elsewhere (Lu *et al.*, 2005). The RGD ligand-targeted nanoparticle for targeting neovasculature has also been studied in ocular neovascularization models (Kim *et al.*, 2004). The results further strengthen the support for this approach as a clinically feasible method for siRNA therapeutics. They also provided a demonstration that siRNA for several genes can be combined in the same nanoparticle to give a better inhibition of disease pathology.

Of course a direct means to translate systemic gene therapy delivery methods to RNAi is through the use of shRNA expression methods. At this time though the advantages and disadvantages of systemic tumor delivery for expression-based RNAi versus siRNA oligonucleotide-based RNAi have yet to be fully elucidated. One clear advantage of siRNA, however, is rapidity with which different siRNA sequences and the matching genes can be studied. Importantly, the siRNA facilitates rapid studies of the interactions of genes, as combination of siRNA oligonucleotides is easy and effective (Kim *et al.*, 2004).

III. *IN VIVO* siRNA APPLICATIONS

A. *In vivo* functional genomics

The traditional approach to identifying genes involved in specific biological processes begins with the determination of genes correlating with the occurrence of that process, i.e., genes that are up- or down-regulated in cells and tissues of the biological system of interest. When applied to pathological processes, this approach generates large pools of up-regulated genes that are candidates as drug targets. However, the success of this approach is limited, as correlation is insufficient for causation or for efficacy. Consequently, a tool that can selectively down-regulate individual genes within the cells and tissues of a pathological process is widely recognized as a valuable means to understand gene function, especially for facilitating the discovery of protein targets for drug intervention. Because siRNA is proving to be a potent and robust means for this important objective, it is being rapidly adopted as the preferred functional genomics tool, at least in cell culture (Buckingham *et al.*, 2004).

One of the key hurdles for drug target discovery or validation directly in animal disease models has been a lack of effective *in vivo* nucleic acid delivery methods. While antisense activity in animals has been limited in general, intravenous administration of aqueous formulations has been used to inhibit liver gene function. Using hydrodynamic delivery of siRNA (Layzer *et al.*, 2004; Lewis *et al.*, 2002; McCaffrey *et al.*, 2002; Sen *et al.*, 2003) is one way to use siRNA gene inhibition for liver gene function. A key element of the challenge is often a large amplitude of inhibition (or overexpression) of the gene for a significant effect on phenotype to be observed. Requirements for large phenotypic effects appear to be a bigger challenge for therapeutics than for gene function studies, particularly for several classes of genes and proteins.

In cancer, the tumorigenesis process is thought to be the result of an abnormal overexpression of oncogenes, growth factors, and mutant tumor suppressors, even though underexpression of other proteins also plays critical roles. Efforts to identify and validate tumorigenic targets have been focused mainly on those targets overexpressed in the tumor tissues and promoting tumorigenesis as a means to enable development of small molecule and antibody anticancer drugs acting through an inhibitor mechanism. There are rapidly increasing numbers of reports demonstrating that siRNA and shRNA are able to knockdown tumorigenic genes *in vitro*, with emerging reports for *in vivo* (Lu et al., 2003b; Xie *et al.*, 2004). Studies designed to reveal whether a gene target plays a tumorigenic role use siRNA duplexes specifically targeting its mRNA sequence to knock down its expression and then observing whether the effect on pathology is a direct and specific effect. One method has been described for a unique target identification approach (Lu *et al.*, 2003a) named Efficacy-First discovery. This method utilizes efficient nucleic acid delivery into xenograft tumors to induce phenotypic effects on tumor growth rate as a method to verify or validate candidate genes for a controlling role in tumor growth behaviors. Using these methods, a pool of gene targets was identified for changes in expression that correlated with changes in tumor growth rate, i.e., their expression correlated with efficacy. Among the well-known cancer cell growth factors, VEGF and VEGF R2 represent two of the most widely recognized and highly validated targets. In these studies, siRNA-mediated downregulation of these endogenous genes in clinically relevant animal models was used as part of a functional genomics study in cancer (Lu *et al.*, 2003a). Marked tumor growth inhibition was observed following repeated delivery of the siRNAs specific to hVEGF and mVEGFR2, which accompanied knockdown of the growth factor at both mRNA and protein levels. The results illustrate the feasibility of using siRNA delivery in animal tumor models for drug target validation according to its ability to achieve efficacy (Lu *et al.*, 2003a). This *in vivo* delivery of siRNA enables differentiation of genes to find out which play a disease-controlling role in tumor growth. This validation process, called disease-control validation, is based on tumor efficacy by siRNA *in vivo* and with clinically relevant xenograft tumor models to give a clear indication of the importance of specific proteins as a drug target (Lu *et al.*, 2003a).

Once siRNA studies in cell culture provide a better understanding of the mechanism of action of candidate gene targets, they lay the foundation for use of the siRNA in a disease model. To be effective, *in vivo* functional genomics requires high efficiency *in vivo* delivery and good activity in a clinically relevant disease model for the clinical indication of interest. Results obtained to date indicate that the gene inhibition of siRNA is not only a sequence-specific effect, but is also sustainable and obtainable with relatively few side effects (low noise level). Importantly, this approach offers an ability to gain insight into the genes and proteins associated with the later stages of pathology, as it can be applied to established disease tissues. Such capabilities to move rapidly from cell culture into clinically relevant disease models promise to revolutionize the speed with which drug targets can be identified and validated for drug discovery. Equally important, the use of siRNA *in vivo* offers the prospects for rapidly obtaining a better understanding of how a particular target can be used to achieve a clinically meaningful therapeutic intervention in disease. This capability is also important for basic research into disease biology, since selective gene inhibition tools and ones that can be constructed quickly and that can operate in animal models will facilitate the rapid expansion of our knowledge. Thus the emerging capability of *in vivo* delivery of siRNA is greatly expanding its power as a tool for functional genomics.

B. Therapeutics

The specificity and potency of siRNA in cell culture and in animal studies suggest that it may be useful as a therapeutic agent (Schiffelers *et al.*, 2004a). However, the development of siRNA as a therapeutic agent faces a number of challenges, especially for systemic routes of administration. The most critical hurdle for *in vivo* delivery is attaining adequate delivery to disease tissue and cells. A growing number of studies are being reported that evaluate the prospects of siRNA for therapeutics, as summarized in Table 6.1.

Surprisingly, *in vivo* stability has not proven to be the major rate-limiting barrier, even for intravenous administration. In fact, rapid excretion of siRNA from blood into the urine occurs before degradation. Nonetheless, efforts are underway to increase its biological stability, primarily via medicinal chemistry originally developed for oligonucleotides. However, chemical stabilization does not address key requirements for better pharmacokinetics and tissue distribution. With conjugation of lipophilic residues to increase serum protein binding, improved pharmacokinetics of the oligonucleotide, alteration of its biodistribution, and reduced urinary excretion were observed (Soutschek *et al.*, 2004). Ultimately, such chemical modification must also address requirements for entering targeted cells, overcoming endosomal and other intracellular barriers, and retaining activity with the cellular RNAi machinery. A different approach to solving these problems using ligand-targeted nanoparticles (Kim *et al.*, 2004; Pardridge, 2004; Schiffelers *et al.*, 2004b) is showing good activity even without chemical stabilization. The fundamental thinking behind this approach is that RNA interference is a natural process involving a complicated

Table 6.1. *In Vivo* Delivery of siRNA for Developing Novel Therapeutics

Therapeutic Applications	Target gene	Model	Delivery vehicle/route	RNAi phenotype and reference
Cancer therapy	VEGF	MCF-7/nude, MDA-MB-435/ nude	Polymer based, intratumoral injection	Reduction of VEGF and inhibition of tumor growth (Lu et al., 2003a,b)
	VEGF R2	N2A/nude	Ligand-targeted nanoparticle, i.v. injection	Tumor growth inhibition (Schiffelers et al., 2004)
	VEGF	PC-3/nude	Atelocollagen, intratumoral injection	Suppressed tumor angiogenesis and tumor growth (Takei et al., 2004)
	VEGF	JT8/nude MDA-MB-231/ SCID	Naked siRNA, i.v., injection	Reduction in VEGF expression and inhibition of tumor growth (Filleur et al., 2003)
	c-raf		Cationic cardiolipin analogue based liposome (CCLA), i.v., injection	Tumor growth inhibition (Chien et al., 2004)
	RRM2	Orthotopic pancreatic/nude	Systemically administration	Suppressed tumor growth, increased tumor apoptosis and inhibition of metastasis through the synergism between RRM2 siRNA and gemcitabine (Duxbury et al., 2004)
Ocular diseases	VEGF	Mice/laser photocoagulation	Saline, local injection	Inhibited choroidal neovascularization (Reich et al., 2003)
	VEGF R1/R2	Mice/HSV induction	Ligand-targeted nanoparticle, i.v. injection	Anti-angiogenesis effect demonstrated by reduction of the neovasculature areas (Kim et al., 2004)
	TGF-beta RII	C57BL6 mice	Polymer-based, subconjunctival injection TransIT-TKO, local injection	Reduced the imflammatory response and matrix deposition, (Nakamura et al., 2004)

Rheumatoid arthritis	TNF-α	Mice	i.a. local injection	Inhibition of collagen-induced arthritis (CIA) (Schiffeleres et al., 2005)
Anti-viral therapy	Influenza A virus genes	C57BL/6 mice	PEI, i.v. administration	Reduced virus production in lungs of infected mice (Ge et al., 2004)
	Influenza A virus genes	BALB/c mice	PBS, hydrodynamic i.v., injection; oligofectamine, intranasal administered.	Reduced lung virus titers in infected mice and protected animals from lethal challenge (Stephen et al., 2004)
	SARS virus genes	Monkey	Intranasal administered	Inhibited SARS virus replication and reduced the SARS-like symptom in infected monkey (Li et al., 2004)
	HBV genes	BALB/c mice	PBS, hydrodynamic i.v. injection	Inhibition in the levels of HBV viral transcripts, viral antigens, and viral DNA detected in the liver and sera (Giladi et al., 2003)
CNS disease	P2X3	Rat models	Saline, intrathecal injection	Diminished P2X3 mRNA expression and P2X3 protein translocation, diminished pain responses, relieved chronic neuropathic pain (Dorn et al., 2004; Ganju et al., 2004)
	DAT	Mice	Saline, ventricular infusion	Down-regulation of DAT mRNA and protein in the brain. Elicited a temporal hyperlocomotor response (Thakker et al., 2004)
	Alpha(2A)-ARs	Rat	Saline	Decreased the levels of both alpha (2A)-AR mRNA and [(3)H] RX821002 binding sites in the brainstem. Decreased anxiety in the adult animals (Shishkina et al., 2004)

(Continues)

Table 6.1. (*Continued*)

Therapeutic Applications	Target gene	Model	Delivery vehicle/route	RNAi phenotype and reference
CNS disease (cont.)	GluR2, Cox-1	Rat	Electroporation	Reduction in the expression levels of both the mRNA and protein of target genes. Treated animals exhibits consistent physiological functions such as glutamate release from presynaptic sites, LTP and LTD (Akaneya et al., 2005)
Others	TNF-α	Mouse	DOTAP, i.p. injection	Inhibited lipopolysaccharide-induced TNF-α expression and development of sepsis (Sorensen et al., 2003; Sioud et al., 2003)
	GAPDH	Mouse	InfaSurf, intranasal administration	Lowered GAPDH protein in lung, heart, and kidney by approximately 50–70 1 and 7 days after siRNA administration (Massaro et al., 2004)
	HO-1	Mouse	Naked, intranasal administration	Enhanced apoptosis, via increased Fas expression and caspase 3 activity, in mouse lung during I-R injury (Zhang et al., 2004)
	ApoB	C57BL/6 mice	Stabilized Chol-siRNA, i.v. injection	Silenced the apoB mRNA in liver and jejunum, decreased apoB protein levels in plasma, and reduced total cholesterol (Soutschek et al., 2004)
	Caspase-8, caspase-3	C57BL/6 mice	10% lipiodol, high-volume portal vein injection	60% reduction in caspase-8 and caspase-3 expression, decrease ischemia/reperfusion injury to the liver (Conttreras et al., 2004)

cellular mechanism where siRNA is the predominant intermediate playing a sequence-specific silencing function. Therefore, preserving the biochemical authenticity of siRNA and improving its *in vivo* delivery efficiency with nanoparticle or other formulations will have the best chance for success in therapeutic application, and, importantly, the best chance to reduce unwanted side effects. Although local and topical delivery of siRNA may fit well for certain disease applications, the systemic delivery of siRNA will have much broader therapeutic applications. The ideal system for siRNA systemic delivery should be a nanoparticle with ligand-directed tissue localization (Fig. 6.3). This three-stage system should be able to first protect the siRNA duplex from excretion and degradation in body fluid such as blood and, at the same time, avoid aggregation and nonspecific binding of the nanoparticles. The system should also have targeting capability to reach disease tissue specifically. When the particle binds and enters the targeted cells, the siRNA content should be released for action. When the siRNA specifically inhibits expression of a disease-causing gene and protein, the clinical benefit will be achieved. As described earlier, a layered nanoplex system has been described that combines neovasculature-targeted

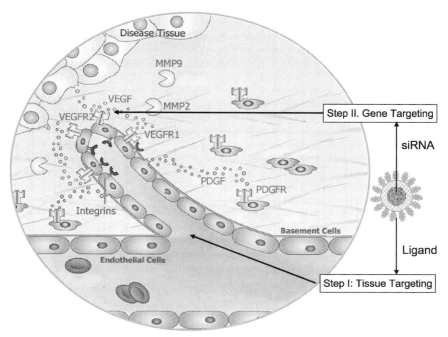

Figure 6.3. Dual-Targeted anti-angiogenesis siRNA systemically delivered using ligand-directed nanoparticle. (See Color Insert.)

nanoparticles giving cytoplasmic delivery with highly potent siRNA active in the cytoplasm. This nanoparticle system appears promising for the therapeutic application of siRNA for metastatic cancer and many other angiogenesis related diseases. The modular design of this nanoparticle should permit incorporation of other ligands applicable for siRNA delivery for other diseases.

Using RNAi for therapy is attractive to directly silencing pathogenic genes or disease-causing mutant genes. As disease mechanisms become increasingly clear, its application can be developed to silence newly identified genes involved in known pathogenic pathways. For example, an obvious siRNA target for the treatment of Alzheimer's disease (AD) is the β-site APP-cleaving enzyme BACE, which is required for the production of Aβ peptide and is present at elevated levels in the cortex of people with AD.

One of the greatest prospects for therapeutics is the ability to develop therapeutic agents that control multiple targets. This is one aspect of siRNA that offers considerable promise relative to the current set of targeted therapeutics such as antibodies. Considerable evidence shows that many types of human diseases result from overexpression of multiple disease-causing genes. Thus the potent and specific properties of siRNA that are controlled solely by changes to their sequence suggest that their combinations will be easily managed. Kim *et al.* (2004) showed the facility with which siRNA can provide combination therapeutics (an siRNA oligo cocktail) targeting multiple genes in the pathology to achieve much better therapeutic efficacy. This approach is based on two important facts: although inhibitory siRNA duplexes are sequence specific, all of them use an identical chemistry (dsRNA oligonucleotides); however, many human diseases are the result of overexpression of multiple endogenous and exogenous disease-causing genes. Using an siRNA oligo cocktail targeting multiple disease-causing genes represents an advantageous therapeutic approach with a synergistic effect. Nevertheless, even as a single agent, siRNA has tremendous therapeutic potential that will be realized when clinically feasible deliveries are developed.

References

Akaneya, Y., Jiang, B., and Tsumoto, T. (2005). RNAi-induced gene silencing by local electroporation in targeting brain region. *J. Neurophysiol.* **93,** 594–602.

Bantounas, I., Phylactou, L. A., and Uney, J. B. (2004). RNA interference and the use of small interfering RNA to study gene function in mammalian systems. *J. Mol. Endocrinol.* **33,** 545–557.

Buckingham, S. D., Esmaeili, B., Wood, M., and Sattelle, D. B. (2004). RNA interference: From model organisms towards therapy for neural and neuromuscular disorders. *Hum. Mol. Genet. Spec.* **No 2,** R275–R288.

Chi, J. T., Chang, H. Y., Wang, N. N., Chang, D. S., Dunphy, N., and Brown, P. O. (2003). Genomewide view of gene silencing by small interfering RNAs. *Proc. Natl. Acad. Sci. USA* **100,** 6364–6369.

Chien, P. Y., Wang, J., Carbonaro, D., Lei, S., Miller, B., Sheikh, S., Ali, S. M., Ahmad, M. U., and Ahmad, I. (2004). Novel cationic cardiolipin analogue-based liposome for efficient DNA and small interfering RNA delivery *in vitro and in vivo*. *Cancer Gene Ther* (2004 Dec 03).

Contreras, J. L., Vilatoba, M., Eckstein, C., Bilbao, G., Anthony Thompson, J., and Eckhoff, D. E. (2004). Caspase-8 and caspase-3 small interfering RNA decreases ischemia/reperfusion injury to the liver in mice. *Surgery* **136,** 390–400.

Davidson, B. L., and Paulson, H. L. (2004). Molecular medicine for the brain: Silencing of disease genes with RNA interference. *Lancet Neurol.* **3**(3), 145–149.

Devroe, E., and Silver, P. A. (2002). Retrovirus-delivered siRNA. *BMC Biotechnol.* **28,** 15.

Ding, Q., Grammer, J. R., Nelson, M. A., Guan, J. L., Stewart, J. E. Jr., and Glodson, C. L. (2004). p27(Kip1) and cyclin D1 are necessary for focal adhesion kinase (FAK) regulation of cell cycle progression in glioblastoma cells propagated *in vitro* and *in vivo* in the scid mouse brain. *J. Biol. Chem.* (2004 Nov 19).

Dorn, G., Patel, S., Wotherspoon, G., Hemmings-Mieszczak, M., Barclay, J., Natt, F. J., Martin, P., Bevan, S., Fox, A., Ganju, P., Wishart, W., and Hall, J. (2004). siRNA relieves chronic neuropathic pain. *Nucleic Acids Res.* **32,** e49.

Dubey, P. K., Mishra, V., Jain, S., Mahor, S., and Vyas, S. P. (2004). Liposomes modified with cyclic RGD peptide for tumor targeting. *J. Drug Target.* **12,** 257–264.

Duxbury, M. S., Ito, H., Zinner, M. J., Ashley, S. W., and Whang, E. E. (2004). RNA interference targeting the M2 subunit of ribonucleotide reductase enhances pancreatic adenocarcinoma chemosensitivity to gemcitabine. *Oncogene* **23,** 1539–1548.

Filleur, S., Courtin, A., Ait-Si-Ali, S., Guglielmi, J., Merle, C., Harel-Bellan, A., Clezardin, P., and Cabon, F. (2003). SiRNA-mediated inhibition of vascular endothelial growth factor severely limits tumor resistance to antiangiogenic thrombospondin-1 and slows tumor vascularization and growth. *Cancer Res.* **63,** 3919–3922.

Ganju, P., and Hall, J. (2004). Potential applications of siRNA for pain therapy. *Expert Opin Biol Ther.* **4,** 531–542.

Ge, Q., Filip, L., Bai, A., Nguyen, T., Eisen, H. N., and Chen, J. (2004). Inhibition of influenza virus production in virus-infected mice by RNA interference. *Proc. Natl. Acad. Sci. USA* **101,** 8676–8681.

Giladi, H., Ketzinel-Gilad, M., Rivkin, L., Felig, Y., Nussbaum, O., and Galun, E. (2004). Small interfering RNA inhibits hepatitis B virus replication in mice. *Mol. Ther.* **8,** 769–776.

Golzio, M., Mazzolini, L., Moller, P., Rols, M. P., and Teissie, J. (2004). Inhibition of gene expression in mice muscle by *in vivo* electrically mediated siRNA delivery. *Gene Ther.* (2004 Dec 09).

Hagstrom, J. E., Hegge, J., Zhang, G., Noble, M., Budker, V., Lewis, D. L., Herweijer, H., and Wolff, J. A. (2004). A facile non-viral method for delivering genes and siRNAs to skeletal muscle of mammalian limbs. *Mol. Ther.* **10**(2), 386–398.

Hassani, Z., Lemkine, G. F., Erbacher, P., Palmier, K., Alfama, G., Giovannangeli, C., Behr, J. P., and Demeneix, B. A. (2004). Lipid-mediated siRNA delivery down-regulates exogenous gene expression in the mouse brain at picomolar levels. *J. Gene Med.* (2004 Oct 28).

Heidel, J. D., Hu, S., Liu, X. F., Triche, T. J., and Davis, M. E. (2004). Lack of interferon response in animals to naked siRNAs. *Nature Biotech.* **22,** 1579–1582.

Hood, J. D., Bednarski, M., Frausto, R., Guccione, S., Reisfeld, R. A., Xiang, R., and Cheresh, D. A. (2002). Tumor regression by targeted gene delivery to the neovasculature. *Science* **296,** 2404–2407.

Jackson, A., Bartz, S. R., Schelter, J., Kobayashi, S. V., Burchard, J., Mao, M., Li, B., Cavet, G., and Linsley, P. S. (2003). Expression profiling reveals off-target gene regulation by RNAi. *Nature Biotechnol.* **21,** 635–637.

Kakizawa, Y., Furukawa, S., and Kataoka, K. (2004). Block copolymer-coated calcium phosphate nanoparticles sensing intracellular environment for oligodeoxynucleotide and siRNA delivery. *J. Control Release* **97,** 345–356.

Kariko, K., Bhuyan, P., Capodici, J., and Weissman, D. (2004). Small interfering RNAs mediate sequence-independent gene suppression and induce immune activation by signaling through toll-like receptor 3. *J. Immunol.* **172,** 6545–6549.

Khan, A., Benboubetra, M., Sayyed, P. Z., Ng, K. W., Fox, S., Beck, G., Benter, I. F., and Akhtar, S. (2004). Sustained polymeric delivery of gene silencing antisense ODNs, siRNA, DNAzymes and ribozymes: *In vitro* and *in vivo* studies. *J. Drug Target.* **12,** 393–404.

Kim, B., Tang, Q., Biswas, P. S., Xu, J., Schiffelers, R. M., Xie, F. Y., Ansari, A. M., Scaria, P. V., Woodle, M. C., Lu, P., and Rouse, B. T. (2004). Inhibition of ocular angiogenesis by siRNA targeting vascular endothelial growth factor pathway genes: Therapeutic strategy for herpetic stromal keratitis. *Am. J. Pathol.* **165,** 2177–2185.

Kobayashi, N., Matsui, Y., Kawase, A., Hirata, K., Miyagishi, M., Taira, K., Nishikawa, M., and Takakura, Y. (2004). Vector-based in vivo RNA interference: Dose- and time-dependent suppression of transgene expression. *J. Pharmacol. Exp. Ther.* **308,** 688–693.

Layzer, J. M., McCaffrey, A. P., Tanner, A. K., Huang, Z., Kay, M. A., and Sullenger, B. A. (2004). *In vivo* activity of nuclease-resistant siRNAs. *RNA* **10,** 766–771.

Lewis, D. L., Hagstrom, J. E., Loomis, A. G., Wolff, J. A., and Herweijer, H. (2002). Efficient delivery of siRNA for inhibition of gene expression in postnatal mice. *Nature Genet.* **32,** 107–108.

Li, B. J., *et al.* (2005). Prophylactic and therapeutic efficacies of siRNA targeting SARS coronavirus in rhesus macaques. Submitted for publication.

Lu, P. Y., Xie, F. Y., and Woodle, M. C. (2002). Tumor inhibition by RNAi-mediated VEGF and VEGFR2 down regulation in xenograft models. *Cancer Gene Ther.* **10**(Suppl. 1), 011.

Lu, P. Y., Xie, F. Y., Scaria, P., and Woodle, M. C. (2003a). From correlation to causation to control: Utilizing preclinical disease models to improve cancer target discovery. *Preclinica* **1,** 31–42.

Lu, P. Y., Xie, F. Y., and Woodle, M. C. (2003b). siRNA-mediated antitumorgenesis for drug target validation and therapeutics. *Curr. Opin. Mol. Ther.* **5,** 225–234.

Lu, P. Y., Xie, F. Y., and Woodle, M. C. (2005). Modulation of angiogenesis with siRNA inhibitors for novel therapeutics. *Trends Mol. Med.* **11,** 104–113.

Massaro, D., Massaro, G. D., and Clerch, L. B. (2004). Noninvasive delivery of small inhibitory RNA and other reagents to pulmonary alveoli in mice. *Am. J. Physiol. Lung Cell Mol. Physiol.* **287,** L1066–L1070.

McCaffrey, A. P., Meuse, L., Pham, T. T., Conklin, D. S., Hannon, G. J., and Kay, M. A. (2002). RNA interference in adult mice. *Nature* **418,** 6893, 38–39.

McManus, M. T., and Sharp, P. A. (2002). Gene silencing in mammals by small interfering RNAs. *Nature Med.* **3,** 737–747.

Minakuchi, Y., Takeshita, F., Kosaka, N., Sosaki, H., Yamamoto, Y., Kouno, M., Honma, K., Nagahara, S., Hanai, K., Sano, A., Kato, A., Terada, M., and Ochiya, T. (2004). Atelocollagen-mediated synthetic small interfering RNA delivery for effective gene silencing in vitro and in vivo. *Nucleic Acids Res.* **32,** e109.

Ogris, M., Walker, G., Blessing, T., Kircheis, R., Wolschek, M., and Wagner, E. (2003). Tumor-targeted gene therapy: Strategies for the preparation of ligand-polyethylene glycol-polyethylenimine/DNA complexes. *J. Controll. Release* **91,** 173–181.

Pardridge, W. M. (2004). Intravenous, non-viral RNAi gene therapy of brain cancer. *Expert Opin. Biol. Ther.* **4,** 1103–1113.

Pun, S. H., Tack, F., Bellocq, N. C., Cheng, J., Grubbs, B. H., Jensen, G. S., Davis, M. E., Brewster, M., Janicot, M., Janssens, B., Floren, W., and Bakker, A. (2004). Targeted delivery of RNA-cleaving DNA enzyme (DNAzyme) to tumor tissue by transferrin-modified, cyclodextrin-based particles. *Cancer Biol. Ther.* **3,** 641–650.

Reich, S. J., Fosnot, J., Kuroki, A., Tang, W., Yang, X., Maguire, A. M., Bennett, J., and Tolentino, M. J. (2003). Small interfering RNA (siRNA) targeting VEGF effectively inhibits ocular neovascularization in a mouse model. *Mol. Vision* **9**, 210–216.

Ren, T., Zhang, G., and Liu, D. (2001). Synthesis of galactosyl compounds for targeted gene delivery. *Bioorg. Med. Chem.* **9**, 2969–2978.

Rutz, S., and Scheffold, A. (2004). Towards *in vivo* application of RNA interference: New toys, old problems. *Arthritis Res. Ther.* **6**, 78–85.

Schiffelers, R. M., Woodle, M. C., and Scaria, P. (2004a). Pharmaceutical prospects for RNA interference. *Pharm. Res.* **21**, 1–7.

Schiffelers, R. M., Ansari, A., Xu, J., Zhou, Q., Tang, Q., Storm, G., Molema, G., Lu, P. Y., Scaria, P. V., and Woodle, M. C. (2004b). Cancer siRNA therapy by tumor selective delivery with ligand-targeted sterically stabilized nanoparticle. *Nucleic Acids Res.* **32**, e149.

Schiffelers, R. M., Xu, J., Storm, G., Woodle, M. C., and Scaria, P. V. (2005). Effects of treatment with small interferring RNA on joint inflammation in mice with collagen-induced arthritis. *Arthritis & Rheumatism* **52**, 1314–1318.

Semizarov, D., Frost, L., Sarthy, A., Kroeger, P., Halbert, D. N., and Fesik, S. W. (2003). Specificity of short interfering RNA determined through gene expression signatures. *Proc. Natl. Acad. Sci. USA* **100**, 6347–6352.

Sen, A., Steele, R., Ghosh, A. K., Basu, A., Ray, R., and Ray, R. B. (2003). Inhibition of hepatitis C virus protein expression by RNA interference. *Virus Res.* **96**, 27–35.

Shishkina, G. T., Kalinina, T. S., and Dygalo, N. N. (2004). Attenuation of alpha (2A)-adrenergic receptor expression in neonatal rat brain by RNA interference or antisense oligonucleotide reduced anxiety in adulthood. *Neuroscience* **129**, 521–528.

Sledz, C. A., Holko, M., de Veer, M. J., Silverman, R. H., and Williams, B. R. (2003). Activation of the interferon system by short-interfering RNAs. *Nat. Cell Biol.* **5**, 834–839.

Song, E., Lee, S. K., Wang, J., Inee, N., Ouyang, N., Min, J., Chen, J., Shankar, P., and Lieberman, J. (2003). RNA interference targeting Fas protects mice from fulminant hepatitism. *Nature Med.* **9**, 347–351.

Sorensen, D. R., Leirdal, M., and Sioud, M. (2003). Gene silencing by systemic delivery of synthetic siRNAs in adult mice. *J. Mol. Biol.* **4**, 327, 761–766.

Soutschek, J., Akinc, A., Bramlage, B., Charisse, K., Constien, R., Donoghue, M., Elbashir, S., Geick, A., Hadwiger, P., Harborth, J., John, M., Kesavan, V., Lavine, G., Pandey, R. K., Racie, T., Rajeev, K. G., Rohl, I., Toudjarska, I., Wang, G., Wuschko, S., Bumcrot, D., Koteliansky, V., Limmer, S., Manoharan, M., and Vornlocher, H. P. (2004). Therapeutic silencing of an endogenous gene by systemic administration of modified siRNAs. *Nature* **432**, 173–178.

Spagnou, S., Miller, A. D., and Keller, M. (2004). Lipidic carriers of siRNA: Differences in the formulation, cellular uptake, and delivery with plasmid DNA. *Biochemistry* **43**, 13348–13356.

Thakker, D. R., Natt, F., Husken, D., Maier, R., Muller, M., van der Putten, H., Hoyer, D., and Cryan, J. F. (2004). Neurochemical and behavioral consequences of widespread gene knockdown in the adult mouse brain by using non-viral RNA interference. *Proc. Natl. Acad. Sci. USA* **101**, 17270–17275.

Tiscornia, G., Singer, O., Ikawa, M., and Verma, I. M. (2003). A general method for gene knockdown in mice by using lentiviral vectors expressing small interfering RNA. *Proc. Natl. Acad. Sci. USA* **100**, 1844–1848.

Tompkins, S. M., Lo, C. Y., Tumpey, T. M., and Epstein, S. L. (2004). Protection against lethal influenza virus challenge by RNA interference *in vivo*. *Proc. Natl. Acad. Sci. USA* **101**, 8682–8686.

Walker, G. F., Fella, C., Pelisek, J., Fahrmeir, J., Boeckle, S., Ogris, M., and Wagner, E. (2005). Toward synthetic viruses: Endosomal pH-triggered deshielding of targeted polyplexes greatly enhances gene transfer *in vitro* and *in vivo*. *Mol. Ther.* **11**(3), 418–425.

Woodle, M. C., Scaria, P., Ganesh, S., Subramanian, K., Titmas, R., Cheng, C., Yang, J., Pan, Y., Weng, K., Gu, C., and Torkelson, S. (2001). Sterically stabilized polyplex: Ligand-mediated activity. *J. Control. Release* **74,** 309–311.

Xia, H., Mao, Q., Paulson, H. L., Davidson, B. L., and Xia, H. (2002). SiRNA-mediated gene silencing *in vitro* and *in vivo. Nature Biotechnol.* **20,** 1006–1010.

Xie, F. Y., Liu, Y., Xu, J., Tang, Q. Q., Scaria, P. V., Zhou, Q., Woodle, M. C., and Lu, P. Y. (2004). Delivering sirna to animal disease models for validation of novel drug targets *in vivo. PharmaGenomics.* July/August, 28–38.

Zender, L., Hutker, S., Liedtke, C., Tillmann, H. L., Zender, S., Mundt, B., Waltemathe, M., Gosling, T., Flemming, P., Malek, N. P., Trautwein, C., Manns, M. P., Kuhnel, F., and Kubicka, S. (2003). Caspase 8 small interfering RNA prevents acute liver failure in mice. *Proc. Natl. Acad. Sci. USA* **100,** 7797–7802.

Zhang, X., Shan, P., Jiang, D., Noble, P. W., Abraham, N. G., Kappas, A., and Lee, P. J. (2003). Small interfering RNA targeting heme oxygenase-1 enhances ischemia-reperfusion-induced lung apoptosis. *J. Biol. Chem.* **279,** 10677–10684.

7 A Novel Gene Expression System: Non-Viral Gene Transfer for Hemophilia as Model Systems

Carol H. Miao

Department of Pediatrics, University of Washington and Children's Hospital and Regional Medical Center, Seattle, Washington 98195

ABSTRACT

It is highly desirable to generate tissue-specific and persistently high-level transgene expression per genomic copy from gene therapy vectors. Such vectors can reduce the cost and preparation of the vectors and reduce possible host immune responses to the vector and potential toxicity. Many gene therapy vectors have failed to produce therapeutic levels of transgene because of inefficient promoters, loss of vector or gene expression from episomal

Advances in Genetics, Vol. 54
0065-2660/05 $35.00
DOI: 10.1016/S0065-2660(05)54007-0

vectors, or a silencing effect of integration sites on integrating vectors. Using *in vivo* screening of vectors incorporating many different combinations of gene regulatory sequences, liver-specific, high-expressing vectors to accommodate factor IX, factor VIII, and other genes for effective gene transfer have been established. Persistent and high levels of factor IX and factor VIII gene expression for treating hemophilia B and A, respectively, were achieved in mouse livers using hydrodynamics-based gene transfer of naked plasmid DNA incorporating these novel gene expression systems. Some other systems to prolong or stabilize the gene expression following gene transfer are also discussed. © 2005, Elsevier Inc.

I. INTRODUCTION

It is the goal of developing gene therapy vectors to achieve persistent and high-level transgene expression with minimal potential toxicity. Long-term expression was previously considered to be possible only with integrating vectors. Most nonintegrating vectors produced transient gene expression. Nevertheless, episomal persistence of DNA has been used by viruses to establish a latent state in the host cells. Some genes from the viral genome confer lifelong expression (Ho and S. 1989; Javier *et al.*, 1988; Steiner *et al.*, 1989). It has also been shown that after gene transfer of naked DNA into muscle (Wolff *et al.*, 1992) or brain cells (Jiao *et al.*, 1992), some plasmids can persist as episomal forms and produce low-level gene expression up to 6 months. Episomal adenoviral vectors that contained large segments of regulatory DNA sequences and untranslated regions have also been shown to persist and produce stable, long-term transgene expression (Balague *et al.*, 2000; Schiedner *et al.*, 1998).

Compared to viral or integrating vectors, non-viral episomal vectors have many advantages, including (1) reduced immunogenicity, (2) low toxicity, (3) ease and low cost of production, (4) avoiding possible recombination event to yield unwanted by-products, (5) avoiding possible oncogenic events by random integration into the host genome, and (6) the ability to deliver large genes. In particular, episomal vectors can avoid the enhancer-dependent insertional mutagenesis events observed in a gene therapy trial for common γ chain-deficient, SCID (γc$-/-$ SCID) patients (Hacein-Bey-Abina *et al.*, 2002). However, non-viral gene transfer in general is much less efficient to transduce cells *in vivo*. Moreover, even though non-viral episomal vectors can persist in the cells (Chen *et al.*, 2001; Hartikka *et al.*, 1996; Miao *et al.*, 2001; Wolff *et al.*, 1992), the vectors still diminish slowly over time, especially in dividing cells. Therefore, developing gene expression cassettes to increase and stabilize the expression of transgene from non-viral vectors is an important key for achieving a therapeutic effect.

II. GENE EXPRESSION SYSTEMS

Tissue-specific and high-level transgene expression per genomic copy from gene transfer vectors are highly desirable features. First, this would allow lower doses of vectors, thus reducing the cost and labor for preparing the vectors as well as the potential risks inherent with their administration and random integration. Second, a lower dose may diminish host immune responses to the vectors and toxicity. Furthermore, restricting transgene expression to target liver cells limits synthesis of transgene in other tissues, lessening further the potential for unregulated gene expression or production of nonfunctional proteins from incomplete processing. Because the transgene would not be expressed in antigen-presenting cells, this may also reduce the potential for inhibitory antibody formation (Pastore et al., 1999).

A. cis-Acting regulatory elements

As a first step toward producing better expression cassettes, various cis-acting regulatory sequences were considered to be incorporated into gene expression cassettes. Our model system is gene transfer of factor IX or VIII into mouse livers for the treatment of hemophilia.

1. Tissue-specific promoter–enhancer

Incorporation of tissue-specific and lineage-specific promoter/enhancers can limit transgene expression in target cells and mediate appropriate levels of gene expression. It was also shown that under certain conditions, use of tissue-specific promoters can modulate the host immune response to the transgene in liver-directed (Pastore et al., 1999) and muscle-directed (Yuasa et al., 2002) gene transfer.

For hepatic gene transfer in our study, a liver-specific promoter from the α_1-antitrypsin gene was used, with or without four copies of the apolipoprotein E (ApoE) enhancer (Shachter et al., 1993). The α_1-antitrypsin (hAAT) promoter was selected because it is a strong liver-specific promoter (Hafenrichter et al., 1994). Furthermore, it was reported that by adding four copies of the strong 154-bp liver-specific ApoE enhancer upstream of the hAAT promoter, expression of the α_1-antitrypsin gene was further increased more than 15-fold over constructs having other promoter–enhancers (Okuyama et al., 1996).

2. Hepatic locus control region (HCR)

The locus control region contains multiple elements of enhancers. It has been introduced into gene therapy vectors to enhance gene expression and to overcome the possible silencing effect by the position of integration in integrating vectors (Ellis and Pannell, 2001; Kowolik et al., 2001). For gene transfer of factor IX into the liver, we elected to incorporate a recently reported locus control region (LCR) for the liver-specific expression of the ApoE gene, ApoE-HCR (Simonet et al., 1993). This hepatic control region is located in the ApoE/CI/CII locus and is localized to a 319-bp region for full functional LCR activities. The enhancer elements contained in the HCR (Dang and Taylor, 1996) include binding sites for HNF3α, TF-LF2, HNF4, C/EBP, and GATA motif that direct high-level gene expression. This element confers copy number-dependent, position-independent gene expression and was shown to exhibit 10-fold higher activity than the ApoE enhancer in a transgenic mouse model (Dang et al., 1995). Because it is unclear how a multimerized LCR would affect gene transcription in vivo, we used a single ApoE-HCR sequence in the plasmid expression cassettes.

HCR incorporated in our hFIX plasmids contributed to the long-term episomal persistence of plasmids in the cells and their persistent, therapeutic level gene expression (Miao et al., 2001). This could be due to a combination of enhanced nuclear retention by the matrix attachment region (MAR) contained in the LCR (Dang et al., 1995) and transcriptionally active nucleosomal structure.

3. Intron

Traditionally, intron sequences have not been used in gene transfer vectors due to their large sizes and lack of dramatic enhancing function on gene expression in cultured cells. However, it has been shown that intron sequences can potentiate gene expression in vivo (Brinster et al., 1988; Liu et al., 1995; Palmiter et al., 1991). In transgenic mice, inclusion of the full-length 6.2-kb or a truncated 1.4-kb fragment of the first intron increased gene expression 40- to 200-fold (Jallat et al., 1990). Kurachi et al. (1995) showed that subregions of the first factor IX intron (intron A) without splicing signals inserted immediately upstream of the hFIX promoter exerted only marginal enhancing or even weakly negative regulatory activities on factor IX gene expression. However, minigene constructs containing further truncated first intron sequences (1.4 and 0.27 kb, respectively) and legitimate splicing sequences (donor, acceptor, and branch sites) increased factor IX gene expression 7- to 9-fold compared to constructs that lacked intronic sequences in cultured hepatoma cells. In addition, a synthetic 5' intron has been shown to augment factor VIII gene expression in cultured hepatic cells (Ill et al., 1997).

Intron elements including the authentic splice donor/acceptor sites have been shown to significantly enhance in vivo gene expression from gene therapy vectors. These include hFIX first intron (Miao et al., 2000; Wang et al., 1996) and minx intron (Hauser et al., 2000). This is attributed to stabilization of the transcript itself or facilitation of synthesis of a mature transcript. Reports have also documented that the incorporation of some introns into gene expression cassettes can be inhibitory to gene expression (Garcia de Veas Lovillo *et al.*, 2003; Notley *et al.*, 2002). The distinct sequence and size of an intron and the position of its insertion are probably very important in determining its functional activity in gene therapy vectors.

4. 3′-Untranslated region (UTR)

3′-Untranslated regions containing a polyadenylation signal regulate transcript cleavage and polyadenylation, as well as its export from the nucleus. Polyadenylation of RNA modulates transcript stability and translational activation (Grzybowska *et al.*, 2001). The bovine growth hormone polyadenylation signal was used in some of our vectors because it contains conserved AAUAAA hexanucleotides, a putative GU element, and a diffuse efficiency element for efficient and accurate polyadenylation (Goodwin and Rottman, 1992).

      Other sequences in 3′-untranslated regions may also affect the stability of mRNAs. Five families with hemophilia B have been described in which a point mutation in the 1.4-kb 3′-UTR of the factor IX gene severely reduced the plasma factor IX concentration to less than 3% of the normal level (Chen *et al.*, 1995; Vielhaber *et al.*, 1993). In addition, a 653-bp deletion located in the 3′-UTR of the factor IX gene was found responsible for mild hemophilia B (de la Salle *et al.*, 1993). Furthermore, in the context of a nonintegrating, episomal recombinant adenoviral vector, limited data suggest that the intron and 3′-untranslated region may have synergistic effects on gene expression *in vivo*, increasing gene expression by 1000-fold (Kaleko *et al.*, 1995). In addition, Kurachi *et al.* (1999) reported that although 3′-UTR had a negative regulatory effect on factor IX gene expression *in vitro*, it substantially increased the steady-state hFIX mRNA and gene expression and is involved in the age-related regulation of factor IX gene *in vivo*. Taken together, these data suggest that the 3′-untranslated region may have an important role in regulation of the expression *in vivo*.

5. Woodchuck hepatitis virus posttranscriptional regulatory element (WPRE) and other posttranscriptional regulatory elements

The WPRE is a viral enhancer element required for the cytoplasmic accumulation of hepatitis viral RNAs . It is believed that WPRE functions by modification of RNA polyadenylation, RNA export, and/or RNA translation (Donello *et al.*, 1998). It has been shown that insertion of WPRE into adenoassociated viral (Loeb *et al.*, 1999; Paterna *et al.*, 2000), lentiviral (Moreau-Gaudry *et al.*, 2001; Ramezani *et al.*, 2000) and retroviral (Schambach *et al.*, 2000; Zufferey *et al.*,

1999), and adenoviral vectors (Xu *et al.*, 2003) increased gene expression *in vivo*. It was also shown that plasmid vectors incorporating WPRE can enhance and stabilize transgene expression following *ex vivo* gene transfer to the central nervous system (Johansen *et al.*, 2003). Since WPRE was thought to function in a promoter- and intron-independent manner, addition of this element in a gene expression cassette with optimized transcriptional regulatory elements may further elevate transgene expression.

Incorporation of other candidate posttranscriptional regulatory elements, such as a translational enhancer, SP163 element (a 163-bp long splice variant derived from the 5′-UTR of vascular endothelial growth factor [VEGF]) (Stein *et al.*, 1998) to further increase gene expression from gene transfer vectors, will need to be further investigated.

6. Insulators

Insulator (or boundary) elements are regulatory sequences that function to define the boundary between differentially regulated loci (reviewed in Bell and Felsenfeld, 1999; Burgess-Beusse *et al.*, 2002; Snyder *et al.*, 1999). Insulators function as chromatin boundary domains, blocking external regulatory influences (e.g., from enhancers, silencers, or locus control regions) and preventing silencing by the encroachment of external condensed chromatin. One of the best studied insulators in vertebrates is the cHS4 element derived from the 5′ end of the chicken β-globin locus. This 1.2-kb fragment has enhancer-blocking activity and can protect expression cassettes in multiple assay systems. Insulator activity maps to a 250-bp core element containing a CTCF, zinc finger DNA-binding protein interaction site, and additional sequences (Burgess-Beusse *et al.*, 2002).

Insulators are usually located between euchromatin (transcriptionally active region) and heterochromatin (transcriptionally inactive region). Deletion of insulators results in spreading of the heterochromatin into the neighboring euchromatin, and consequently the silencing of the gene (Noma *et al.*, 2001). Insulators have been considered important to be incorporated into integrating vectors to protect the transgene expression cassettes from silencing by adjacent chromatin sequences, as well as preventing the insertional mutagenesis event from viral enhancers exerting an enhancing effect on the adjacent genes. However, bacterial DNA in non-viral episomal vectors are transcriptionally inactive in mammalian cells, whereas the transgene expression cassettes directed by eukaryotic promoter–enhancer sequences are active. The inert bacterial DNA sequences are similar to the heterochromatin in eukaryotic cells. It has been shown that the bacterial backbone may be inhibitory to transgene expression *in vivo* due to spreading of its inactivity to downstream eukaryotic promoter/enhancers (Chen *et al.*, 2004). If this is the case, incorporation of insulator elements such as the cHS4 element into the plasmid vector may overcome this silencing effect as an alternative to deletion of bacterial DNA sequences.

7. EBV oriP and EBNA-1 and other replication signals

It was found that following gene transfer of plasmid vectors, the plasmids are retained mostly as episomal forms inside the cells (Miao et al., 2001). These episomal vectors can still persist in the cells (Chen et al., 2001; Hartikka et al., 1996; Miao et al., 2001; Wolff et al., 1992), yet still diminish slowly over time, especially in dividing cells. If the plasmid vectors could replicate intracellularly, thereby increasing the relative copy numbers after transduction, this might permit the initial use of less efficient gene delivery systems to achieve a therapeutic effect. In addition, even though some cells such as hepatocytes replicate relatively slowly, it may be beneficial to include a self-replication signal in the plasmid vectors so that gene expression is maintained following either tissue damage or target cell division.

Plasmid vectors incorporating the origin of replicaion, oriP, of the Epstein–Barr virus (EBV) and EBV nuclear antigen-1 (EBNA-1) gene sequence replicate in cell culture (Krysan et al., 1989; Mucke et al., 2000). It was also demonstrated that plasmid vectors carrying EBV sequences and a 19-kb genomic fragment (SERPINA1) encoding the human α_1-antitrypsin gene produced a high level (>300 μg/ml) of AAT following hydrodynamics-based gene delivery and persisted for the >9-month duration of the experiment (Stoll et al., 2001). Nevertheless, replication and retention of EBV ori have not been demonstrated in primary cells or in animals. Furthermore, the EBV system is limited by its dependence on a viral protein and its host cell range for replication. It has been reported that plasmid containing an SV-40 ORI and a SAR/MAR can confer replication in CHO cells without the presence of large T-antigen (Piechaczek et al., 1999). Because there is a MAR sequence in the ApoE-HCR region, we have incorporated a SV40 origin of replication into the human factor IX plasmid (hFIX) in combination with ApoE-HCR. It was found that a SV-40 ori alone (without the large T-antigen) (C. H. Miao, unpublished results) or a MAR sequence (Cossons et al., 1997; Miao et al., 2001) is capable of facilitating the nuclear retention of vectors, generating a persistent and higher level of gene expression. Nevertheless no replication of the plasmids was detected.

Several putative human origins have been identified (Aladjem et al., 1998; Pelletier et al., 1999; Taira et al., 1994; Wu et al., 1993). These sequences permit autonomous replication of plasmids transfected into human cells (Krysan et al., 1989; Mucke et al., 2000; Nielsen et al., 2000). It will be of interest to test if plasmid vectors incorporating human origins can replicate autonomously in vivo.

8. Other genomic sequences

Using an episomal adenoviral vector with all viral-coding sequences deleted and containing the complete hAAT (PI) locus, high levels of very stable expression of hAAT were produced for more than 10 months (Balague et al., 2000; Schiedner et al., 1998). Prolonged and enhanced gene expression was also observed in plasmid vectors containing the whole gene of hAAT in combination with EBV sequences as mentioned previously. Another example is from gene transfer with a helper virus-independent adnoviral vector carrying the genomic human apo A-1 sequence which produced stable gene expression of human apo A-1 for 6 months (Van Linthout et al., 2002). These results indicate that there may be additional enhancer sequences in genomic sequences encoding the whole gene as well as sequences or structures in genomic DNA that resist silencing of gene expression.

B. Establishment of high-level and tissue-specific gene expression cassette

We used hemophilia as our gene transfer model system. Hemophilia B is a bleeding disorder that results from a deficiency of functional factor IX in plasma. Different gene therapy techniques are being developed for the treatment of hemophilia (reviewed in Nathwani et al., 2004). Developing high-expressing gene transfer vectors is important in achieving therapeutic levels of transgene expression. In order to optimize the production of human FIX in hepatocytes, gene expression was evaluated by incorporating various combinations of cis-acting regulatory sequences, including a tissue-specific promoter–enhancer, hepatic locus control region, intron A, and 3'-untranslated region into gene transfer vectors.

Studies have suggest that vector-mediated gene expression from the liver did not always correlate with that obtained from tissue culture studies (Kay et al., 1992; Schmidt et al., 1990), suggesting that the evaluation of the gene expression vectors must be established in vivo. Two different methods for plasmid-mediated gene delivery to mouse livers have been described. First, plasmid DNAs were delivered in hypertonic solution into the portal vein of mice whose hepatic vein was transiently occluded. The intraportal delivery of naked DNA is at least a thousand times more efficient than interstitial delivery into mouse liver (Budker et al., 1996). High levels of luciferase expression and β-galactosidase expression in 1% of the hepatocytes throughout the entire liver were obtained using 100 μg of the plasmid. Second, a hydrodynamic transfection method using the rapid infusion of plasmids in aqueous solutions via the tail vein has been developed to achieve even higher levels ($\sim$40%) of transgene expression in the liver (Liu et al., 1999; Zhang et al., 1999).

We started to evaluate the series of constructs with different combinations of regulatory elements in cell culture system, and subsequently *in vivo* by using the two direct plasmid injection methods. The results obtained from the cell culture system did not correlate well with the *in vivo* studies, indicating the importance of studying expression cassettes *in vivo* before implementation into gene therapy applications. The direct injection method through the portal vein was more labor intensive and had much higher mortality rates (~50% vs ~2%) than the hydrodynamics-based method via tail vein injection. Furthermore, it was demonstrated that by using the *in vivo* screening method in C57Bl/6 mice, the plasmid incorporating the ApoE-HCR element and hAAT promoter in combination with the hFIX minigene sequence (hFIXmg) and a bovine growth hormone polyadenylation signal (bpA) produced the highest gene expression level: ~67-fold higher (0.7–1.5 μg/ml) than the basal level obtained from LX-FIX with a retroviral promoter 2 days after plasmid injection. More importantly, at 8 weeks of age, mice injected with constructs containing an intron and 3'-UTR sequences still produced considerable levels of FIX, whereas mice infused with the other plasmids did not.

These results suggested that some of the cis DNA elements may have been responsible for differences in the persistence in gene expression. Systematic studies were then undertaken to attempt to determine which elements were responsible for this finding. Eight plasmids with a Bluescript vector backbone were injected into eight groups of mice ($n = 6$/group) by hydrodynamics-based plasmid injection. From comparing the results obtained from the eight plasmids (Fig. 7.1), it was determined that inclusion of an intron and one polyadenylation signal increased and prolonged transgene expression in the context of a plasmid after direct injection and, in combination with a locus control region (ApoE-HCR), resulted in persistent as well as therapeutic levels of hFIX gene expression (Miao *et al.*, 2000).

The mechanism for persistent gene expression from some of the plasmid DNAs is currently unknown. The ApoE-HCR contains a matrix attachment region (MAR) as well as liver-specific enhancer elements, which may augment gene expression in the plasmid. The MAR and intronic sequences together may interact with the nuclear matrix and may be primarily responsible for the persistence of gene expression from the plasmid DNA. Furthermore, it may be necessary and important to include at least one intron in gene therapy vectors so that the resulting transcripts can be assembled more efficiently into splicesome complexes and better protected in the nucleus from random degradation. Furthermore, from deletion analyses with or without an external polyadenylation signal, it was demonstrated that while only one polyadenylation signal from either hFIX 3'-UTR or bpA was essential for high-level gene expression, the hFIX 3'-UTR enhanced hFIX gene expression approximately twofold more than bpA alone. This could be due to specific functional sequences

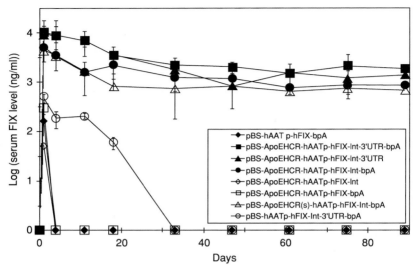

Figure 7.1. Delineation of essential elements required for persistent hFIX gene expression by testing different plasmids in mouse experiments. Twenty micrograms of the respective Bluescript in 2 ml 0.9% saline solution was injected into the tail vein of C57Bl/6 mice ($n = 6$ per group) over 6–8 s. Serum or plasma was analyzed for human factor IX levels by ELISA at different time points. Different symbols represent expression levels from mice injected with different plasmids: filled diamonds, pBS-hAATp-bpA; filled squares, pBS-ApoEHCR-hAATp-hFIX-Int-3'UTR-bpA; filled triangles, pBS-ApoEHCR-hAATp-hFIX-Int-3'UTR; filled circles; pBS-ApoEHCR-hAATp-hFIX-Int-bpA; open diamonds, pBS-ApoEHCR-hAATp-hFIX-Int; open squares, pBS-ApoEHCR-hAATp-hFIX-bpA; open triangles, pBS-ApoEHCR(s)-hAATp-hFIX-Int-bpA; open circles, pBS-hAATp-hFIX-Int-3'UTR-bpA.

in the 3'-UTR region. There may be enhancer elements contained in the 3'-UTR region to potentiate transcription of the gene or regions that are needed to stabilize the mRNA or elements required for posttranscriptional or posttranslational regulation.

These studies established the importance of studying expression cassettes *in vivo* before implementation into gene therapy applications. Because enhancers can potentiate gene transcription in a position-dependent manner, it is of interest to test if combining multimerized liver-specific transcription factor-binding sites such as HNF4, HNF3, HNF1, or C/EBP with the hepatic locus control region and a liver-specific promoter, can produce higher levels of gene expression. Furthermore, it needs to be determined if the cis sequences described here can function in a similar manner with heterologous cDNAs. It is also of importance to investigate the mechanism leading to persistent gene expression, which should aid in developing a non-viral gene transfer strategy for human factor IX deficiency. While our original goal was to use this approach to screen for sequences that might enhance gene expression in gene transfer vectors *in vivo*, the unexpected results related to prolonged gene expression allowed us to pursue specific cis DNA elements that may be useful in achieving long-term as well as therapeutic gene expression from non-viral vectors *in vivo*.

C. Persistence of episomal plasmid DNA and its gene expression

1. Persistent gene human FIX gene expression after direct plasmid injection into mouse livers

In order to explore long-term gene expression of hFIX by non-viral delivery of plasmid DNA, we performed a series of analyses at different time points with C57Bl/6 mice infused with a high-expressing plasmid (pBS-HCRHP-FIXIA abbreviated from pBS-ApoE-HCR-hAATp-hFIXmg-bpA, Fig. 7.2A) and a low-expressing plasmid (pBS-HP-FIXA abbreviated from pBS-hAATp-hFIX-bpA, Fig. 7.2A), respectively. As shown in Fig. 7.2B–D, mice infused with pBS-HP-FIXA always produced transient, low-level factor IX gene expression, whereas mice infused with pBS-HCRHP-FIXIA produced initially high-level factor IX gene expression, averaged to 10 μg/ml of hFIX protein (normal = 5 μg/ml), which decreased and stabilized at 0.5–2 μg/ml up to 550 days (duration of the experiments). This concentration of hFIX is in the therapeutic range for treating hemophilia B.

2. Episomal persistence of the vector DNA over time

To examine the correlation of vector DNA and gene expression levels, we investigated the vector status in different organs by Southern blot analyses. It was found that the majority of the DNA was taken up by the liver (Fig. 7.3A). Very low levels of vector DNA were observed in lung, spleen, and heart, but none were detectable in brain, kidney, and testes. This is probably because rapid delivery of the large volume of plasmid solution resulted in volume overload and back-flow into the liver associated with high-pressure delivery. For either plasmid, the amount of vector DNA retained in the cells peaked at 1 day, then declined and stabilized at the plateau level from 30 up to 240 days. In addition, total cellular DNA isolated from the livers in plasmid-treated mice sacrificed at different time points were subjected to restriction digestion and Southern blot analyses. Similar patterns of bands in samples from plasmid-treated mice and episomal plasmid controls were observed. The average copy numbers of vectors calculated from groups of mice were comparable between digested and undigested samples, suggesting that most of the input vectors remained episomal.

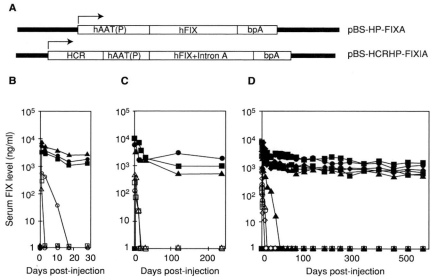

Figure 7.2. (A) The structure of constructs containing different hFIX expression cassettes. The plasmid
pBS-HCRHP-FIXIA is abbreviated from pBS-ApoEHCR-hAATp-hFIX+1IntA-bpA
(12), and pBS-HP-FIXA is abbreviated from pBS-hAATp-hFIX-bpA. HCR, hepatic
locus control region from the ApoE gene locus (771 bp); hAAT(P), human α1-anti-
trypsin promoter (408 bp); hFIX, human factor IX cDNA (1.4 kb); Intron A, truncated
human factor IX first intron (1.4 kb); bpA, bovine growth hormone polyadenylation
signal (265 bp). (B–D) Expression Levels of hFIX over time after naked DNA transfer
into mouse livers by two different plasmids shown in (A). Mice were rapidly (5–8 s)
infused with 20 μg of plasmid in 2 ml saline into the tail vein. Groups of mice (n = 3/
group) treated with each plasmid were sacrificed at four different time points
postinjection, 1 day (data not shown), 3 days (data not shown), (B) 30 days, and (C)
240 days. (D) Human FIX levels were followed for the rest of the treated mice for the
duration of the experiments (550 days). Filled symbols, pBS-HCRHP-FIXIA; open
symbols, pBS-HP-FIXA.

3. Transgene expression is tissue specific and its level is dependent on expression cassettes

Reverse transcription-polymerase chain reaction (RT-PCR) was performed to
examine the status of the transcripts in different tissues over time. Transcripts
were only observed in the liver (Fig. 7.3B), which is as predicted from the tissue-
specific promoter incorporated in the expression cassette. For high-expressing
plasmid pBS-HCRHP-FIXIA, the amount of transcript was highest 1 day postinjection and decreased gradually to a steady-state level over time. For low-
expressing plasmid pBS-HP-FIXA, a low level of transcript was observed 1 day

after plasmid injection and no transcript can be detected afterward. Relative levels of mRNA were positively associated with protein expression levels over time. This suggested that although retention of the plasmids in the nucleus is necessary for stable expression of the transgene, the expression level was further controlled by transcriptional activation and processing to the stable transcript.

4. Transient and acute liver damage was induced by the rapid tail vein injection procedure

Transient and acute liver damage was observed following gene transfer as indicated by the transient elevation of ALT and AST levels (Table 7.1). The toxicity in the mouse livers caused by the injection procedure was examined. The damage was related to rapid injection of a large volume, as the levels elevated were comparable in saline and plasmid-injected animals. Histological examinations showed that the amount and pattern of tissue damage at different time points correlated with the transaminase levels found in the different series of mice. Focal coagulative necrosis and hemorrhage are confined primarily to the subcapisular region (Fig. 7.4). The total extent of damage represented less than 5% of the total liver with the majority of the hepatic parenchyma showing no significant histologic abnormality. Considerable reparative changes were observed in mice 10 days postplasmid injection and the focal lesions had resolved afterward, as no significant histologic abnormalities were detected in liver sections from mice 180 days postinjection. There was no significant fibrosis or inflammatory infiltrate in these long-term mice, indicating complete recovery from the transient hepatic damage.

5. Possible mechanism

The high-expressing vector DNA persisted at approximately 2.3 copies per diploid genome. Taking into account that $\sim$20% of the liver cells were transduced when a Bluescript plasmid encoding β-galactosidase gene was injected rapidly through the tail vein of mice, each transduced liver cell would contain approximately 11 copies of the plasmid. It has been shown that when a plasmid first enters the nucleus, the DNA is immediately condensed into a nucleosome-like structure by histones and other molecules (Cereghini and Yaniv, 1984; Jeong and Stein, 1994). The rapid decline in the amount of DNA and transgene expression in the first few days may be due to a combination of elimination of transduced cells damaged in the infusion process and degradation of any unstable plasmid DNA in the nucleus. The remaining plasmids probably persist in the nucleosome-like structure and are associated with a certain compartment of the nucleus. However, only plasmids with certain

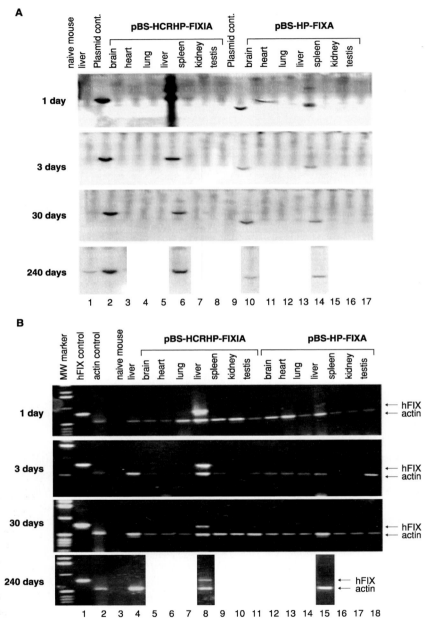

Figure 7.3. DNA and RNA analyses. (A) Southern blot analyses. Total cellular DNA was isolated from different organs of four groups of mice ($n = 3$, one representative animal shown

regulatory DNA sequences were able to produce prolonged high-level gene expression. It is possible that the established high-expressing hFIX cassettes meet the requirement for episomal persistence of plasmids and preservation of active promoters. The hepatic locus control region (HCR) from the ApoE gene locus (Simonet *et al.*, 1993) contains DNase I hypersensitive sites and is

Table 7.1. Liver Toxicity after Direct Plasmid Injection[a]

	Day 1	Day 3	Day 10	Day 180
ALT level (U/liter)				
Normal	29 ± 7	—	—	—
Saline treated	809 ± 459	89 ± 38	28 ± 8	27 ± 1
Plasmid treated	1447 ± 829	44 ± 21	35 ± 6	27 ± 5
AST level (U/liter)				
Normal	36 ± 3	—	—	
Saline treated	395 ± 203	93 ± 54	34 ± 5	30 ± 2
Plasmid treated	736 ± 424	44 ± 21	35 ± 6	27 ± 5
Billirubin level (mg/dl)				
Normal	0.15 ± 0.02	—	—	—
Saline treated	0.32 ± 0.15	0.45 ± 0.39	0.16 ± 0.01	0.10 ± 0.01
Plasmid treated	0.12 ± 0.07	0.22 ± 0.13	0.12 ± 0.02	0.14 ± 0.05

[a]Data represent means (± standard deviation) of six mice per group. Mice were untreated (normal) or treated with either saline or plasmid pBS-HCRHP-FIX-IA. ALT, alanine amino transferase; AST, aspartate amino transferase.

from each group) that had received plasmid DNAs, pBS-HP-FIXA or pBS-HCRHP-FIXIA, as described in Fig. 1. Each group was killed at a different time point (day 1, 3, 30, or 240). DNA was then digested with *Pst*I and analyzed by Southern blot with a radiolabeled human factor IX probe. Lane 1, liver DNA from a naïve mouse; lane 2, 100 pg of pBS-HCRHP-FIXIA; lanes 3–9, DNA from different organs as marked on the graph from pBS-HCRHP-FIXIA-treated mouse; lane 10, 100 pg of pBS-HP-FIXA; lanes 11–17, DNA from different organs as marked from pBS-HCRHP-FIXIA-treated mouse. (B) RT-PCR analyses. The mRNA was isolated from different tissues of four groups of mice (n = 3 mice/group, one representative animal shown) that received plasmid pBS-HP-FIXA or pBS-HCRHP-FIXIA. The isolated transcripts were subjected to RT-PCR analysis using primers complementary to sequences in exons 2–8 of the hFIX cDNA spanning 810 bp (top) and primers complementary to β-actin mRNA (bottom) to confirm that the mRNA was intact in all samples. The PCR products were electrophoresed on a 1% ethidium bromide gel. Lane 1, positive control, DNA-based PCR using the plasmid as a template; lane 2, actin control; lane 3, liver RNA treated in an identical manner without reverse transcriptase; lane 4, liver RNA from a naive mouse; lanes 5–11, mRNA from different organs as marked from pBS-HCRHP-FIXIA-treated mouse; lanes12–18, mRNA from different organs as marked from pBS-HP-FIXA-treated mouse.

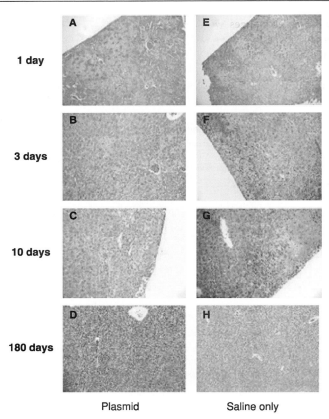

Plasmid   Saline only

Figure 7.4. Histology of the liver sections. Livers isolated from four groups of mice that had received 20 μg of plasmid pBS-HCRHP-FIXIA or saline-only control and killed at different time points after injection were sectioned and examined by routine hematoxylin and eosin staining. (A, B, C, D) From livers of plasmid-treated mice, (A) day 1, (B) day 3, (C) day 10, (D) day 180. (E, F, G, H) From livers of saline-treated mice, (E) day 1, (F) day 3, (G) day 10, (H) day 180. (A, B, C, E, F, and G) are with high-power magnification to show the lesion sites, whereas (D and H) are with low-power magnification. Arrows point to some of the focal regions of hemorrhage. (See Color Insert.)

primarily responsible for keeping the chromatin transcriptionally active. Furthermore, the ApoE-HCR contains a matrix attachment region, as well as liver-specific enhancer elements, which may augment gene expression in the plasmid (Dang *et al.*, 1995). The MAR and intronic sequences together may interact with the nuclear matrix (Gindullis and Meier, 1999; Meissner *et al.*, 2000; Wei *et al.*, 1999) and may be primarily responsible for the persistence of gene expression from the plasmid DNA.

An alternative possibility is that some of the vector genomes have integrated into the cellular chromosome and contributed to the long-term gene expression. Although this cannot be ruled out at this time, our results argue against significant genomic integration. First, liver regeneration after partial hepatectomy resulted in a substantial decline in the hFIX level, suggesting that regeneration and cell replication were detrimental to plasmid maintenance. Second, Southern blot analyses showed that the average copy numbers of low molecular weight vectors calculated from groups of mice were comparable between digested and undigested samples and that most of the vector DNA had restriction digest fragments identical to those seen in the plasmid control. These results agree with those obtained by Wolff (Jiao et al., 1992) and others (Monthorpe et al., 1993) who did not identify any integration events of plasmid DNAs in muscle cells after naked DNA transfer.

In conclusion, it was demonstrated that hFIX was sustained at therapeutic levels in plasma for more than one and a half years from targeted liver expression by tail vein injection of naked plasmids (Miao et al., 2001). No long-term toxicity or persistent immune response was observed. It is also important to note that our experimental results were obtained in groups of several animals and were reproducible, as opposed to many other non-viral protocols. This study establishes the basis to develop clinically feasible strategies into the liver for the treatment of hemophilia using the non-viral delivery of plasmids containing high-expressing, liver-specific expression cassettes.

D. Non-viral high-expressing gene therapy vectors to accommodate heterologous genes

1. Construction of liver-specific vectors

In order to develop vectors that can accommodate heterologous genes for efficient liver gene transfer, we have incorporated the same and/or additional cis-regulatory sequence used in pBS-HCRHP-FIXIA into the Bluescript vector. In particular, the promoter of the α_1-antitrypsin gene in conjunction with its natural noncoding exon sequence was connected with a downstream heterologous intron with authentic splice donor/acceptor sites, followed by a multiple cloning site for the insertion of cDNA (Fig. 7.5A). The noncoding exon was included to construct an authentic splice donor site for the heterologous intron and an efficient transcriptional start site. This arrangement was made to enhance the stability and to ensure correct processing for the mature transcript.

In summary, five DNA elements were incorporated into the Bluescript vectors: (1) a hepatic locus control region from the ApoE gene (HCR), (2) a liver-specific α_1-antitrypsin promoter plus its native noncoding exon (HP), (3) a 1.4-kb truncated hFIX first intron (I) or a 0.3-kb synthetic minx intron (mI),

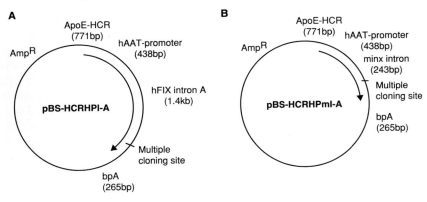

Multiple Cloning Site Sequence: 5'- CATATGGTCGACGAATTCAAGCTTGATATC-3'
* Unique sites in the vector   NdeI*   SalI*   ECoRI   HindIII   ECoRV*

Figure 7.5. Liver-specific, high-expressing constructs. The expression cassettes contain a combination of *cis*-regulatory elements and a multiple cloning site. (A) pBS-HCRHPI-A with a 1.4-kb truncated factor IX intron. (B) pBS-HCRHPmI-A with a synthetic minx intron. *Cis*-regulatory sequence in addition to a heterologous intron, includes a hepatic locus control region from the ApoE gene (ApoE-HCR), a liver-specific α1-antitrypsin promoter (HP), and a bovine growth hormone polyadenylation signal (bpA).

(4) a multiple cloning site for inserting cDNA sequences, and (5) a bovine growth hormone polyadenylation signal to generate pBS-HCRHPI-A (Fig. 7.5A) or pBS-HCRHPmI-A (Fig. 7.5B). Although the hFIX first intron has been shown to promote hFIX gene expression *in vivo* (Miao *et al.*, 2001), we contructed a second vector with a minx intron to take advantage of its much smaller size so that a smaller gene expression cassette can be established. The minx intron (Niwa *et al.*, 1990) has been used successfully in an adenoviral vector to enhance transgene expression (Hartigan-O'Connor *et al.*, 2001; Hauser *et al.*, 2000).

2. Evaluation of the vector systems using hFIX and GFP as reporter genes

These vectors were first tested by inserting hFIX and GFP as reporter genes. hFIX cDNA was inserted into pBS-HCRHPI-A and pBS-HCRHPmI-A to generate pBS-HCRHPI-FVIXA and pBS-HCRHPmI-FIXA, respectively. Comparable hFIX gene expression levels (0.5–5 μg/ml) were observed in mice treated with constructs pBS-HCRHPI-FIXA and pBS-HCRHP-FIXIA, whereas about twofold lower levels were observed with pBS-HCRHPmI-FIXA over a period of

6 months (experimental duration). This could be due to the size of the intron fragment or the presence of an enhancer element present in the hFIX first intron. In addition, RT-PCR and Western analyses showed that transcripts are currently processed to produce functional hFIX. Furthermore, 15–30% of the hepatocytes were positive for GFP expression in plasmid pBS-HCRHPI-GFPA-treated mice 3 days after injection. The positive cells dispersed evenly throughout the entire liver, which is consistent with our previous observation of hFIX expression after naked plasmid transfer by immunohistochemical staining (Ye *et al.*, 2003).

3. Evaluation of HFVIII gene expression

Next, a partial B domain-deleted hFVIII cDNA (Bi *et al.*, 1995) was inserted into the multiple cloning site in the liver-specific plasmid pBS-HCRHPmI-A or pBS-HCRHPI-A as described earlier to make pBS-HCRHPmI-VIIIA or pBS-HCRHPI-FVIIIA, respectively (Fig. 7.6A). A control plasmid, pBS-HP-FVIIIA, with the α_1-antitrypsin promoter, B domain-deleted human FVIII fragment, and bpA was also made. As shown in Fig. 7.6B, we found that following hydrodynamics-based gene delivery into immunodeficient Rag 2−/− mice, high-level gene expression (20–310 ng/ml, normal.7 = 100 ng/ml) persisted for at least 200 days (duration of the experiments) from the two liver-specific constructs pBS-HCRHPmI-VIIIA or pBS-HCRHPI-FVIIIA as compared to only brief, transient, and low-level hFVIII expression from the control construct pBS-HP-FVIIIA. Furthermore, construct pBS-HCRHPI-FVIIIA with HFIX intron A persistently produced approximately twofold higher HFVIII gene expression than the construct pBS-HCRHPmI-FVIIIA with minx intron (Fig. 7.6C). The HFVIII protein circulating in mouse plasma was functional as measured by a chromogenic assay (COATEST, measuring factor Xa generation; chromogenix AB, Sweden) and a modified clotting assay utilizing a modified activated partial thromboplastin time (aPTT)reagent and FVIII-deficient plasma and was composed of an 80-kDa light chain and a 90-kDa heavy chain lacking the B domain shown in the Western blot analysis.

The development of these non-viral constructs circumvents the limitations of accommodating the much larger cDNA such as factor VIII and its additional regulatory elements. This study demonstrates that persistent therapeutic to supraphysiological levels of human factor VIII can be achieved with established liver-specific vectors in an efficient non-viral approach *in vivo* (Miao *et al.*, 2003). Furthermore, these vectors will be able to deliver other heterologous genes for high-level transgene expression in the liver, such as those for treating metabolic diseases, or to study the roles of exogenous proteins, such as viral proteins with unknown functions in the liver. The resulting gene expression cassettes can also be readily inserted into other viral or non-viral vectors for achieving effective hepatic gene therapy.

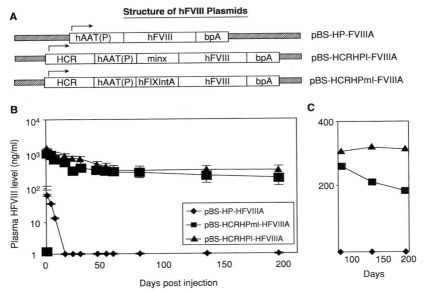

Figure 7.6. Gene expression level of hFIX in C57BL/6 mice after naked plasmid transfer of pBS-HCRHP-FIXIA, pBS-HCRHPIFIXA, and pBS-HCRHPmI-FIXA. (A) Three hFIX constructs were used to evaluate gene expression from liver-specific vectors. (B) hFIX levels over time in C57BL/6 mice after treatment. Fifty micrograms of plasmid in 2 ml of saline solution was injected into the tail vein of mice ($n = 5$) in 5–8 sec. Mice were then bled at regular intervals. Circulating hFIX levels in plasma were evaluated by ELISA. (C) Comparison of long-term hFIX expression levels (note change to linear scale for hFIX levels).

E. Other improvements

1. Elimination of bacterial sequence

Studies have shown that enhanced levels of transgene expression can be obtained from a bacterial DNA-free expression cassette either formed from a fragmented plasmid in mouse liver (Chen et al., 2001) or delivered as a mini-circle vector (Chen et al., 2003). It was suggested that bacterial DNA sequences played a role in episomal transgene silencing (Chen et al., 2004). The molecular mechanism underlining the bacterial DNA silencing is currently unknown. However, it was suggested the silencing is not the consequence of a direct inhibition, but is more likely a result from an event occurring at bacterial DNA sequences such as the inactive heterochromatin structure or *de novo* CpG methylation and then spreading to the downstream eukaryotic promoter directing the transgene expression cassette, resulting in its inactivation (Kass *et al.*, 1997).

2. Elimination of CpG motifs

It was found that delivery of plasmid vectors into some tissues or organs such as muscle that an inflammatory reaction occurs in response to unmethylated CpG sequences present in the bacterial part of plasmid DNA, leading to the induction of B- and T-cell-mediated immune responses. Such CpG motifs are common in bacterial genomes but are absent or methylated in mammalian DNA. Unmethylated CpG motifs interact with the Toll-like receptor (TLR)-9 in cells of the innate immune system. This interaction triggers an inflammatory reaction that in turn drives the adaptive response to the vector-encoded protein (Chuang *et al.*, 2002; Krug *et al.*, 2001).

Several approaches have been used to decrease the immune responses induced against the transgene product, including those induced by CpG motifs present in bacterial plasmid. Global immunosuppressive regimens (Dai *et al.*, 1995) or blockade of costimulatory interactions such as B7/CD28 (Kay *et al.*, 1995) and CD40/CD40L (Yang *et al.*, 1996) has inhibited the induction of immune responses and allowed long-term gene expression. However, these treatments lack antigen specificity and exert global effects on the immune system. Cumulative reductions (Yew *et al.*, 1999, 2000), elimination (Hodges *et al.*, 2004; Reyes-Sandoval and Ertl, 2004; Yew *et al.*, 2002), or methylation (Hong *et al.*, 2001) of immunostimulatory CpG sequences in plasmid vectors has been carried out to prevent the induction of immune responses, resulting in higher and more sustained levels of gene expression than their CpG-replete analogs.

III. NON-VIRAL GENE TRASNFER FOR HEMOPHILIA AS MODEL SYSTEMS

Hemophilia A and B result from a deficiency of blood clotting factors VIII and IX, respectively. Hemophilia has long been used as a model system for gene therapy due to extensive previous work, well-established assays for factor activity and antigenic levels, and the available animal models. Progress in the successful gene transfer of factor IX or factor VIII into animals by viral vectors (Balague *et al.*, 2000; Herzog *et al.*, 1999; Snyder *et al.*, 1999; VandenDriessche *et al.*, 1999) has led to clinical trials for hemophilia B and A patients (Greengard and Jolly, 1999; High, 2001; Kay *et al.*, 2000). Despite the limited success of these protocols, the high titers of virus required for therapy are very labor-intensive. Furthermore, immunogenicity toward the transgene or vector and the possibility of long-term toxicity pose major obstacles for clinical gene therapy protocols. Non-viral gene therapy, however, provides an alternative strategy that avoids the potentially harmful effects of viral gene therapy.

A. Persistent gene expression of factor IX in hemophilia B mice

1. Complete and sustained phenotypic correction of hemophilia B in mice

As described in Section II, it was demonstrated that injecting either circular or linear naked DNA containing a liver-specific gene expression cassette into the liver results in long-term, therapeutic levels of hFIX (0.5–2 μg/ml) in normal C57BL/6 mouse plasma (Chen et al., 2001; Miao et al., 2001). The optimal plasmid, pBS-HCRHPI-FIXA, was subsequently delivered into hemophilia B mice using the hydrodynamics-based gene delivery method. To our surprise, the technique was well tolerated in hemophilia B mice and the liver damage and recovery were comparable to normal mice. Hemophilia B mice produced an average of 5 μg/ml hFIX 1 day after injection, and hFIX levels of 1–5 μg/ml persisted for more than 1 year. A clotting assay indicated that FIX activity increased to 20–100% of normal. Moreover, the bleeding time measured by transaction of the tail tips of the treated hemophilia B mice was restored to that of normal mice (3–5 min), indicating complete correction of the phenotype. No bleeding episode was ever observed during or after the plasmid infusion.

2. Repeated administration boosts hFIX expression levels

Although we obtained long-term, therapeutic-level hFIX gene expression in mouse livers, the levels still fell gradually over time due to the loss of the episomal plasmids. It is not totally clear how long the half-life of turnover of hepatocytes is. It is estimated around 2 years for human hepatocytes. Therefore, repeated infusions of plasmid DNA would probably be needed to maintain therapeutic levels of hFIX over a lifetime for human gene therapy. Repeated injections of liver-specific plasmids were performed three times at 8-week intervals into hemophilia B mice. Human FIX expression levels were elevated after each injection, and elevated levels of 10–15 μg/ml were maintained for at least 8 weeks (duration of the experiments) (Fig. 7.7A). A functional clotting assay using FIX-deficient plasma corroborated with the elevated antigen levels and indicated 200–300% hFIX activity in plasma from hemophilia B mice that had received multiple injections of naked DNA (Fig. 7.7B). A Southern blot analysis showed that there were increased plasmid copies in the liver after repeated injections. These results suggest that lifelong gene expression of hFIX can be achieved with repeated infusions of hFIX plasmids if needed.

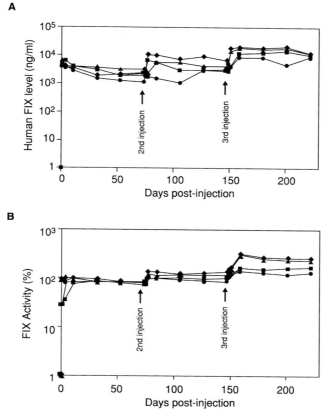

Figure 7.7. Human FIX antigen and activity, and anti-hFIX antibody levels in hemophilia B mice after repeated gene transfer of pBS-HCRHP-FIXIA. (A) Human FIX levels over time in hemophilia B mice after repeated gene transfer. The plasmids were infused three times at 75-day intervals into hemophilia B mice (n = 4) the same way as described in the legend of Fig. 1(A). (B) Clotting activity of hFIX over time in the hemophilia B mice after repeated treatment. (C) The anti-hFIX antibody titer after repeated gene transfer. Antibody titer in μg IgGmL^{-1} was determined by an antigen-specific ELISA. Each line represents an individual animal.

3. Exclusive and diffusively dispersed expression of hFIX in hepatocytes

The number and distribution of hepatocytes expressing hFIX following tail vein injection were examined by staining the liver sections with affinity-purified antihuman FIX antisera. The expression of hFIX was detected in hepatocytes from livers isolated 4 h after tail vein injection (Fig. 7.8A and B). Our Southern

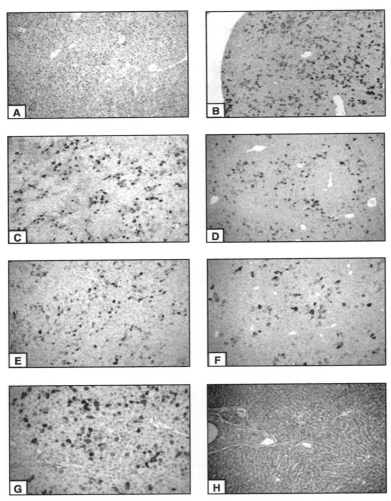

Figure 7.8. Immunohistochemical analysis of hepatic hFIX gene expression following tail vein injection [immunoperoxidase stain for hFIX with hematoxylin counterstain, original magnification × 10 (A) and (C–F), original magnification × 4 (B)]. Mouse livers were harvested 1 h (A), 4 h (B), 1 day (C), 3 days (D), 10 days (E), and 1 year (F) post rapid tail-vein injection with pBS-HCRHP-FIXIA. Note the lack of staining 1 h after injection is similar to that observed in normal control and saline treated mice (not shown). Higher magnification (20×) of liver section harvested 4 h postinjection (G). Sections of human liver show a low level diffuse staining for hFIX (H). (See Color Insert.)

blot analyses of the genomic DNA isolated from hepatic nuclei showed that significant amounts of plasmid DNA had entered the nucleus as early as 15 min postinfusion (data not shown). These results are consistent with our expression data, as there is a time lag from plasmid entry to gene expression. Approximately 30% of the hepatocytes showed strong expression for hFIX 4 to 24 h postinjection (Fig. 7.8B and C), then declined, and remained constant at approximately 10–15% from 30 days to 1 year (Fig. 7.8D–F). Human FIX was expressed exclusively in hepatocytes, as no hFIX antigen was observed in splenocytes, endothelial cells, or bile duct epithelial cells (Fig. 7.8G). Hepatocytes expressing hFIX were widely distributed throughout all the sections examined with no predilection for either portal or central venous zones of the hepatic architecture (Fig. 7.8G). Furthermore, the hFIX staining in transduced hepatocytes is much more intense than in control human hepatocytes (Fig. 7.8H). This agrees well with our previous DNA analyses (Miao et al., 2001) that there is an average of 10 copies of vector DNA per transduced cell as compared with a single copy of hFIX gene per normal human hepatic cell.

This study establishes that complete phenotypic correction of hemophilia B can be achieved by transferring a plasmid, pBS-HCRHP-FIXIA, into mouse hepatocytes (Ye et al., 2003). This is the first case of long-term correction of hemophilia B by non-viral gene transfer methods.

B. High-level gene expression of factor VIII in hemophilia A mice

1. Initial high-level factor VIII gene expression was achieved in hemophilia A mice

In order to test if non-viral gene transfer can be used to correct the hemophilia A phenotype in factor VIII knockout mice, the plasmid pBS-HCRHPI-FVIIIA was used to transduce hemophilia A mice (Liu et al., 1995). Initial high levels of human factor VIII (3–12 IU/ml) were observed in immunocompetent, FVIII-deficient animals of 129/sv × C57Bl/6 mixed genetic background within 3 days after naked plasmid transfer into mouse livers. The human factor VIII protein circulating in mouse plasma was fully functional, as measured by both COATEST and a modified aPTT assay using factor VIII-deficient plasma.

2. Humoral response against factor VIII limited the success of gene transfer

The level of functional factor VIII protein, however, declined rapidly to undetectable levels within 2–4 weeks postgene delivery (Fig. 7.9A). This decline occurred in parallel with the formation of inhibitory antibodies against human factor VIII (Fig. 7.9B). Inhibitory antibody titers remained high over a follow-up

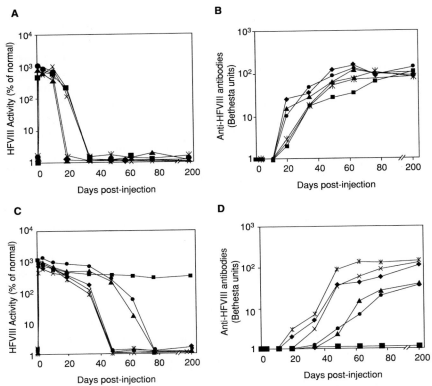

Figure 7.9. Naked plasmid transfer of hFVIII plasmids into hemophilia A mice with and without transient immunosuppression. Fifty micrograms of the plasmid, pBSHCRHPI-FVIIIA in 2 ml of saline solution was injected into the tail vein of mice (n = 6) in 5 – 8 s. Mice were then bled at regular intervals. Circulating hFVIII activities in plasma were evaluated by COATEST and a modified clotting assay. (A) Human FVIII levels over time in hemophilia A mice after plasmid treatment. (B) Inhibitor antibody formation against hFVIII over time. (C) Human FVIII levels over time in hemophilia A mice after plasmid treatment and transient immunosuppression with cyclophosphamide. Cyclophosphamide (50 mg/kg) was administered intraperitoneally at the time of vector injection and 14 and 28 days postinjection. (D) Inhibitor antibody formation against hFVIII over time with plasmid and cyclophosphamide treatment.

period of more than 200 days. This finding strongly suggests that exogenous gene delivery led to a persistent humoral immune response that was likely mediated by persistent transgene expression.

Consistent with this, we confirmed persistent episomal plasmid maintenance and continued synthesis of FVIII protein in the liver of treated animals as demonstrated by Southern blot analysis of the liver DNA, Western blot analysis of liver cell lysates, and immunohistochemical staining of FVIII protein in the liver section. In addition, we found very few CD8$^+$ T cells being recruited into the liver at day 1 or other later time points. Taken together, these observations suggest that the loss in functional protein was not mediated by activation of a cytotoxic T-cell response directed at factor VIII producing cells. Thus, after naked DNA transfer of a liver-specific plasmid, the phenotype of hemophilia A mice was completely corrected initially; however, a humoral immune response that followed completely blocked the function of the transgene product, despite persistent expression of factor VIII gene in the mouse liver.

It was also found that the immune response to exogenous factor VIII is unlikely to be mediated by species-specific sequence differences (Ye et al., 2004a). Inhibitory antibodies generated against factor VIII were characterized. The majority of anti-hFVIII IgG was the IgG1 isotype over the entire experimental period. Smaller amounts of anti-FVIII IgG2a, IgG2b, and IgG3 isotype antibodies were also detectable. This antibody response is similar to that reported in hemophilia A mice treated with repeated human factor VIII protein infusions (Wu et al., 2001) and to that present in hemophilia A patients (Reding et al., 2001). These observations suggest that transgene-specific antibodies were predominantly generated in response to Th2-induced activation signals.

3. Transient immunomodulation strategies to prolong transgene expression

In order to reduce or eliminate the production of anti-FVIII inhibitory antibodies, we utilized the immunosuppressive agent cyclophosphamide to induce tolerance in hemophilia A mice. The mice were treated simultaneously with a plasmid injection (tail vein) and 50 mg/kg cyclophosphamide intraperitoneally, followed by the same dosage given biweekly for a total of 4 weeks (three doses). As shown in Fig. 7.9C, prolonged high-level expression of functional human factor VIII was observed in all the mice compared to mice without cyclophosphamide treatment. However, five out of the six mice eventually developed high-level inhibitory antibodies (Fig. 7.9D). One mouse exhibited persistent circulating functional human factor VIII for at least 6 months and never developed detectable inhibitory antibodies.

In this study, we demonstrated that in immunocompetent hemophilia A mice, naked DNA transfer of a high-expressing, liver-specific factor VIII plasmid produced initial high levels of factor VIII gene expression (Ye et al., 2004a). The initial very high levels of functional protein, however, fell gradually to an undetectable level within 2–3 weeks and their disappearance correlated with the generation of high-titer, inhibitory antifactor VIII antibodies. Transient immunomodulation by cyclophosphamide significantly delayed or abolished inhibitory antibody formation against the transgene.

IV. CONCLUSIONS AND FUTURE PROSPECTS

We have successfully developed an effective gene expression system for delivering factor IX or factor VIII into mouse livers. Nevertheless, while the current method of hydrodynamics-based delivery resulted in gene transfer in about 40% of hepatocytes, this method caused transient right-sided heart failure and acute liver injury (Liu *et al.*, 1999; Zhang *et al.*, 2000) and therefore is not suitable for clinical applications in its current form. Development of catheter-mediated hydrodynamic delivery of plasmid DNA via the hepatic vein to the liver produced promising preliminary results for enhanced transgene expression in rabbits (Eastman *et al.*, 2002). Gene transfer of naked plasmid DNA mediated by physical methods, such as mechanical massage (Liu and Huang, (2002); Liu *et al.*, 2004), electroporation (Fewell *et al.*, 2001), ultrasound (Miao *et al.*, 2005), and laser (Tirlapur and Konig, 2002), are also under development. It is anticipated that some of these technologies alone or in combination with other delivery vehicles may provide a clinically feasible and safe non-viral gene delivery method for human gene therapy.

Furthermore, in collaboration with Dr. Shuwha Lin's group, we also have preliminary *in vivo* data showing that the functional factor IX activity can be further enhanced by the mutation of three amino acid sequences in the factor IX protein (S. Lin and C. L. Miao, unpublished results). This approach of delivering a gene encoding a more active protein will allow further reduction of gene transfer vectors needed or the use of a less efficient gene delivery methodology to achieve a therapeutic effect.

Finally, it was demonstrated that a transgene-specific immune response to factor VIII was induced in hemophilia A mice following non-viral gene transfer. The development of antibodies to a previously unexpressed protein product may limit the success of human gene therapy approaches. Development of ways to induce long-term tolerance against a specific antigen will be very important to assure the success of gene therapy. We are currently exploring transient immunomodulation therapies. It was demonstrated that long-term tolerance to factor VIII was achieved by transient immunosuppression by CTLA4Ig and anti-CD40L without inducing a lasting immunodeficiency in hemophilia A mice following naked DNA transfer (Ye *et al.*, 2004b). Other clinically relevant immunomodulation regimens are also currently under investigation.

Acknowledgments

The contributions of Drs. Xin Ye, Peiqing Ye, and Zhenping Shen to the research from our laboratory are greatly appreciated. We also acknowledge our collaborators for their help to this work, including Drs. Mark Kay, Arthur Thompson, Keith Loeb, Hans Ochs, David Rawlings, Darrel

Stafford, Rita Sarkar, Shuwha Lin, David Lillicrap, and Randal Kaufman. We are also thankful for many helpful discussions with Drs. Earl Davie, Anthony Blau, and Steven Hauschka. This work was supported by a Scientist Development Grant from the American Heart Association, a grant-in-aid from the Hemophilia Association of New York, a Career Development Grant from the National Hemophilia Foundation, and NIH Grant HL69409-02.

References

Aladjem, M. I., Rodewald, L. W., Kolman, J. L., and Wahl, G. M. (1998). Genetic dissection of a mammalian replicator in the human beta-globin locus. *Science* **281**, 1005–1009.

Balague, C., Zhou, J., Dai, Y., Alemany, R., Josephs, S. F., Andreason, G., Hariharan, M., Sethi, E., Prokopenko, E., jan, H. Y., Lou, Y. C., Hubert-Leslie, D., Ruiz, L., and Zhang, W. W. (2000). Sustained high-level expression of full-length human factor VIII and restoration of clotting activity in hemophilic mice using a minimal adenovirus vector. *Blood* **95**, 820–828.

Bell, A. C., and Felsenfeld, G. (1999). Stopped at the border: Boundaries and insulators. *Curr. Opin. Genet. Dev.* **9**, 191–198.

Bi, L., Lawler, A. M., Antonarakis, S. E., High, K. A., Gearhart, J. D., and Kazazian, H. H. J. (1995). Targeted disruption of the mouse factor VIII gene produces a model of haemophilia A. *Nature Genet.* **10**, 119–121.

Brinster, R. L., Allen, J. M., Behringer, R. R., Gelinas, R. E., and Palmiter, R. D. (1988). Introns increase transcriptional efficiency in transgenic mice. *Proc. Natl. Acad. Sci. USA* **85**, 836–840.

Budker, V., Zhang, G., Knechtle, S., and Wolff, J. A. (1996). Naked DNA delivered intraportally expresses efficiently in hepatocytes. *Gene Ther.* **3**, 593–598.

Burgess-Beusse, B., Farrell, C., Gaszner, M., Litt, M., Mutskov, V., Recillas-Targa, F., Simpson, M., West, A., and Felsenfeld, G. (2002). The insulation of genes from external enhancers and silencing chromatin. *Proc. Natl. Acad, Sci. USA* **99**(Suppl. 4), 16433–16437.

Cereghini, S., and Yaniv, M. (1984). Assembly of transfected DNA into chromatin: Structural changes in the origin-promoter-enhancer region upon replication. *EMBO J.* **3**, 1243–1253.

Chen, S. H., Schoof, J. M., Weinmann, A. F., and Thompson, A. R. (1995). Heteroduplex screening for molecular defects in factor IX genes from haemophilia B families. *Br. J. Haematol.* **89**, 409–412.

Chen, Z. Y., He, C. Y., Ehrhardt, A., and Kay, M. A. (2003). Minicircle DNA vectors devoid of bacterial DNA result in persistent and high-level transgene expression *in vivo*. *Mol. Ther.* **8**, 495–500.

Chen, Z. Y., He, C. Y., Meuse, L., and Kay, M. A. (2004). Silencing of episomal transgene expression by plasmid bacterial DNA elements *in vivo*. *Gene Ther.* **11**, 856–864.

Chen, Z. Y., Yant, S. R., He, C. Y., Meuse, L., Shen, S., and Kay, M. A. (2001). Linear DNAs concatemerize *in vivo* and result in sustained transgene expression in mouse liver. *Mol. Ther.* **3**, 403–410.

Chuang, T. H., Lee, J., Kline, L., Mathison, J. C., and Ulevitch, R. J. (2002). Toll-like receptor 9 mediates CpG-DNA signaling. *J. Leukocyte Biol.* **71**, 538–544.

Cossons, N., Nielsen, T. O., Dini, C., Tomilin, N., Young, D. B., Riabowol, K. T., Rattner, J. B., Johnston, R. N., Zannis-Hadjopoulos, M., and Price, G. B. (1997). Circular YAC vectors containing a small mammalian origin sequence can associate with the nuclear matrix. *J. Cell Biochem.* **67**, 439–450.

Dai, Y., Schwarz, E. M., Gu, D., Zhang, W. W., Sarvetnick, N., and Verma, I. M. (1995). Cellular and humoral immune responses to adenoviral vectors containing factor IX gene: Tolerization of factor IX and vector antigens allows for long-term expression. *Proc. Natl. Acad. Sci. USA* **92**, 1401–1405.

Dang, Q., and Taylor, J. (1996). *In vivo* footprinting analysis of the hepatic control region of the human apolipoprotein E/C-I/C-IV/C-II gene locus. *J. Biol. Chem.* **271,** 28667–28676.

Dang, Q., Walker, D., Taylor, S., Allan, C., Chin, P., Fan, J., and Taylor, J. (1995). Structure of the hepatic control region of the human apolipoprotein E/C-I gene locus. *J. Biol. Chem.* **270,** 22577–22585.

de la Salle, C., Charmantier, J. L., Baas, M. J., Schwartz, A., Wiesel, M. L., Grunebaum, L., and Cazenave, J. P. (1993). A deletion located in the 3' non translated part of the factor IX gene responsible for mild haemophilia B. *Thromb. Haemost.* **70,** 370–371.

Donello, J. E., Loeb, J. E., and Hope, T. J. (1998). Woodchuck hepatitis virus contains a tripartite posttranscriptional regulatory element. *J. Virol.* **72,** 5085–5092.

Eastman, S. J., Baskin, K. M., Hodges, B. L., Chu, Q., Gates, A., Dreusicke, R., Anderson, S., and Scheule, R. K. (2002). Development of catheter-based procedures for transducing the isolated rabbit liver with plasmid DNA. *Hum. Gene Ther.* **13,** 2065–2077.

Ellis, J., and Pannell, D. (2001). The beta-globin locus control region versus gene therapy vectors: A struggle for expression. *Clin. Genet.* **59,** 17–24.

Fewell, J. G., MacLaughlin, F., Mehta, V., Gondo, M., Nicol, F., Wilson, E., and Smith, L. C. (2001). Gene therapy for the treatment of hemophilia B using PINC-formulated plasmid delivered to muscle with electroporation. *Mol. Ther.* **3,** 574–583.

Garcia de Veas Lovillo, R. M., Ruijter, J. M., Labruyere, W. T., Hakvoort, T. B., and Lamers, W. H. (2003). Upstream and intronic regulatory sequences interact in the activation of the glutamine synthetase promoter. *Eur. J. Biochem.* **270,** 206–212.

Gindullis, F., and Meier, I. (1999). Matrix attachment region binding protein MFP1 is localized in domains at the nuclear envelope. *Plant Cell* **11,** 1117–1128.

Goodwin, E. C., and Rottman, F. M. (1992). The 3'-flanking sequence of the bovine growth hormone gene contains novel elements required for efficient and accurate polyadenylation. *J. Biol. Chem.* **267,** 16330–16334.

Greengard, J. S., and Jolly, P. J. (1999). Animal testing of retroviral-mediated gene therapy for factor VIII deficiency. *Thromb. Haemost.* **82,** 555–561.

Grzybowska, E. A., Wilczynska, A., and Siedlecki, J. A. (2001). Regulatory functions of 3'UTRs. *Biochem. Biophys. Res. Commun.* **288,** 291–295.

Hacein-Bey-Abina, S., Le Deist, F., Carlier, F., *et al.* (2002). Sustained correction of X-linked severe combined immunodeficiency by *ex vivo* gene therapy. *N. Engl. J. Med.* **346,** 1185–1193.

Hafenrichter, D. G., Wu, X., Rettinger, S. D., Kennedy, S. C., Flye, M. W., and Ponder, K. P. (1994). Quantitative evaluation of liver-specific promoters from retroviral vectors after *in vivo* transduction of hepatocytes. *Blood* **84,** 3394–3404.

Hartigan-O'Connor, D., Kirk, C. J., Crawford, R., Mule, J. J., and Chamberlain, J. S. (2001). Immune evasion by muscle-specific gene expression in dystrophic muscle. *Mol. Ther.* **4,** 525–533.

Hartikka, J., Sawdey, M., Cornefert-Jensen, F., Margalith, M., Barnhart, K., Nolasco, M., Vahlsing, H. L., Meek, J., Marquet, M., Hobart, P., Norman, J., and Manthorpe, M. (1996). An improved plasmid DNA expression vector for direct injection into skeletal muscle. *Hum. Gene Ther.* **7,** 1205–1217.

Hauser, M. A., Robinson, A., Hartigan-O'Connor, D., Williams-Gregory, D. A., Buskin, J. N., Apone, S., Kirk, C. J., Hardy, S., Hauschka, S. D., and Chamberlain, J. S. (2000). Analysis of muscle creatine kinase regulatory elements in recombinant adenoviral vectors. *Mol. Ther.* **2,** 16–25.

Herzog, R. W., Yang, E. Y., Couto, L. B., Hagstrom, J. N., Elwell, D., Fields, P. A., Burton, M., Bellinger, D. A., Read, M. S., Brinkhous, K. M., Podsakoff, G. M., Nichols, T. C., Kurtzman, G. J., and High, K. A. (1999). Long-term correction of canine hemophilia B by gene transfer of blood coagulation factor IX mediated by adeno-associated viral vector. *Nature Med.* **5,** 56–63.

High, K. A. (2001). Gene therapy: A 2001 perspective. *Haemophilia* **7**(Suppl. 1), 23–27.

Ho, D. Y., and Mocarski, E. S. (1989). Herpes simplex virus latent RNA (LAT) is not required for latent infection in the mouse. *Proc. Natl. Acad. Sci. USA* **86**, 7596–7600.

Hodges, B. L., Taylor, K. M., Joseph, M. F., Bourgeois, S. A., and Scheule, R. K. (2004). Long-term transgene expression from plasmid DNA gene therapy vectors is negatively affected by CpG dinucleotides. *Mol. Ther.* **10**, 269–278.

Hong, K., Sherley, J., and Lauffenburger, D. A. (2001). Methylation of episomal plasmids as a barrier to transient gene expression via a synthetic delivery vector. *Biomol. Eng.* **18**, 185–192.

Ill, C. R., Yang, C. Q., Bidlingmaier, S. M., Gonzales, J. N., Burns, D. S., Bartholomew, R. M., and Scuderi, P. (1997). Optimization of the human factor VIII complementary DNA expression plasmid for gene therapy of hemophilia A. *Blood Coagul. Fibrinolysis* **8**(Suppl. 2), S23–S30.

Jallat, S., Perraud, F., Dalemans, W., Balland, A., Dieterle, A., Faure, T., Meulien, P., and Pavirani, A. (1990). Characterization of recombinant human factor IX expressed in transgenic mice and in derived trans-immortalized hepatic cell lines. *EMBO J.* **9**, 3295–3301.

Javier, R. T., Stevens, J. G., Dissette, V. B., and Wagner, E. K. (1988). A herpes simplex virus transcript abundant in latently infected neurons is dispensable for establishment of the latent state. *Virology* **166**, 254–257.

Jeong, S., and Stein, A. (1994). Micrococcal nuclease digestion of nuclei reveals extended nucleosome ladders having anomalous DNA lengths fro chromatin assembled on non-replacting plasmids in transfected cells. *Nucleic Acids Res.* **22**, 370–375.

Jiao, S., Acsadi, G., Jani, A., Felgner, P. L., and Wolff, J. A. (1992). Persistence of plasmid DNA and expression in rat brain cells *in vivo*. *Exp. Neurol.* **115**, 400–413.

Johansen, J., Tornoe, J., Moller, A., and Johansen, T. E. (2003). Increased *in vitro* and *in vivo* transgene expression levels mediated through cis-acting elements. *J. Gene Med.* **5**, 1080–1089.

Kaleko, M., Kayda, D., Sakhuja, K., Mehaffey, M., and McClelland, A. (1995). Genomic sequences increase adenoviral vector-mediated factor IX expression 1,900 fold and enable sustained expression *in vivo*. *J. Cell. Biochem.* **21**, A366.

Kass, S. U., Landsberger, N., and Wolffe, A. P. (1997). DNA methylation directs a time-dependent repression of transcription initiation. *Curr. Biol.* **7**, 157–165.

Kay, M. A., Baley, P., Rothenberg, S., Leland, F., Fleming, L., Ponder, K. P., Liu, T., Finegold, M., Darlington, G., Pokorny, W., and Woo, S. L. C. (1992). Expression of human alpha 1-antitrypsin in dogs after autologous transplantation of retroviral transduced hepatocytes. *Proc. Natl. Acad. Sci. USA* **89**, 89–93.

Kay, M. A., Holterman, A. X., Meuse, L., Gown, A., Ochs, H. D., Linsley, P. S., and Wilson, C. B. (1995). Long-term hepatic adenovirus-mediated gene expression in mice following CTLA4Ig administration. *Nature Genet.* **11**, 191–197.

Kay, M. A., Manno, C. S., Ragni, M. V., Larson, P. J., Cuoto, L. B., McClelland, A., Glader, B., Chew, A. J., Tai, S. J., Herzog, R. W., Arruda, V., Johnson, F., Scallan, C., Skarsgard, E., Flake, A. W., and High, K. A. (2000). Evidence for gene transfer and expression of factor IX in haemophilia B patients treated with an AAV vector. *Nature Genet.* **24**, 257–261.

Kowolik, C. M., Hu, J., and Yee, J. K. (2001). Locus control region of the human CD2 gene in a lentivirus vector confers position-independent transgene expression. *J. Virol.* **75**, 4641–4648.

Krug, A., Towarowski, A., Britsch, S., Rothenfusser, S., Hornung, V., Bals, R., Giese, T., Engelmann, H., Endres, S., Krieg, A. M., and Hartmann, G. (2001). Toll-like receptor expression reveals CpG DNA as a unique microbial stimulus for plasmacytoid dendritic cells which synergizes with CD40 ligand to induce high amounts of IL-12. *Eur. J. Immunol.* **31**, 3026–3037.

Krysan, P. J., Haase, S. B., and Calos, M. P. (1989). Isolation of human sequences that replicate autonomously in human cells. *Mol. Cell. Biol.* **9**, 1026–1033.

Kurachi, S., Deyashiki, Y., Takeshita, J., and Kurachi, K. (1999). Genetic mechanisms of age regulation of human blood coagulation factor IX. *Science* **285**, 739–743.

Kurachi, S., Hitomi, Y., Furukawa, M., and Kurachi, K. (1995). Role of intron I in expression of the human factor IX gene. *J. Biol. Chem.* **270,** 5276–5281.

Liu, F., and Huang, L. (2002). Noninvasive gene delivery to the liver by mechanical massage. *Hepatology* **35,** 1314–1319.

Liu, F., Lei, J., Vollmer, R., and Huang, L. (2004). Mechanism of liver gene transfer by mechanical massage. *Mol. Ther.* **9,** 452–457.

Liu, F., Song, Y., and Liu, D. (1999). Hydrodynamics-based transfection in animals by systemic administration of plasmid DNA. *Gene Ther.* **6,** 1258–1266.

Liu, K., Sandgren, E. P., Palmiter, R. D., and Stein, A. (1995). Rat growth hormone gene introns stimulate nucleosome alignment *in vitro* and in transgenic mice. *Proc. Natl. Acad. Sci. USA* **92,** 7724–7728.

Loeb, J. E., Cordier, W. S., Harris, M. E., Weitzman, M. D., and Hope, T. J. (1999). Enhanced expression of transgenes from adeno-associated virus vectors with the woodchuck hepatitis virus posttranscriptional regulatory element: Implications for gene therapy. *Hum. Gene Ther.* **10,** 2295–2305.

Meissner, M. D. T., Gerner, C., Grimm, R., Foisner, R., and Sauermann, G. (2000). Differential nuclear localization and nuclear matrix association splicing factors PSF and PTB. *J. Cell Biochem.* **76,** 559–566.

Miao, C. H., Brayman, A. A., Loeb, K. R., Ye, P., Zhou, L., Mourad, P., and Crum, L. A. (2005). "Enhancement of non-viral gene transfer of factor IX into mouse livers by ultrasound." *Human Gene Therapy,* in press.

Miao, C. H., Ohashi, K., Patijn, G. A., Meuse, L., Ye, X., Thompson, A. R., and Kay, M. A. (2000). Inclusion of the hepatic locus control region, an intron, and untranslated region increases and stabilizes hepatic factor IX gene expression *in vivo* but not *in vitro*. *Mol. Ther.* **1,** 522–532.

Miao, C. H., Thompson, A. R., Loeb, K., and Ye, X. (2001). Long-term and therapeutic-level hepatic gene expression of human factor IX after naked plasmid transfer *in vivo*. *Mol. Ther.* **3,** 947–957.

Miao, C. H., Ye, X., and Thompson, A. R. (2003). High-level factor VIII gene expression *in vivo* achieved by non-viral liver-specific gene therapy vectors. *Hum. Gene Ther.* **14,** 1297–1305.

Monthorpe, M., Cornefert-Jensen, F., Hartikki, J., Felgner, J., Rundell, A., Margalith, M., and Dwarki, V. (1993). Gene therapy by intramuscular injection of plasmid DNA: Studies on firefly luciferase gene expression in mice. *Hum. Gene Ther.* **4,** 419–431.

Moreau-Gaudry, F., Xia, P., Jiang, G., Perelman, N. P., Bauer, G., Ellis, J., Surinya, K. H., Mavilio, F., Shen, C. K., and Malik, P. (2001). High-level erythroid-specific gene expression in primary human and murine hematopoietic cells with self-inactivating lentiviral vectors. *Blood* **98,** 2664–2672.

Mucke, S., Draube, A., Polack, A., Pawlita, M., Massoudi, N., Staratschek-Jox, A., Bohlen, H., Bornkamm, G., Diehl, V., and Wolf, J. (2000). Suppression of the tumorigenic growth of Burkitt's lymphoma cells in immunodeficient mice by cytokine gene transfer using EBV-derived episomal expression vectors. *Int. J. Cancer* **86,** 301–306.

Nathwani, A. C., Davidoff, A. M., and Tuddenham, E. G. (2004). Prospects for gene therapy of haemophilia. *Haemophilia* **10,** 309–318.

Nielsen, T. O., Cossons, N. H., Zannis-Hadjopoulos, M., and Price, G. B. (2000). Circular YAC vectors containing short mammalian origin sequences are maintained under selection as HeLa episomes. *J. Cell Biochem.* **76,** 674–685.

Niwa, M., Rose, S. D., and Berget, S. M. (1990). *In vitro* polyadenylation is stimulated by the presence of an upstream intron. *Genes Dev.* **4,** 1552–1559.

Noma, K., Allis, C. D., and Grewal, S. I. (2001). Transitions in distinct histone H3 methylation patterns at the heterochromatin domain boundaries. *Science* **293,** 1150–1155.

Notley, C., Killoran, A., Cameron, C., Wynd, K., Hough, C., and Lillicrap, D. (2002). The canine factor VIII 3′-untranslated region and a concatemeric hepatocyte nuclear factor 1 regulatory element enhance factor VIII transgene expression *in vivo*. *Hum. Gene Ther.* **13**, 1583–1593.

Okuyama, T., Huber, R. M., Bowling, W., Pearline, R., Kennedy, S. C., Flye, M. W., and Ponder, K. P. (1996). Liver-directed gene therapy: A retroviral vector with a complete LTR and the ApoE enhancer-alpha 1-antitrypsin promoter dramatically increases expression of human alpha 1-antitrypsin *in vivo*. *Hum. Gene Ther.* **7**, 637–645.

Palmiter, R. D., Sandgren, E. P., Avarbock, M. R., Allen, D. D., and Brinster, R. L. (1991). Heterologous introns can enhance expression of transgenes in mice. *Proc. Natl. Acad. Sci. USA* **88**, 478–482.

Pastore, L., Morral, N., Zhou, H., Garcia, R., Parks, R. J., Kochanek, S., Graham, F. L., Lee, B., and Beaudet, A. L. (1999). Use of a liver-specific promoter reduces immune response to the transgene in adenoviral vectors. *Hum. Gene Ther.* **10**, 1773–1781.

Paterna, J. C., Moccetti, T., Mura, A., Feldon, J., and Bueler, H. (2000). Influence of promoter and WHV post-transcriptional regulatory element on AAV-mediated transgene expression in the rat brain. *Gene Ther.* **7**, 1304–1311.

Pelletier, R., Price, G. B., and Zannis-Hadjopoulos, M. (1999). Functional genomic mapping of an early-activated centromeric mammalian origin of DNA replication. *J. Cell Biochem.* **74**, 562–575.

Piechaczek, C., Fetzer, C., Baiker, A., Bode, J., and Lipps, H. J. (1999). A vector based on the SV40 origin of replication and chromosomal S/MARs replicates episomally in CHO cells. *Nucleic Acids Res.* **27**, 426–428.

Ramezani, A., Hawley, T. S., and Hawley, R. G. (2000). Lentiviral vectors for enhanced gene expression in human hematopoietic cells. *Mol. Ther.* **2**, 458–469.

Reding, M. T., Wu, H., Krampf, M., Okita, D. K., Diethelm-Okita, B. M., Key, N. S., and Conti-Fine, B. M. (2001). CD4+ T cells specific for factor VIII as a target for specific suppression of inhibitor production. *Adv. Exp. Med. Biol.* **489**, 119–134.

Reyes-Sandoval, A., and Ertl, H. C. (2004). CpG methylation of a plasmid vector results in extended transgene product expression by circumventing induction of immune responses. *Mol. Ther.* **9**, 249–261.

Schambach, A., Wodrich, H., Hildinger, M., Bohne, J., Krausslich, H. G., and Baum, C. (2000). Context dependence of different modules for posttranscriptional enhancement of gene expression from retroviral vectors. *Mol. Ther.* **2**, 435–445.

Schiedner, G., Morral, N., Parks, R. J., Wu, Y., Koopmans, S. C., Langston, C., Graham, F. L., Beaudet, A. L., and Kochanek, S. (1998). Genomic DNA transfer with a high-capacity adenovirus vector results in improved *in vivo* gene expression and decreased toxicity. *Nature Genet.* **18**, 180–183.

Schmidt, E. V., Christoph, G., Zeller, R., and Leder, P. (1990). The cytomegalovirus enhancer: A pan-active control element in transgenic mice. *Mol. Cell. Biol.* **10**, 4406–4411.

Shachter, N. S., Zhu, Y., Walsh, A., Breslow, J. L., and Smith, J. D. (1993). Localization of a liver-specific enhancer in the apolipoprotein E/C-I/C-II gene locus. *J. Lipid Res.* **34**, 1699–1707.

Simonet, W. S., Bucay, N., Lauer, S. J., and Taylor, J. M. (1993). A far-downstream hepatocyte-specific control region directs expression of the linked human apolipoprotein E and C-I genes in transgenic mice. *J. Biol. Chem.* **268**, 8221–8229.

Snyder, R. O., Miao, C., Meuse, L., Tubb, J., Donahue, B. A., Lin, H. F., Stafford, D. W., Patel, S., Thompson, A. R., Nichols, T., Read, M. S., Bellinger, D. A., Brinkhous, K. M., and Kay, M. A. (1999). Correction of hemophilia B in canine and murine models using recombinant adeno-associated viral vectors. *Nature Med.* **5**, 64–70.

Stein, I., Itin, A., Einat, P., Skaliter, R., Grossman, Z., and Keshet, E. (1998). Translation of vascular endothelial growth factor mRNA by internal ribosome entry: Implications for translation under hypoxia. *Mol. Cell. Biol.* **18**, 3112–3119.

Steiner, I., Spivack, J. G., Lirette, R. P., Brown, S. M., MacLean, A. R., Subad-Shapre, J. H., and Fraser, N. W. (1989). Herpes simplex virus type 1 latency-associated transcripts are evidentlyt not essential for latent infection. *EMBO J.* **8,** 505–511.

Stoll, S. M., Sclimenti, C. R., Baba, E. J., Meuse, L., Kay, M. A., and Calos, M. P. (2001). Epstein-Barr virus/human vector provides high-level, long-term expression of alpha1-antitrypsin in mice. *Mol. Ther.* **4,** 122–129.

Taira, T., Iguchi-Ariga, S. M., and Ariga, H. (1994). A novel DNA replication origin identified in the human heat shock protein 70 gene promoter. *Mol. Cell. Biol.* **14,** 6386–6397.

Tirlapur, U. K., and Konig, K. (2002). Targeted transfection by femtosecond laser. *Nature* **418,** 290–291.

Van Linthout, S., Lusky, M., Collen, D., and De Geest, B. (2002). Persistent hepatic expression of human apo A-I after transfer with a helper-virus independent adenoviral vector. *Gene Ther.* **9,** 1520–1528.

VandenDriessche, T., Vanslembrouck, V., Goovaerts, I., Zwinnen, H., Vanderhaeghen, M. L., Collen, D., and Chuah, M. K. (1999). Long-term expression of human coagulation factor VIII and correction of hemophilia A after *in vivo* retroviral gene transfer in factor VIII-deficient mice. *Proc. Natl. Acad. Sci. USA* **96,** 10379–10384.

Vielhaber, E., Jacobson, D. P., Ketterling, R. P., Liu, J. Z., and Sommer, S. S. (1993). A mutation in the 3′ untranslated region of the factor IX gene in four families with hemophilia B. *Hum. Mol. Genet.* **2,** 1309–1310.

Wang, J. M., Zheng, H., Sugahara, Y., Tan, J., Yao, S. N., Olson, E., and Kurachi, K. (1996). Construction of human factor IX expression vectors in retroviral vector frames optimized for muscle cells. *Hum. Gene Ther.* **7,** 1743–1756.

Wei, X., Somanathan, S., Samarabandu, J., and Berezney, R. (1999). Three-dimensional visualization of transcription sites and their association with splicing factor-rich nuclear speckles. *J. Cell Biol.* **146,** 543–558.

Wolff, J. A., Ludtke, J. J., Acsadi, G., Williams, P., and Jani, A. (1992). Long-term persistence of plasmid DNA and foreign gene expression in mouse muscle. *Hum. Mol. Genet.* **1,** 363–369.

Wu, C., Zannis-Hadjopoulos, M., and Price, G. B. (1993). *In vivo* activity for initiation of DNA replication resides in a transcribed region of the human genome. *Biochim. Biophys. Acta* **1174,** 258–266.

Wu, H., Reding, M., Qian, J., Okita, D. K., Parker, E., Lollar, P., Hoyer, L. W., and Conti-Fine, B. M. (2001). Mechanism of the immune response to human factor VIII in murine hemophilia A. *Thromb. Haemost.* **85,** 125–133.

Xu, Z. L., Mizuguchi, H., Mayumi, T., and Hayakawa, T. (2003). Woodchuck hepatitis virus post-transcriptional regulation element enhances transgene expression from adenovirus vectors. *Biochim. Biophys. Acta* **1621,** 266–271.

Yang, Y., Su, Q., Grewal, I. S., Schilz, R., Flavell, R. A., and Wilson, J. M. (1996). Transient subversion of CD40 ligand function diminishes immune responses to adenovirus vectors in mouse liver and lung tissues. *J. Virol.* **70,** 6370–6377.

Ye, P., Thompson, A. R., Sarkar, R., Shen, Z., Lillicrap, D. P., Kaufman, R. J., Ochs, H. D., Rawlings, D. J., and Miao, C. H. (2004a). Naked DNA transfer of factor VIII induced transgene-specific, species-independent immune response in hemophilia A mice. *Mol. Ther.* **10,** 117–126.

Ye, P., Zhou, L., Shen, Z., Thompson, A. R., Rawlings, D. J., Ochs, H. D., and Miao, C. H. (2004b). Transient immuno-suppression by CTLA4Ig and anti-CD40L promotes long-term tolerance specifically against hFVIII in hemophilia A mice following naked DNA transfer. American Society of Gene Therapy, 7th annual meeting. [Abstract 158].

Ye, X., Loeb, K. R., Stafford, D. W., Thompson, A. R., and Miao, C. H. (2003). Complete and sustained phenotypic correction of hemophilia B in mice following hepatic gene transfer of a high-expressing human factor IX plasmid. *J. Thromb. Haemost.* **1,** 103–111.

Yew, N. S., Wang, K. X., Przybylska, M., Bagley, R. G., Stedman, M., Marshall, J., Scheule, R. K., and Cheng, S. H. (1999). Contribution of plasmid DNA to inflammation in the lung after administration of cationic lipid: pDNA complexes. *Hum. Gene Ther.* **10,** 223–234.

Yew, N. S., Zhao, H., Przybylska, M., Wu, I. H., Tousignant, J. D., Scheule, R. K., and Cheng, S. H. (2002). CpG-depleted plasmid DNA vectors with enhanced safety and long-term gene expression *in vivo. Mol. Ther.* **5,** 731–738.

Yew, N. S., Zhao, H., Wu, I. H., Song, A., Tousignant, J. D., Przybylska, M., and Cheng, S. H. (2000). Reduced inflammatory response to plasmid DNA vectors by elimination and inhibition of immunostimulatory CpG motifs. *Mol. Ther.* **1,** 255–262.

Yuasa, K., Sakamoto, M., Miyagoe-Suzuki, Y., Tanouchi, A., Yamamoto, H., Li, J., Chamberlain, J. S., Xiao, X., and Takeda, S. (2002). Adeno-associated virus vector-mediated gene transfer into dystrophin-deficient skeletal muscles evokes enhanced immune response against the transgene product. *Gene Ther.* **9,** 1576–1588.

Zhang, G., Budker, V., and Wolff, J. A. (1999). High levels of foreign gene expression in hepatocytes after tail vein injections of naked plasmid DNA. *Hum. Gene Ther.* **10,** 1735–1737.

Zhang, G., Song, Y. K., and Liu, D. (2000). Long-term expression of human alpha1-antitrypsin gene in mouse liver achieved by intravenous administration of plasmid DNA using a hydrodynamics-based procedure. *Gene Ther.* **7,** 1344–1349.

Zufferey, R., Donello, J. E., Trono, D., and Hope, T. J. (1999). Woodchuck hepatitis virus posttranscriptional regulatory element enhances expression of transgenes delivered by retroviral vectors. *J. Virol.* **73,** 2886–2892.

8

Site-Specific Integration with ϕC31 Integrase for Prolonged Expression of Therapeutic Genes

Daniel S. Ginsburg and Michele P. Calos

Department of Genetics
Stanford University School of Medicine
Stanford, California 94305

ABSTRACT

Need of a site-specific integrating vector in gene therapy has become pressing, as recent work has shown that many of the current integrating vectors used preferentially integrate in the vicinity of genes. A site-specific integrating vector would reduce the risk of insertional mutagenesis posed by randomly integrating vectors, and a non-viral vector would reduce the safety and immunogenicity problems associated with viral vectors. The ϕC31 integrase is a protein from *Streptomyces* phage ϕC31 that has been developed as a non-viral site-specific gene therapy vector. The ϕC31 integrase catalyzes the integration of a plasmid containing *attB* into pseudo *attP* sites in mammalian genomes. It has been shown to function in tissue culture cells as well as in mice. Vectors based on the ϕC31 integrase were able to treat tyrosinemia type I in a mouse model and two forms of epidermolysis bullosa in keratinocytes from patients, demonstrating its effectiveness as a gene therapy vector. Development of ϕC31 integrase-based vectors is

Advances in Genetics, Vol. 54
0065-2660/05 $35.00
DOI: 10.1016/S0065-2660(05)54008-2

still underway, but it has already been shown to provide long-term expression through site-specific integration.   © 2005, Elsevier Inc.

I. INTRODUCTION

The goal of gene therapy is to provide permanent correction of a disease through genomic modification of patient cells. Gene therapies need to overcome three major hurdles to be successful. First, they should be safe and not cause harm to the patient. Second, they need to deliver the therapeutic gene into the appropriate cells. Third, the appropriate level and duration of expression of the therapeutic gene must be provided. Current gene therapy vectors meet these challenges with varying effectiveness. Viral vectors are efficient vehicles for delivering DNA or RNA into cells, but have many limitations. For example, they often provoke an immune response that causes danger to the patient and can eliminate corrected cells. Non-viral vectors are safer and less immunogenic than viruses, but have had difficulty achieving delivery to the appropriate cells and sustaining a useful level of gene expression.

For most gene therapies, the desired duration of expression is the lifetime of the patient. Long-term expression can be achieved in both viral and non-viral vectors by genomic integration of therapeutic DNA, i.e., covalent linkage of the vector DNA with chromosomal DNA. While genomic integration prevents the therapeutic DNA from being lost during cell division, integration raises its own safety issues. Most current methods for genomic integration insert the vector DNA in a quasi-random fashion into the genome. These randomly integrating vectors often show a preference for locations in or near genes. Murine leukemia virus (MLV)-based vectors preferentially integrate just upstream of genes (Wu et al., 2003). HIV-1 (Schroder et al., 2002) and adeno-associated virus (AAV) (Nakai et al., 2003) based vectors preferentially integrate within genes. The risk of insertional mutagenesis from retroviral vectors was shown to be real, when two patients in a clinical trial treating X-linked severe combined immunodeficiency (SCID) developed leukemia partly because of vector integration in the vicinity of the oncogene LMO2, resulting in its activation (Hacein-Bey-Abina et al., 2003).

In contrast to retroviral vectors, most non-viral gene therapy vectors to date have been nonintegrating. While such episomal vectors do not pose a threat of insertional mutagenesis, they are lost over time in dividing cells. Some tissues that are targets for gene therapy, such as brain, liver, and skeletal muscle, contain largely quiescent cells. Episomal vectors are retained in these tissues, but expression often decreases over time, probably due to gene silencing. Special sequence arrangements in the vector, such as inclusion of Epstein-Barr virus (EBV) sequences (Sclimenti et al., 2003; Stoll and Calos, 2002) or exclusion of bacterial sequences (Chen *et al.*, 2004), can improve long-term expression, probably by counteracting gene silencing. Other tissue targets for gene therapy, such as stem cells, are dividing, and integration provides a straightforward route to long-term expression, particularly in these situations.

A number of different integration systems are being developed for use as non-viral vectors. Some of these, including the Sleeping Beauty transposon (Yant *et al.*, 2000), achieve long-term expression through chromosomal integration. These vectors correct the transient expression problem of most non-viral vectors, but they integrate randomly (Vigdal *et al.*, 2002). It is likely that they will share similar insertional mutagenesis problems with randomly integrating viruses. In addition, expression from randomly integrating vectors can be low due to integration into transcriptionally inactive regions of the genome.

This review focuses on the φC31 integrase. This integration system achieves a high degree of site-specific vector integration in the genome. Aside from the challenge of delivery to cells, the φC31 integrase system has advantages for gene therapy because it may lower the insertional mutagenesis risk compared to randomly integrating vectors and it may overcome the loss of gene expression of nonintegrating vectors.

II. φC31 INTEGRASE

The φC31 integrase is an integrase enzyme present in the *Streptomyces* phage φC31 (Kuhstoss and Rao, 1991). The enzyme is a 613 amino acid, 67-kDa member of the serine site-specific recombinase family (Thorpe and Smith, 1998). The φC31 integrase catalyzes recombination of two attachment, or *att*, sites, *attB* and *attP* (Fig. 8.1). The *attP* site is found in the φC31 phage genome, and *attB* is found in the host *Streptomyces* genome. Recombination of *attB* and *attP* allows the phage to enter the lysogenic phase of the life cycle, integrating its genome into the bacterial chromosome. Recombination of *attB* and *attP* results in hybrid sites *attL* and *attR*. Recombination of *attL* and *attR* mediated by the φC31 integrase is inefficient without the presence of an excisionase protein (Thorpe and Smith, 1998). Therefore, the integration reaction is essentially

attB   GTGCCAGGGCGTGCCC TT GGGCTCCCCGGGCGCG

attP   CCCCAACTGGGGTAACCT TT GAGTTCTCTCAGTTGGGGG

psA   TAAGTACTTGGGTTTCCC TT GGTGTCCCCATGGAGATTT

Figure 8.1. Sequences of φC31 attachment sites and human pseudo *attP* site psA. Sequences of the minimal 34-bp *attB* and 39-bp *attP* and the human pseudo *attP* site psA are shown. Bases involved in the imperfect inverted repeat are underlined, and the central TT core where recombination occurs is indicated. The psA pseudo *attP* site has 44% identity with *attP*.

unidirectional. The minimal lengths of the ϕC31 attB and attP sites have been shown to be 34 and 39 bp, respectively (Groth et al., 2000). The att sites share a 2-bp TT identical core at which recombination takes place (Smith and Till, 2004), and the arm sequences on either side of the core contain an imperfect inverted repeat (Fig. 8.1).

The recombination mechanism of the ϕC31 integrase is thought to be similar to that of $\gamma\delta$ resolvase (Stark et al., 1992). By this scheme, a pair of integrase subunits binds to each att site. Staggered cuts of 2 bp are made simultaneously in each att site by nucleophilic attack of the catalytic serine hydroxyl group. One pair of integrase subunits rotates 180° relative to the other. Strand exchange completes the recombination reaction to form hybrid sites attL and attR (Smith et al., 2004).

III. ϕC31 INTEGRASE FUNCTION IN MAMMALIAN CELLS

The purified ϕC31 integrase protein was shown to catalyze recombination of attB and attP in vitro without the addition of any cofactors (Thorpe and Smith, 1998). This result suggested that the integrase might function in human cells (Groth et al., 2000). To test this hypothesis, a transient transfection assay was done in human 293 cells. Cells were transfected with an integrase-expressing plasmid and an assay plasmid containing the att sites flanking a marker gene. In this assay, the integrase gave an intramolecular excision frequency of 50%. An extrachromosomal intermolecular integration reaction occurred at a frequency of 7.5% (Groth et al., 2000). The ϕC31 integrase was also shown to catalyze integration into an attP site placed into the chromosomes of 293 cells at an integration frequency of approximately 1% (Thyagarajan et al., 2001). Chromosomal excision between attB and attP was observed at a frequency of 51% (Hollis et al., 2003).

More interesting than its ability to catalyze recombination with its own att sites in the chromosome was the ability of the ϕC31 integrase to integrate a plasmid containing attB into endogenous sites in the human genome (Fig. 8.2). In a chromosomal integration experiment in human cells in which an attP site was inserted, only 15% of integrants occurred in the placed attP. The other 85% of integrations occurred at pseudo attP sites in the genome (Thyagarajan et al., 2001). Analysis of these integration sites revealed preferential integration into a small number of locations. From an initial experiment in 293 cells, a commonly used ϕC31 integration site in the human genome, termed human pseudo site A (psA), was found on chromosome 8p22. The sequence of psA is 44% identical to wild-type attP (Fig. 8.1). Site-specific integration at psA has been observed at frequencies of 5–50% in various studies, but may typically be lower (Ortiz-Urda et al., 2002, 2003; Sclimenti et al., 2001).

Figure 8.2. Unidirectional integration mediated by φC31 integrase. A plasmid containing a therapeutic transgene (white arrow) and *attB* (gray arrow) is integrated into a pseudo *attP* site in the chromosome (black arrow). After integration, the therapeutic gene is flanked by hybrid *att* sites that are not acted on by the integrase. Therefore, the integration reaction is unidirectional.

Genomic integration mediated by φC31 has been shown to provide long-term transgene expression in many different systems. For example, integrase-mediated transgene expression was first tested in tissue culture. 293 cells were transfected with a luciferase-*attB* donor plasmid and a CMV-integrase plasmid or a control empty plasmid. Luciferase levels were 10-fold higher in cells receiving the integrase plasmid compared to cells that received the control plasmid, reflecting genomic integration and robust expression of the integrated gene (Thyagarajan *et al.*, 2001).

The φC31 integrase was also shown to be effective *in vivo* in mice, where it provided long-term expression of integrated genes (Olivares *et al.*, 2002). Mice were treated with a donor plasmid containing a human factor IX (hFIX) minigene and *attB* and a CMV-integrase plasmid or a control plasmid. The vectors were delivered via hydrodynamic tail vein injection to obtain liver transduction. A group of mice was given partial hepatectomies, removing two-thirds of the liver, 100 days after injection to cause the hepatocytes to divide, reducing the amount of unintegrated DNA. Levels of hFIX in the integrase-treated mice remained stable at approximately 4 μg/ml after 8 months, even in those mice given a partial hepatectomy. Factor IX levels without integrase were 40-fold lower and dropped to near background after partial hepatectomy. Approximately 2% of liver cells contained an integration at a mouse pseudo *attP* site termed mpsL1, which accounted for the vast majority of integration events (Olivares *et al.*, 2002). Therefore, in the mouse liver system, φC31 integrase was effective in providing a good frequency of integration, strong long-term expression of the transgene, and highly site-specific integration.

In addition to mouse liver, the utility of φC31 integrase has been demonstrated in the rat retina (Chalberg *et al.*, 2004). The eye is a desirable target for gene therapy because of genetic diseases such as retinitis pigmentosa, for which there are no current therapies. A donor vector containing *attB* and a luciferase-GFP fusion protein, along with an integrase expression vector or a control vector, were delivered to rat retinal pigment epithelium by subretinal injection followed by electroporation. Animals receiving the φC31 integrase plasmid expressed luciferase in the eye at levels 85-fold above expression levels in animals that did not receive the integrase plasmid. This expression was maintained for the 4.5-month time course of the experiment and provides encouragement for the utility of the integrase system in the eye (Chalberg et al., 2005).

The experiments just described in mouse liver and rat retina showed that the ϕC31 integrase was effective for gene therapy in these tissues and organisms. To show that integrase-based vectors could cure a genetic disease, a mouse model of hereditary tyrosinemia type I (HT1) was used (Held et al., 2005). Mice lacking functional fumarylacetoacetate (FAH) were kept alive with the drug NTBC. They were treated with a donor plasmid containing attB and the FAH gene, as well as an integrase expression plasmid or a control empty vector. Plasmids were delivered via hydrodynamic tail vein injection. The total integration frequency in the livers of mice treated with integrase was 3.6%. All mice examined contained integrations at the mouse pseudo attP site mpsL1 seen previously in the hFIX experiment (Olivares et al., 2002). Seven other mouse pseudo sites were found, but none were used as often as mpsL1 (Held et al., 2005).

The ϕC31 integrase was effective in curing the FAH-deficient mice. Corrected cells have a selective advantage in the mouse model and grow to form nodules that then repopulate the liver of the animal. Integrase-treated animals formed hepatocyte nodules at a frequency 26-fold greater than in animals that did not receive integrase. Serial transplantation of hepatocytes from treated animals into untreated animals was able to correct the FAH deficiency in the transplant recipients. Thus, FAH expression from ϕC31 integrase vectors was stable through many cell generations (Held et al., 2005).

Not only was the ϕC31 integrase effective in treating an animal model of HT1, but it was also able to correct recessive dystrophic epidermolysis bullosa (RDEB) in human cells (Ortiz-Urda et al., 2002). RDEB is caused by a deficiency of collagen VII due to mutation of the COL7A1 gene. Primary keratinocytes obtained from RDEB patients were transfected with a donor plasmid containing the 8.9-kb COL7A1 gene and attB, as well as an integrase-expressing plasmid or a control plasmid. A short drug treatment was used to enrich the corrected keratinocyte population. The corrected cells were then grafted onto immune-deficient mice. Grafts containing integrase-treated cells showed phenotypic correction for the entire 14-week duration of the experiment. Cells that received the empty vector instead of a ϕC31 integrase-expressing plasmid showed no phenotypic correction. Integration sites were analyzed in treated keratinocytes plated at a single cell dilution to obtain clones, with the result that the majority of the clones contained an integration at the psA pseudo attP site (Ortiz-Urda et al., 2002).

In similar experiments, epidermal cells from junctional epidermolysis bullosa (JEB) patients were transfected with a donor plasmid containing attB and the *LAMB3* gene encoding the laminin 5 β3 subunit, as well as an integrase expression plasmid or empty vector (Ortiz-Urda et al., 2003). Total integration frequency in these cells was approximately 17%. Corrected cells were enriched by a short drug treatment and grafted onto immune-deficient mice. The grafts showed normal morphology for the 16-week course of the experiment. Cells that did not receive integrase showed no phenotypic correction. Analysis of integration sites revealed that 4 of 10 clones contained an integration at pseudo site psA (Ortiz-Urda et al., 2003).

While the wild-type ϕC31 integrase effectively catalyzes integration into mammalian chromosomes, there is work underway to improve both the frequency and the specificity of the recombination reaction. DNA shuffling was used to evolve the wild-type ϕC31 integrase to improve its reaction at human pseudo *attP* site psA (Sclimenti et al., 2001). A shuffled library of mutated integrase genes was screened in *Escherichia coli* for improved recombination between *attB* and the psA pseudo *attP* site. Shuffled integrases that were improved in the *E. coli* assay were tested individually in 293 cells in both a selection assay for overall integration frequency and a quantitative polymerase chain reaction assay for integration at psA. The mutant integrase with the highest frequency of integration at psA, 11C2, gave a total integration frequency 10-fold lower than that of the wild-type integrase, but 30% of its integrations occurred at psA compared to 5% for the wild-type integrase (Sclimenti et al., 2001). Therefore, DNA shuffling and screening appeared to be capable of modest improvement in the specificity of the integration reaction.

The ϕC31 integrase, at 67 kDa, is expected to be too large to achieve nuclear entry by passive diffusion through nuclear pores. The integrase obviously enters the nucleus, as shown by the success in chromosomal integration in the experiments described earlier. The Cre recombinase has been shown to contain an endogenous nuclear localization sequence (NLS) (Le et al., 1999). Addition of another NLS to Cre does not improve recombination in mammalian cells. It is likely that the ϕC31 integrase also contains an endogenous NLS. Addition of an exogenous NLS to the ϕC31 integrase was tested to see if it could improve recombination in mammalian cells (Andreas et al., 2002). Addition of a C-terminal SV40 large T-antigen NLS increased the activity of the ϕC31 integrase by 75% in an extrachromosomal excision reaction. Intrachromosomal excision was increased two- to threefold by the NLS (Andreas et al., 2002).

IV. CONCLUSIONS

The problems of random integration and immunogenicity in gene therapy vectors are being addressed with the development of new site-specific non-viral vectors. Especially promising are site-specific non-viral vectors based on the

ϕC31 integrase. The ϕC31 integrase has been successful in achieving long-term therapeutic gene expression through integration in tissue culture cells and in animals. In addition, it has been shown to correct hereditary tyrosinemia in a mouse model as well as two forms of epidermolysis bullosa in keratinocytes from patients. The ϕC31 integrase system catalyzes a unidirectional reaction without the need for any host-specific cofactors. It is not expected to be limited by vector DNA size. The genome of the ϕC31 phage is approximately 45 kb and is successfully recombined by the integrase. The ϕC31 integrase mediates integration preferentially into a small number of sites in the human genome. These features are the reasons for the success thus far of the ϕC31 integrase in gene therapy studies. Ongoing work to improve the efficiency and specificity of the ϕC31 integration system has the potential to increase further the utility of an already useful non-viral gene transfer system.

Acknowledgment

This work was supported by NIH Grants DK58187 and HL68112 to MPC.

References

Andreas, S., Schwenk, F., Kuter-Luks, B., Faust, N., and Kuhn, R. (2002). Enhanced efficiency through nuclear localization signal fusion on phage PhiC31-integrase: Activity comparison with Cre and FLPe recombinase in mammalian cells. *Nucleic Acids Res.* **30**, 2299–2306.

Chalberg, T. W., Genise, H. L., Vollrath, D., and Calos, M. P. (2005). ϕC31 integrase confers site-specific integration and long-term expression in rat retina. *Invest. Ophthal. Vis. Sci.* **46**, in press.

Chen, Z. Y., He, C. Y., Meuse, L., and Kay, M. A. (2004). Silencing of episomal transgene expression by plasmid bacterial DNA elements *in vivo*. *Gene Ther.* **11**, 856–864.

Groth, A. C., Olivares, E. C., Thyagarajan, B., and Calos, M. P. (2000). A phage integrase directs efficient site-specific integration in human cells. *Proc. Natl. Acad. Sci. USA* **97**, 5995–6000.

Hacein-Bey-Abina, S., Von Kalle, C., Schmidt, M., McCormack, M. P., Wulffraat, N., Leboulch, P., Lim, A., Osborne, C. S., Pawliuk, R., Morillon, E., *et al.* (2003). LMO2-associated clonal T cell proliferation in two patients after gene therapy for SCID-X1. *Science* **302**, 415–419.

Held, P. K., Olivares, E. C., Aguilar, C. P., Finegold, M., Calos, M. P., and Grompe, M. (2005). *In vivo* correction of murine hereditary tyrosinemia type I by ϕC31 integrase mediated gene delivery. *Mol. Ther.* **11**, 399–408.

Hollis, R. P., Stoll, S. M., Sclimenti, C. R., Lin, J., Chen-Tsai, Y., and Calos, M. P. (2003). Phage integrases for the construction and manipulation of transgenic mammals. *Reprod. Biol. Endocrinol.* **1**, 79.

Kuhstoss, S., and Rao, R. N. (1991). Analysis of the integration function of the streptomycete bacteriophage phi C31. *J. Mol. Biol.* **222**, 897–908.

Le, Y., Gagneten, S., Tombaccini, D., Bethke, B., and Sauer, B. (1999). Nuclear targeting determinants of the phage P1 cre DNA recombinase. *Nucleic Acids Res.* **27**, 4703–4709.

Nakai, H., Montini, E., Fuess, S., Storm, T. A., Grompe, M., and Kay, M. A. (2003). AAV serotype 2 vectors preferentially integrate into active genes in mice. *Nature Genet.* **34**, 297–302.

Olivares, E. C., Hollis, R. P., Chalberg, T. W., Meuse, L., Kay, M. A., and Calos, M. P. (2002). Site-specific genomic integration produces therapeutic Factor IX levels in mice. *Nature Biotechnol.* **20,** 1124–1128.

Ortiz-Urda, S., Thyagarajan, B., Keene, D. R., Lin, Q., Calos, M. P., and Khavari, P. A. (2003). PhiC31 integrase-mediated non-viral genetic correction of junctional epidermolysis bullosa. *Hum. Gene Ther.* **14,** 923–928.

Ortiz-Urda, S., Thyagarajan, B., Keene, D. R., Lin, Q., Fang, M., Calos, M. P., and Khavari, P. A. (2002). Stable non-viral genetic correction of inherited human skin disease. *Nature Med.* **8,** 1166–1170.

Schroder, A. R., Shinn, P., Chen, H., Berry, C., Ecker, J. R., and Bushman, F. (2002). HIV-1 integration in the human genome favors active genes and local hotspots. *Cell* **110,** 521–529.

Sclimenti, C. R., Neviaser, A. S., Baba, E. J., Meuse, L., Kay, M. A., and Calos, M. P. (2003). Epstein-Barr virus vectors provide prolonged robust factor IX expression in mice. *Biotechnol. Prog.* **19,** 144–151.

Sclimenti, C. R., Thyagarajan, B., and Calos, M. P. (2001). Directed evolution of a recombinase for improved genomic integration at a native human sequence. *Nucleic Acids Res.* **29,** 5044–5051.

Smith, M. C., and Till, R. (2004). Switching the polarity of a bacteriophage integration system. *Mol. Microbiol.* **51,** 1719–1728.

Smith, M. C., Till, R., Brady, K., Soultanas, P., and Thorpe, H. (2004). Synapsis and DNA cleavage in phiC31 integrase-mediated site-specific recombination. *Nucleic Acids Res.* **32,** 2607–2617.

Stark, W. M., Boocock, M. R., and Sherratt, D. J. (1992). Catalysis by site-specific recombinases. *Trends Genet.* **8,** 432–439.

Stoll, S. M., and Calos, M. P. (2002). Extrachromosomal plasmid vectors for gene therapy. *Curr. Opin. Mol. Ther.* **4,** 299–305.

Thorpe, H. M., and Smith, M. C. (1998). *In vitro* site-specific integration of bacteriophage DNA catalyzed by a recombinase of the resolvase/invertase family. *Proc. Natl. Acad. Sci. USA* **95,** 5505–5510.

Thyagarajan, B., Olivares, E. C., Hollis, R. P., Ginsburg, D. S., and Calos, M. P. (2001). Site-specific genomic integration in mammalian cells mediated by phage phiC31 integrase. *Mol. Cell. Biol.* **21,** 3926–3934.

Vigdal, T. J., Kaufman, C. D., Izsvak, Z., Voytas, D. F., and Ivics, Z. (2002). Common physical properties of DNA affecting target site selection of sleeping beauty and other Tc1/mariner transposable elements. *J. Mol. Biol.* **323,** 441–452.

Wu, X., Li, Y., Crise, B., and Burgess, S. M. (2003). Transcription start regions in the human genome are favored targets for MLV integration. *Science* **300,** 1749–1751.

Yant, S. R., Meuse, L., Chiu, W., Ivics, Z., Izsvak, Z., and Kay, M. A. (2000). Somatic integration and long-term transgene expression in normal and haemophilic mice using a DNA transposon system. *Nature Genet.* **25,** 35–41.

9 *Sleeping Beauty* Transposon-Mediated Gene Therapy for Prolonged Expression

Perry B. Hackett, Stephen C. Ekker, David A. Largaespada, and R. Scott McIvor

Department of Genetics, Cell Biology and Development
Arnold and Mabel Beckman Center for Transposon Research
University of Minnesota
Minneapolis, Minnesota 55455 and
Discovery Genomics, Inc.
Minneapolis, Minnesota 55413

I. Introduction
    A. Non-viral vectors for gene therapy: potential utility and current limitations
    B. Transposons: non-viral vectors that deliver long-term gene expression
II. The *Sleeping Beauty* Transposon System
    A. The transposition process
    B. Origin and development of the SB transposon system
    C. Characteristics of the SB transposon system
    D. Integration-site preferences of SB transposons
    E. Applications of the SB transposons for gene discovery
III. Applications of *Sleeping Beauty* Transposons for Gene Therapy
    A. SB-mediated gene expression in the liver
    B. SB-mediated gene expression in the lung
    C. SB-mediated gene expression in hematopoietic cells
    D. SB-mediated gene expression in tumors
    E. Safety issues for transposon-mediated gene therapy
IV. Future Directions
    A. Delivery methods for the SB transposon system
    B. Efficiency and evolution of the SB transposon system

Advances in Genetics, Vol. 54
0065-2660/05 $35.00
DOI: 10.1016/S0065-2660(05)54009-4

ABSTRACT

The *Sleeping Beauty* (SB) transposon system represents a new vector for non-viral gene transfer that melds advantages of viruses and other forms of naked DNA transfer. The transposon itself is comprised of two inverted terminal repeats of about 340 base pairs each. The SB system directs precise transfer of specific constructs from a donor plasmid into a mammalian chromosome. The excision of the transposon from a donor plasmid and integration into a chromosomal site is mediated by *Sleeping Beauty* transposase, which can be delivered to cells vita its gene or its mRNA. As a result of its integration in chromosomes, and its lack of viral sequences that are often detected by poorly understood cellular defense mechanisms, a gene in a chromosomally integrated transposon can be expressed over the lifetime of a cell. SB transposons integrate nearly randomly into chromosomes at TA-dinucleotide base pairs although the sequences flanking the TAs can influence the probability of integration at a given site. Although random integration of vectors into human genomes is often thought to raise significant safety issues, evidence to date does not indicate that random insertions of SB transposons represent risks that are equal to those of viral vectors. Here we review the activities of the SB system in mice used as a model for human gene therapy, methods of delivery of the SB system, and its efficacy in ameliorating disorders that model human disease. © 2005, Elsevier Inc.

I. INTRODUCTION

In the 21st century we can expect a revolution in the delivery of therapeutics. We can expect genetic medicines that will confer permanent solutions to chronic and acute ailments. How these genetic medicines will be delivered and controlled, without adverse side effects, are the pressing issues facing modern medicine. Gene therapy theoretically represents the best form of treatment for some medical disorders because natural biological products instead of chemicals are employed for their natural function. Delivery of the therapeutic is relatively constant at a physiologically effective level instead of cycles of high and low concentrations that result from the introduction of therapeutics at periodic intervals. Conceptually, gene therapy has the potential to provide a marked clinical and economic improvement over infused recombinant protein used in protein-replacement therapies. The essential goal of gene therapy is to provide what all patients want, an improved quality of life. For these reasons, gene therapy will become the treatment of choice for disorders such as hemophilia (Mannucci and Tuddenham, 2001). Gene therapy is applicable to both genetic and acquired diseases. This chapter reviews a new vector for non-viral gene therapy, the *Sleeping Beauty* transposon system. This vector combines the advantages of viral vectors and directed integration of single copies of a therapeutic gene, with the advantage of non-viral vectors, the absence of protein factors that can elicit adverse reactions.

A. Non-viral vectors for gene therapy: potential utility and current limitations

Nature uses two devices for introducing new genetic material into chromosomes of all organisms. The first is viruses, which have evolved elaborate strategies for efficiently introducing their genomes into cells and occasionally into the chromosomes of infected cells. Because of the high number of potential viruses in the environment and the deleterious aspects of viral infection, most animals have defensive systems to protect their chromosomes from outside intruders. Defenses include acquired immune responses against viral proteins and innate immune responses against selected motifs of viral genomes and/or their transcripts. Nonetheless, due to their efficiencies in gene delivery to cells, they have been used in about 70% of the approximate 1000 gene therapy trials through 2003 (www.wiley.co.uk/genmed/clinical/). The second method is transposons, which have evolved the means to enter chromosomes over such long evolutionary periods that there are few if any host defenses. However, unless facilitated by artificial laboratory techniques, random fragments of DNA that are not transposons, "naked DNA," enter genomes at low rates. Delivery of nontransposon DNA by a variety of methods has been the basis of about 30% of gene therapy trials. Thus, the use of either naked DNA or viruses for gene therapy has serious drawbacks. Here is why.

A fundamental component of any gene therapy strategy is the vehicle used for the delivery of genes into a cell and into its nucleus for appropriate expression. There are five major barriers in the delivery of genetic material to cells: (i) stability of the transgene in the extracellular environment, (ii) transfer of genetic material across the cell membrane, (iii) delivery of the genetic material to the nucleus without intracellular degradation, (iv) integration of the transgenic material into chromosomes so that it can be replicated, and (v) reliable expression of the transgene following integration into a genome. Many viruses are good at penetrating some or all of these barriers, but as mentioned, they have other problems. The problems of surmounting these barriers with non-viral DNA have been reviewed (Niidome and Huang, 2002; Nishikawa and Huang, 2001) and are discussed briefly later.

Non-viral, DNA-mediated gene transfer has been explored extensively as a means of expressing new genes in cells and tissues and constitutes an alternative with several potential advantages over viral delivery systems. (i) Viral vector preparations from cultured mammalian cells come with the risk of contamination by a variety of different infectious agents, including replication-competent virus generated by recombination between virus vector and packaging functions (Kay *et al.*, 2001). In addition, the viral particle itself can be toxic, depending on the dose and site of administration. The risks of DNA-mediated delivery, by comparison, are limited to those associated with plasmid preparation from bacterial extracts (endotoxin, etc.) and whatever chemical component is conjugated with the DNA for the purpose of delivery. (ii) Viral vector preparations are likely to be more highly immunogenic than DNA-based delivery systems. The best example of this is the acute immune/inflammatory response brought about by adenovirus vector administration and transduction in the liver (Lozier *et al.*, 1997). (iii) DNA-mediated delivery is not constrained by many of the biophysical and genetic limitations of viral vectors, such as genome size and elements required for regulation of expression and replication. (iv) DNA-mediated delivery systems are likely to be less expensive, more stable than viral vector preparations, and more amenable to pharmaceutical formulation. (v) A further complication in the use of retroviruses (Mitchell *et al.*, 2004; Wu *et al.*, 2003), lentiviruses (Schroder *et al.*, 2002), and adeno-associated viruses (AAV) (Nakai *et al.*, 2003) may come from their preference for integrating in or near promoters and transcriptional units, where they may have increased chances of causing adverse effects (Baum *et al.*, 2004; Dave' *et al.*, 2004; Lehrman, 1999; Thomas *et al.*, 2003).

DNA-mediated gene transfer presents a superior alternative to viral vectors for gene therapy. *In vivo* DNA-mediated gene transfer into a variety of different target sites has been studied extensively. Naked DNA can provide long-term expression in muscle, albeit after injection of relatively large quantities of DNA (Wolff *et al.*, 1990, 1992). DNA-mediated gene transfer has also been characterized in liver (Miao *et al.*, 2000, 2001; Zhang *et al.*, 2000a), heart (Acsadi *et al.*, 1991; Kleiman *et al.*, 2003; Li and Huang, 1997; Morishita, 2002), lung (Ferkol *et al.*, 1995; Ferrari *et al.*, 1997; Liu *et al.*, 2004b; Yoshimura *et al.*, 1992), brain (Lam and Breakefield, 2001), and endothelial cells (Liu *et al.*, 1999b; Nabel *et al.*, 1989; Zhu *et al.*, 1993) when administered in association with various cationic lipids, polycations, and other conjugating substances (Przybylska *et al.*, 2004; Svahn *et al.*, 2004). However, the primary limitation of DNA-mediated gene transfer in these systems is the relatively short duration of gene expression. The "long-term" gene expression that has been observed in muscle and in liver is associated with persistence of the newly introduced DNA in an extrachromosomal form (Miao *et al.*, 2001; Wolff *et al.*, 1990, 1992; Zhang *et al.*, 2000a). The stability of newly introduced DNA sequences can be improved greatly by integration into the host cell chromosome. However, stable integration in tissues after DNA-mediated gene transfer occurs rarely and primarily by random (illegitimate) recombination.

B. Transposons: non-viral vectors that deliver long-term gene expression

We have developed a new means to achieve stable integration of DNA sequences in vertebrates using the *Sleeping Beauty* (SB) transposon system (Ivics *et al.*, 1997). Since its creation in 1997, *Sleeping Beauty* has been shown to mediate transposition in different cultured cell types (Converse *et al.*, 2004; Izsvak *et al.*, 2000) as well as in zebrafish embryos (Balciunas *et al.*, 2004; Clark *et al.*, 2004; Davidson *et al.*, 2003; Hackett *et al.*, 2004; Wadman *et al.*, 2005), mouse embryos (Dupuy *et al.*, 2002), mouse embryonic stem cells (Luo *et al.*, 1998), mouse germ cells (Carlson *et al.*, 2003; Dupuy *et al.*, 2001; Fischer *et al.*, 2001; Horie *et al.*, 2001, 2003), and mouse somatic tissues (Belur *et al.*, 2003; He *et al.*, 2004; Kren *et al.*, 2003; Liu *et al.*, 2004a,c; Montini *et al.*, 2002; Ohlfest *et al.*, 2004, 2005; Ortiz *et al.*, 2003; Yant *et al.*, 2000, 2005). *Sleeping Beauty* thus provides a means of achieving chromosomal integration and long-term expression both *in vitro* and in experimental animals, thereby circumventing a primary limitation of nontransposon, DNA-mediated gene delivery for human therapy (Ivics and Izsvak, 2004). The success of the SB system has led to the development of other transposon and transposon-like vector systems, including *Frog Prince* (Miskey *et al.*, 2003), *Tol2* (Kawakami and Shima, 1999; Koga *et al.*, 2003), ΦC31 (Olivares *et al.*, 2002), and the retrotransposon L1 (Farley *et al.*, 2004; Han *et al.*, 2004; Luning Prak and Kazazian, 2000). The following sections review the current status of SB transposons for gene therapy.

This review concentrates on the *Sleeping Beauty* transposon system. SB transposons represent a type of mobile element that belongs to the *Tc1/mariner* class of transposons that transpose via movement of a DNA element. *Tc1/mariner*-type transposons comprise almost 3% of the human genome (Lander *et al.*, 2001; Venter, 2001) and therefore are a minority class of transposon species in human and other vertebrate genomes—retrotransposons comprise most transposons in vertebrate genomes, of which the LINE and SINE families comprise the largest subfraction, approximately 33% of the genome (Lander *et al.*, 2001; Venter, 2001). DNA transposons move in a simple, cut-and-paste manner (Fig. 9.1) in which a precise DNA segment is excised from one DNA molecule and moved to another site in the same or different DNA molecule (Plasterk, 1993). The protein that catalyzes this reaction, the transposase, is encoded within the transposon for an autonomous element or

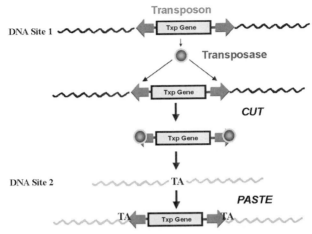

Figure 9.1. The cut-and-paste mechanism of transposition for a DNA transposon with an active transposase (Txp) gene. SB transposons integrate only into TA base pairs. After duplication of the TA site during transposition, TA sequences are formed on each end of the transposon. The inverted red arrows, representing 230 bp each, are the only DNA sequences required by the transposase enzyme for transposition. For gene therapy, the transposase gene is replaced by a DNA sequence that encodes a therapeutic product that could be either a protein or an RNA molecule. (See Color Insert.)

can be supplied in *trans* by another source for a nonautonomous element. *Tc1/mariner*-type transposases require a TA dinucleotide base pair for an integration site, a sequence that is duplicated during the integration process. The *Tc1/mariner*-type SB transposon system consists of two components: (i) a transposon, made up of a gene of interest flanked by inverted repeats [IRs, shown as arrowheads (IR-DR) in Fig. 9.1] and (ii) a source of transposase. During *Sleeping Beauty*-mediated transposition, the SB transposase recognizes the ends of the IRs, excises the transposon from the delivered plasmid DNA, and then inserts the transposon into another DNA site. The transposon structure shown in Fig. 9.1 is representative of the class of *autogenous* transposons, i.e., a transposon that encodes an active transposase that directs the movement of the transposon with transposase. To date, no active *Tc1/mariner*-type or SB-like transposase gene has been found in any vertebrate genome, although thousands of highly mutated transposase genes have been found in genome sequencing projects. Consequently, all of the ca. 20,000 *Tc1/mariner*-type transposons that reside in human genomes are stable. In contrast, some retroelements are active and do hop occasionally in humans (Kazazian and Goodier, 2002; Prak and Kazazian, 2000).

Tc1/mariner-type transposable elements are ubiquitous in animal genomes and generally can be mobilized in cell-free systems in the presence of their respective transposase enzymes made in *Escherichia coli* (Lampe *et al.*, 1996; Vos *et al.*, 1996), suggesting that they require few, if any, species-specific host factors. The presumed simplicity of this form of transposon made them attractive candidates for use in human gene therapy. However, transposition of SB transposons in cell-free systems has not been demonstrated; it appears that there are host factors that play roles in the transposition process for some *Tc1/mariner*-type elements (Izsvak *et al.*, 2004; Yant and Kay, 2003; Zayed *et al.*, 2003). This difference between the SB transposon system and the *Tc1* and *mariner* transposons has not interfered with using SB transposons for gene delivery to vertebrate genomes.

II. THE *SLEEPING BEAUTY* TRANSPOSON SYSTEM

For the purposes of human gene therapy there are several important facets of using the SB transposon system as a vector that need to be appreciated. (1) The SB transposase directs the integration of precisely defined, single copies of a DNA sequence into chromatin (Fig. 9.2). (2) The integrated gene is stable with respect to expression as a result of the integration, providing long-lasting expression of a therapeutic gene. (3) The transposase elevates the frequency of integration of a desired gene by about 100-fold or more, depending on the transposon, the transposase, the target cells, and method of delivery of the system. (4) The SB system is *binary*, meaning that the transposon is *not* autonomous or able to transpose on its own. An appropriate transposase-encoding sequence must be supplied either in *trans* on a second vector [shown in Fig. 9.2 (Ivics *et al.*, 1997)] or in *cis* on the same DNA molecule as the transposon (Converse *et al.*, 2004; Harris *et al.*, 2002; Mikkelsen *et al.*, 2003). Consequently, there is not a "containment issue" for use of this vector system. (5) The SB transposase is synthetic—it was derived from sequences found in fish genomes (more on this in Section II.B.2) and therefore does not bind measurably to transposons in human or other mammalian cells (Ivics *et al.*, 1997; Izsvak *et al.*, 1997). (6) Transposons per se are not constrained rigorously by the genes they carry or their size, although the efficiency of transposition and passage of the plasmids that carry them through cellular membranes (plasma and/or nuclear, as discussed earlier in Section I.A) appears to decrease with size (Geurts *et al.*, 2003; Izsvak *et al.*, 2000; Karsi *et al.*, 2001). (7) The transposase has nuclear localization sequences that may enhance translocation of the transposon from the cytoplasm to the nucleus of nondividing cells (Zanta *et al.*, 1997). Figure 9.2 schematizes the use of SB transposons for gene therapy.

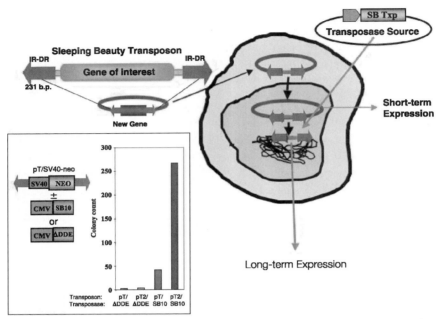

Figure 9.2. SB transposons for gene therapy. Delivery of a transposon with a transposase-encoding
sequence, shown here on two plasmid vectors, can provide long-term expression of the
transposed gene compared with shorter durations of expression when the gene remains
as an episome in transformed cells *in vivo*. (Inset) The increase in levels of gene
expression as a result of transposition in the presence of SB transposase compared to
levels observed when a defective form (ΔDDE) of the enzyme is supplied (Ivics *et al.*,
1997). T is the original transposon and T2 is an improved version (Cui *et al.*, 2002) as
measured by the frequency of G418-resistant HeLa cell colony formation following
transposition of an SV40-Neo construct in either of the two transposons and with or
without active SB transposase. The numbers of colonies obtained in the ΔDDE
experiments are about equivalent to the levels found following delivery without any
transposase and represent random, illegitimate recombination into chromosomes. (See
Color Insert.)

A. The transposition process

The transposition–integration process is shown in more detail in Fig. 9.3. The
SB transposase cleaves the transposon-donor site at the flanking TA dinucleotide base pairs in a staggered manner such that three bases, GTC, extend at each
3′ end of the transposon (lines 2 and 3 in Fig. 9.3). The 3′ ends of the excised
transposon invade the target DNA molecule (indicated by the gold ellipses) at
the pair of TA sequences that extend from the 5′ ends that are produced when
SB transposase cleaves the target site at a TA dinucleotide base pair (third line

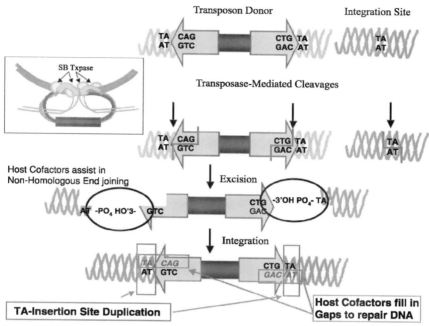

Figure 9.3. SB-mediated transposition from a donor site (green lines) to an integration site (purple lines). Two SB transposase molecules, shown as yellow circles in the boxed insert on the left, bind on each of the inverted terminal repeat (arrows) to introduce three cleaves: two flanking the transposon (pink structure with inverted arrows representing the inverted terminal repeats) and one in the target integration site (second line). The insert emphasizes that SB transposase molecules act in concert in a complex of transposon donor and target integration site. The excision step is shown on the third line with integration occurring by the invasion of the 3′ ends of the transposon joining the exposed TA nucleotides at the integration site (shown in the ellipsoids in the third line). Following ligation of the single strands on each side, DNA repair enzymes fill in the remaining five nucleotide gaps (shown in red in the fourth line). TA target site duplication is indicated in the last line by the boxed TA dinucleotide base pairs. The transposon-donor sequence is resealed and the single base mismatch is repaired by cellular enzymes (lower right corner). (See Color Insert.)

in Fig. 9.3). This process is called "nonhomologous end joining" and is mediated by cellular cofactors (Plasterk *et al.*, 1999; Yant and Kay, 2003). Integration is completed by repair of the five base gaps on both strands. Note that the original TA dinucleotide base pair target sequence is duplicated on both flanks of the transposon following integration (boxes in line 4 of Fig. 9.3), the left and right ends of the donor DNA. CAG overhangs on their 3′ ends are brought together with a single A–A pairing in the center that is resolved by one or the other A's being replaced by a T during DNA repair. As a result, in the most common case, the original **TA** in the donor site is modified to contain a "footprint" **TAC** (A or T)GTA. Sometimes the repair process introduces more alterations in the donor site, with greater losses of sequence (Plasterk, 1993; Plasterk *et al.*, 1999). It should be noted that transposase is not like a restriction enzyme that cleaves DNA molecules into fragments that freely diffuse; rather, the excision–integration reactions are highly coordinated events such that the paste step of transposition follows directly from the cleavage step.

The model in Fig. 9.3 raises an obvious question—are the excision and integration steps always coupled? A related question is whether a single transposase molecule can cleave DNA even though it takes four proteins to form an integration complex. The answers are not established for the SB system and there do not appear to be hard rules for transposons. For instance, whereas Mu and Tn7 select a target site prior to cleavage, Tn10 synaptic complexes can integrate into a target DNA after excision (Sakai and Kleckner, 1997; Williams *et al.*, 1999). Our current knowledge of SB transposition is based on separate measurements of excision and integration. The excision step can be semiquantified by polymerase chain reaction (PCR) by what is called an *excision assay* (Dupuy *et al.*, 2002; Liu *et al.*, 2004a). Essentially, PCR primers flanking a transposon will direct the synthesis of a predictable fragment following the removal of several thousand base pairs from the donor site as shown in Fig. 9.4; sometimes, however, the repair process is not exact and does not produce the canonical footprint shown in the Fig. 9.4. In contrast, the complete transposition process can be estimated by genetic selection experiments. When the two assays were combined to study the effects of various mutations in the transposon, it was found that transposition correlated with excision, as expected from the model shown in Fig. 9.3 (Liu *et al.*, 2004a).

There have been no studies with SB transposase to determine if cleavage can occur by monomeric SB enzymes. However, in a related process mediated by the λ phage integrase, it was found that a single recombinase protein can cleave DNA, but that the enzyme tends to form multimeric complexes in physiological salt solutions, as does SB transposase (Z. Cui and P. Hackett, unpublished result), that inhibit activity by single proteins (Lee *et al.*, 2004). Thus, at this point the evidence strongly suggests a coupling of excision and integration. Currently, the transposition events in organs of multicellular animals have been evaluated by sequencing insertion sites in chromosomes of treated animals because each cell in which the transposon integrates is a separate event that is nonclonal. This procedure is difficult and is not quantitative. However, the excision assay allows a rapid evaluation of relative level of transposase activity *in vivo* when several methods of gene delivery are used.

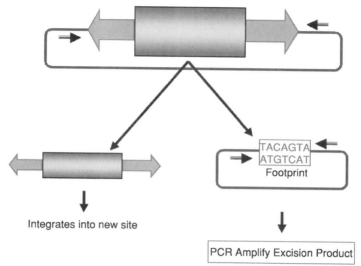

Figure 9.4. The excision assay for quantifying one step in the transposition process. A pair of primers (black arrowheads) is designed to flank a transposon in a donor molecule, shown here in a plasmid vector (blue line). Following excision of the plasmid, a resealed vector is produced that will be considerably smaller than the donor DNA and often will have a precise, canonical footprint, shown here in the box. The middle A-T base pair shown could be T-A as well. Following delivery of transposons to a multicellular tissue, each of the integration sites will be unique but the remaining plasmid will be the same so that a single set of primers can record all excision (and correlated integration) events. (See Color Insert.)

B. Origin and development of the SB transposon system

As noted earlier, *Tc1/mariner* transposons are found in just about every animal genome in which they have been sought, although in most animals in general and in vertebrates in particular their transposase genes appear to be defective (Hartl *et al.*, 1997a). Although the use of these transposons seemed trivial at first, when we and others examined the abilities of known, active transposons to move in vertebrate cells, including those from zebrafish and humans, we found marginal transposition activity (Fadool *et al.*, 1998; Gibbs *et al.*, 1994; Raz *et al.*, 1998; Zhang *et al.*, 1998) (Z. Ivics and Z. Izsvák, unpublished result). Consequently, the SB transposon system was constructed based on phylogenetic principles in a 10-step process of site-specific mutagenesis of a salmonid transposase gene that became evolutionarily dormant more than 10 million years ago (Ivics *et al.*, 1996; Izsvak *et al.*, 1995; Radice *et al.*, 1994). The *awakened* transposase was named *Sleeping Beauty* (Ivics *et al.*, 1997). The *T* transposon plus SB transposase comprise the SB transposon system. For gene therapy, both components of the SB system are delivered to cells in plasmids, the transposon, and the transposase gene that when expressed can cut the transposon out of the plasmid carrier for reinsertion into a chromosome. The SB transposase gene may be on the same plasmid carrier as the transposon (more on this later). As shown in the insert in Fig. 9.2, the original SB transposase is able to improve integration from 20- to 40-fold in cultured mammalian cells and about 20-fold in zebrafish embryos (Izsvak *et al.*, 1997, 2000). In a head-to-head competition, Fischer *et al.* (2001) showed that the rate of transposition mediated by the SB transposon system was nearly an order of magnitude higher than those observed for a variety of transposons from nematodes (Schouten *et al.*, 1998) and flies (Klinakis *et al.*, 2000) in cultured human HeLa cells. In all of these experiments, *nonautonomous* transposons were used, i.e., transposons in which the transposase gene was replaced with alternative genetic cargo. The transposase was generally supplied by another plasmid carrying the transposase gene or by mRNA encoding the transposase.

A significant difference exists between the SB transposon and other commonly used members of the *Tc1/mariner* transposon family. SB transposons, called Tn [where n is the version of transposon, the original was simply *T* (Ivics *et al.*, 1997)], contain two "repeats" within each inverted terminal repeat (called IR-DRs for inverted repeats containing direct repeats) compared with other transposons that contain a single inverted repeat (Fig. 9.5). Significantly, the original *T* transposon (Ivics *et al.*, 1997) contained DR sequences that varied depending on their position. More recent changes in specific base pairs within the DR sequences have led to further development of the efficiency of SB transposons with significantly higher transpositional activities (Cui *et al.*, 2002; Zayed *et al.*, 2004), the first being T2, which has consensus inner (Li and Ri) and outer (Lo and Ro) DR sequences.

1. Improvements in SB transposons

Analyses in our laboratories of the DR sequences identified several aspects of the SB transposon system that were unexpected (Cui *et al.*, 2002; Liu *et al.*, 2004a). First, differences between the inner and the outer DR sequences are important—there does not appear to be a universal DR sequence that can be used effectively for the four sites in a complete transposon. Second, SB transposase binds more tightly to the inner DR sequences than to the outer DR sequences. Third, when the outer DR sequences are altered to increase the binding affinity of SB transposase, transposition rates decrease dramatically, suggesting that improvements in transposon sequences cannot be evaluated simply on the basis of transposase DR-binding energies. Fourth, the 170-bp spacer sequences between the DRs within an inverted terminal repeat are important—altering them often

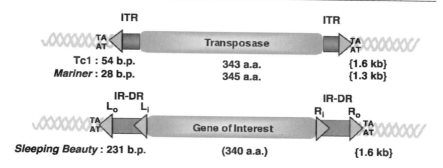

Figure 9.5. Comparative structures of *Tc1/mariner*-like transposons and SB transposons. ITR, inverted terminal repeat sequence; IR-DR, inverted repeat containing direct repeated sequences. The lengths of the repeat sequences are noted for each transposon as well as the transposase size (or reconstructed size in the case of SB—the transposase is never supplied from a gene flanked by two IR-DRs). Consensus sizes of the transposons are shown in the brackets on the right. The specific sequences of the four DRs in the original SB transposon, *T*, are shown at the bottom along with the consensus sequences that were used to build an improved transposon called T2. The SB footprint refers to the portion of the DR sequence that is protected from DNase hydrolysis when bound by SB transposase (Ivics *et al.*, 1997). (See Color Insert.)

leads to depression of transposition activity. A "transpositional enhancer" has been identified in the inter-DR sequence (Izsvak *et al.*, 2002) that may facilitate the DNA bending and pairing that occurs between the two ends of the transposon after four transposase molecules have bound (see inset in Fig. 9.3). A nucleic acid-binding protein, HMGB1, has also been shown to elevate rates of transposition, possibly by facilitating binding of SB transposase to the inner DRs and bending the inverted terminal repeat sequences to a form compatible with transposition (Zayed *et al.*, 2003). All of these data suggest that the process of SB-mediated transposition involves the interactions of several cellular cofactors that together contort the transposon, and probably the new integration site, in a complex manner in concert with four SB transposase molecules (Ivics *et al.*, 2004; Izsvak *et al.*, 2004). Our findings that maximal binding of transposase molecules to the DNA sites can be deleterious to function is similar to results obtained from analyses of transcriptional regulators. For instance, control and appropriate function are lost when binding to all sites is too tight at *lac* operator sites (Kalodimos *et al.*, 2004; Oehler *et al.*, 1994) and other cases where flexibility and dynamic activity on DNA are required (von Hippel, 2004). Two versions of a highly improved SB transposon vector, T2 and T/SA, are now readily employed in the field (Cui *et al.*, 2002; Zayed *et al.*, 2004) and more are under development.

2. Improvements in SB transposase

Several studies have been conducted to improve the activity of the SB transposase enzyme (Geurts *et al.*, 2003; Yant *et al.*, 2004; Zayed *et al.*, 2004), the other half of the SB system. Figure 9.6 shows the functional domains of the 360 amino acid protein. The first one-third represents a domain with two functions that were thought to be mediated by a leucine zipper motif (Ivics *et al.*, 1996, 1997), although this structure is not predicted by a commonly used algorithm (Lupas *et al.*, 1991) and the corresponding region of a related *mariner* family transposase does not form a coiled-coil, as determined by the crystal structure of the

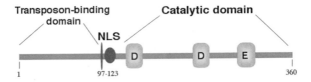

Transposon-Binding Domain
· 1 Transposase molecule binds to each DR sequence
· Coiled-coil structure
· Dimerization activity between two SB transposase molecules

Nuclear Localization Sequence (NLS)
· Bipartite structure

Catalytic Domain
· Mediates cleavage at the ends of the Inverted Terminal Repeats
· Recognizes integration sites at TA dinucleotide basepairs
· Mediates non-homologous end-joining - the "paste" step

Figure 9.6. Diagram of SB transposase. The three functional domains are identified at the top; numbers below the structure are the approximate amino acid residue boundaries of the domains. The transposon-binding domain, often called the DNA-binding domain, binds to DRs and can protect the nested sequence from degradation by DNase I as identified in Fig. 9.4. The transposon-binding domain is also responsible for dimerization of transposase molecules to form the complex shown in Fig. 9.3. The NLS sequence comprises amino acids 79–123 and has two clusters of basic amino acids separated by a 10 amino acid spacer. The catalytic domain characterized by the DDE motif is commonly found in all cut-and-paste recombination enzymes. (See Color Insert.)

N-terminal region of the Tc3 transposase (van Pouderoyen *et al.*, 1997). The first is to bind specifically to the DR sequences in the inverted terminal repeats of the transposon. The second role is to bring the two ends of the transposon together to form a synaptic complex (Fig. 9.3) that invades the target site. The middle, relatively short motif is a nuclear localization sequence (NLS) that is composed of two clusters of basic sequences, thereby making it a bipartite NLS. The bipartite structure of nuclear localizing sequences in transposons kept them from being identified until the one present in SB transposase was uncovered (Ivics *et al.*, 1996). The third, catalytic domain comprises the carboxy-terminal half of the transposase. Like the N-terminal sequence, it has two functions: one to identify TA insertion sites and the other to catalyze the three cleavages and one paste reaction of the complete transposition reaction (Fig. 9.3). The catalytic domain is characterized by the DDE motif, which represents two aspartic acids (D) and one glutamic acid (E) and which is found in all cut-and-paste recombination enzymes, such as retroviral integrases, transposases, and phage integrases (Craig, 1995, 1997). Commonly, there are 35 amino acids separating the second D and the E residues. An exception to the DDE rule is the *Mos1* transposase, which has a DDD motif with 34 amino acids separating the second and third aspartic acids; substitution of an E for the third D abrogates enzyme activity (Lohe *et al.*, 1997).

The combination of functions of the DNA recognition domain and the catalytic domain present problems for directed mutagenesis to improve the transposase. This is common in transposases—X-ray crystallography of the Tn5 transposon synaptic intermediate complex showed that the domains of Tn5 transposase do not have discrete functions; rather, the amino-terminal domain, catalytic, and carboxy-terminal domains all participate in DNA binding; each plays a role in forming the structure of the synaptic complex (Davies *et al.*, 2000). As noted earlier, one strategy for the improvement of SB transposition based on increasing the binding strength of SB transposase to DRs is unlikely to succeed due to the necessity of maintaining the dynamic structures that the transposon passes through during the transposition process. As a result, most improvements have come from further efforts to derive a more accurate consensus sequence based on selected, defunct transposase genes (Geurts *et al.*, 2003; Zayed *et al.*, 2004) or by systematically substituting a leucine for many amino acids in the polypeptide to determine the effect on transposition (Yant *et al.*, 2004). Both procedures found several sites that when mutated gave enhanced activity, and both strategies found that combinations of mutations that enhanced transposition alone did not always act in an additive manner; indeed some combinations canceled each other. All of the assays for enhancement of SB transposase activity were conducted in HeLa cells using the frequency of G418-resistant colony formation as a measure (Fig. 9.2 inset). Assays in other cell types may be useful to detect alternative mutations that lead to higher activity in specific tissues because various

SB transposases may not have equal activities in all cells. Consideration of the presumed complex topography of charges in the catalytic center of SB transposase suggests that any simple and directed modification to alter the TA recognition site has a large probability of altering (reducing) the enzymatic process at the same time. Consequently, development of a site-specific SB transposase (Kaminski *et al.*, 2002) does not appear likely in the near term.

C. Characteristics of the SB transposon system

Features of the SB transposon system important for gene therapy can be divided into two broad categories: limitations on delivery of the complete transposon system into cells of target tissues and limitations on the transposition process. Limitations on delivery of the transposon system are based on the transposon carrier. Most experiments have used plasmids, but one report discusses delivery of the SB system using an adenovirus as the carrier (Yant *et al.*, 2002). Considerations on methods of delivery to cells of specific tissues are discussed in Section III. Limitations on transposition include (1) the size of the transgene in the transposon, (2) the size of the transposon-donor plasmid, (3) the ratio of two components of the SB system, the transposase and the transposon, (4) methylation state of the transposon, and (5) the "state" of the target cells.

     The effect of the size of the transgenic construct, comprising the gene and transcriptional regulatory components, is an obvious concern. Size effects occur at the plasma and nuclear membranes as well as transposition into chromosomes. Three studies to examine the effects of size have been conducted using transposons that carried a selectable marker plus various "spacer" DNA sequences (Geurts *et al.*, 2003; Izsvak *et al.*, 2000; Karsi *et al.*, 2001). The first two studies employed prokaryotic sequences and the third used DNA that flanked the carp β-actin gene. In all cases there was a nearly linear decrease of transposition as a function of transposon size, but the decrease was significantly less when the spacer was vertebrate DNA compared to prokaryotic DNA. The results are reconcilable if the prokaryotic DNA was more apt to be methylated and thereby transcriptionally silenced than the eukaryotic DNA spacer that had a lower GC content. Thus, the rates of transposition are likely to be higher than indicated by selection for gene expression. The Geurts *et al.* (2003) study indicated that transpositional activity was reduced about 50% when the size of the transgenic construct was about 6 kbp, a size that would accommodate about 85% of cDNAs made to known mRNAs. However, if a transposon is flanked by two complete DR elements in an inverted orientation, the "sandwiched" transposon can be mobilized with transgenic constructs as large as 10 kbp (Zayed *et al.*, 2004).

     A second influence on transposition is the size of the transposon-donor plasmid. Plasmid size will affect uptake and transport of the plasmid into and through the cell to the chromosomes. The size also has an effect on the effective separation of the termini of the transposons. The transposon-containing plasmid shown in Fig. 9.4 is an arrangement where the DNA sequence through the transposon is shorter than the DNA sequence of the plasmid carrier. However, when larger genes are put into the transposon, the DNA sequence length of the plasmid may be shorter than that of the transposon so that the effective separation of the transposon ends is defined by the plasmid. Izsvak *et al.* (2000) demonstrated that indeed the closer the transposon ends were, either by engineering the transposon or the plasmid vector, the more effective the transposition.

Four transposase molecules are required to bring the two ends of a transposon together for the transposition reaction (Fig. 9.3). As noted earlier (Fig. 9.6), the transposase molecules can form dimers and tetramers to form the synaptic complex. As a consequence, overexpression of SB transposase leads to inhibition of transposition by quenching the reaction; extra transposase molecules can dimerize with those bound to the DRs to prevent their interactions. Overexpression inhibition was found earlier in *mariner* transposons (Hartl *et al.*, 1997a) and has been shown to occur for SB transposons as well (Geurts *et al.*, 2003). Gene therapy applications of SB transposons must take into consideration overexpression inhibition, which can be accomplished in two ways. The first is to use various ratios of plasmids with either the transposon or the transposase gene. The ratio of transposase to transposon will vary depending on the cell type, the method of delivery, the effects of size on the uptake of the two different plasmids, and the strength of the promoter driving the SB gene. Alternatively, if a *cis* configuration of transposon plus transposase gene is used, then the primary way to vary the ratio of the two components is by choice of the promoter for the SB transposase gene (Mikkelsen *et al.*, 2003; Score *et al.*, 2003). Thus, different conditions must be tested for different vectors and different target tissues (Montini *et al.*, 2002; Ortiz *et al.*, 2003; Yant *et al.*, 2000).

The methylation state of the transposon appears to be important. The frequency of transposition from one site in chromosomal DNA to another location is more than 100- to 1000-fold higher in mouse germ cells (Dupuy *et al.*, 2001, 2002; Horie *et al.*, 2001, 2003) than in mouse embryonic stem cells (Luo *et al.*, 1998). Apparently, CpG methylation of the transposon, but not necessarily its cargo, enhances SB transposition (Yusa *et al.*, 2004). Moreover, removal of CpG sites in plasmids and their accompanying methylation appears to reduce innate immune responses directed against unmethylated prokaryotic plasmids, thereby extending the lifetime of transgene's expression and presence of its encoded polypeptide product (Chevalier-Mariette *et al.*, 2003; Reyes-Sandoval and Ertl, 2004). These findings led to several questions, including (1) Do those few transposons that integrate do so as a result of being methylated in the nucleus prior to integration? (2) Are there effective ways of methylating the transposon portion of a vector but not its cargo to enhance transposition without inhibiting expression of the transgene? (3) Will methylation of the transposon reduce promoter activities inside the transposon? Answers to these questions may provide means for raising the efficiencies of transposition.

DNA methylation has been invoked as a method for suppression of transposon hopping within genomes (Jones and Talai, 2001; Martienssen, 1998) based on the findings in plants that most transposons are methylated (Yoder *et al.*, 1997) and that abolishing DNA methylation resulted in the activation of transposition (Bird, 2002; Miura *et al.*, 2001). In most cases this appears to involve the suppression of retrotransposons. However, methylation is also regulated by double-stranded (ds) RNA-mediated mechanisms and, via these activities, may serve to regulate the expression of transposase genes as well as the methylation state of transposons (Lippman *et al.*, 2004; Montgomery, 2004; Vastenhous and Plasterk, 2004). Because SB transposons do not contain their cognate transposase, dsRNA events to regulate their expression probably do not play direct roles in transposon silencing. However, depending on the insertion site, some, but not all, SB transposons would be silenced as a result of heterochromatin spread (Richards and Elgin, 2002).

A major question that has been unanswered since the first report on the SB transposon system is why relatively few cells that take up the transposon and express it within a few days actually support transposition. For example, in our common assays with HeLa cells, about 10^6 to 10^7 transposon molecules are delivered per cell and about 80% of cells take up and express the DNA, yet at most only 3% of the cells can be recovered following genetic screening for expression of the transgene (Ivics *et al.*, 1997) (J. Bell *et al.*, unpublished result). In mouse tissues the reduction seems about the same, even though the ratios of transposon plasmid to transposase plasmid may be as much as 20:1, depending on the experiment (Belur *et al.*, 2003; Montini *et al.*, 2002; Yant *et al.*, 2000) (E. Aronovich, unpublished result). Because a large fraction of cells transiently express the transgene contained in the transposon, it is clear that SB transposons can enter the nuclei of most cells, as can transposase with its NLS sequences, so that nuclear entry does not appear to be a key barrier. Factors mentioned earlier, including methylation status and effective ratios of transposon to transposase, which change as the number of transposase transcripts accumulate in a cell, may affect the overall rates of transposition. Nevertheless, on a per vector basis, transposition is a highly improbable process, opening the possibility that the "state" of the target cells may be important. At present, despite its clear importance, we have little understanding of what "state" really means and how it might vary in the cells from one tissue to another. More work is needed to understand this issue. It may involve subtleties of the cell cycle beyond simply the disappearance of the nuclear membrane, which does not appear to be a critical barrier for entry of SB transposons into nuclei.

D. Integration-site preferences of SB transposons

Based on mapping of nearly 2000 transposon–integration events in either mouse or human genomes from cultured cells and tissues in mice, transposons integrate almost randomly, equally into exons, introns, and intergenic sequences as a function of their lengths (Carlson et al., 2003; Dupuy et al., 2001, 2002; Horie et al., 2001, 2003; Roberg-Perez et al., 2003; Vigdal et al., 2002; Yant et al., 2005), unlike retroviral (Mitchell et al., 2004; Wu et al., 2003), lentiviral (Mitchell et al., 2004; Schroder et al., 2002), and AAV (Nakai et al., 2003) vectors, which preferentially integrate nearby promoter elements and transcriptional units (Fig. 9.7A). Although on a macroscale SB transposons seem to integrate in a nearly random fashion, some Tc1/mariner-type transposons are known to prefer "hot spots" that consist of particular sequences flanking the invariant TA integration site (Ketting et al., 1997). Consequently, the flanking sequences of integration sites for SB transposons have revealed a consensus AYATATRT or $(AT)_4$ simple sequence palindrome (Carlson et al., 2003; Vigdal et al., 2002; Yant et al., 2005). One feature of such AT-rich sequences is that they are highly deformable (Vigdal et al., 2002). However, in a genome like that of the mouse that is 58% (A + T), roughly 30% of the dinucleotide base pairs may be TA, corresponding to about 2×10^9 integration sites per diploid genome. With so many sites, it can be difficult to find preferential integration sites that might vary from the TA palindrome. Consequently, Liu et al. (2005) looked for preferential sites within limited sequences such as plasmids. They found that there were highly favored sites that did not match the consensus but did have high deformation coefficients (Fig. 9.7C). In particular, they found that these sites exhibited characteristic alternating higher and lower angles of rotation between adjacent base pairs, a greater spacing between the TA base pairs, and a tilting of the base pairs at the TA integration site. These same characteristics are found in the TA palindromes. In these analyses, there was little concern for relative accessibility of the DNA sequences to transposition, a consideration that probably plays a role in genomes with variations in the degrees of sequence condensation between active and transcribed chromosomal regions.

Although a full description of preferred sites is ongoing, the following can be concluded at this stage. (1) By and large, transposition appears to be far more random for transposons than for many viral vectors. (2) SB transposase does not appear to have a distinct preference for exons, introns, or transcriptional regulatory motifs such as enhancers and promoters. (3) There are preferred sites, but they appear to be fairly common and are not simply sequence based. In a sense this is a bit surprising because one might expect that given the millions of potential AT integration sites, the first sites encountered by a transposon–transposase complex might result in integration. It has been suggested that heterochromatin is preferentially located at the periphery of

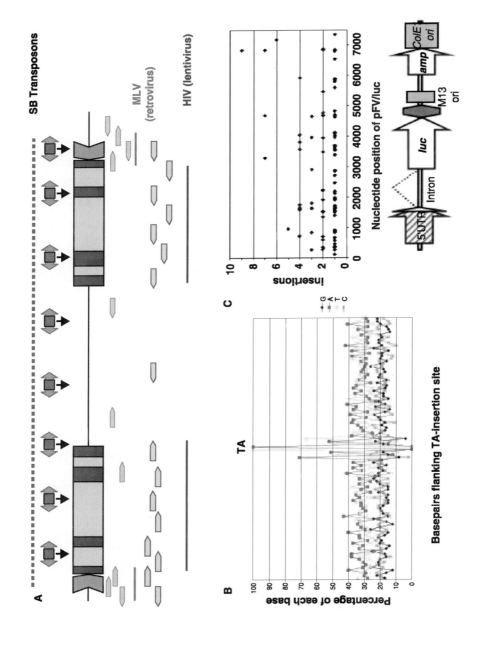

A

SB Transposons

MLV (retrovirus)

HIV (lentivirus)

B

G
A
T
C

TA

Percentage of each base

100
90
80
70
60
50
40
30
20
10
0

Basepairs flanking TA-insertion site

C

Insertions

10
8
6
4
2
0

0   1000   2000   3000   4000   5000   6000   7000

Nucleotide position of pFV/luc

5′UTR   Intron   luc   M13 ori   amp   ColE ori

mammalian nuclei and that a transcriptionally active chromatin is located more toward the interior (Misteli, 2004). As transposon integration does not occur preferentially (or nonpreferentially) in heterochromatin, other factors such as chromatin structure and assorted factors affect the probabilities of integration. Over all, it is possible that the lack of integration preference by transposons may cause fewer adverse effects due to insertional mutagenesis than by viruses. However, the random nature of SB integration makes this vector a useful agent for insertional mutagenesis, as discussed next.

E. Applications of the SB transposons for gene discovery

SB transposons have been used for insertional mutagenesis in mice and fish. In zebrafish, SB transposons have been used for functional genetic studies following their introductions into early embryos (Balciunas et al., 2004; Clark et al., 2004; Davidson et al., 2003). Transposons also have been used in medaka for the same purposes (Grabher and Wittbrodt, 2004; Grabher et al., 2003). In mice the most productive random screens have employed strains that express SB transposase and have resident SB transposons that can be remobilized in the germ line to produce animals with new insertions in every cell. The reported rates of hopping vary from about 0.1 new insertions per offspring to about 2 new insertions per offspring, depending on the numbers and locations of the resident transposons (Carlson et al., 2003; Dupuy et al., 2001, 2002; Fischer et al., 2001; Horie et al., 2001, 2003). All of the studies have shown that most, but not all, of the remobilized transposons tend to hop to a site relatively close (10 Mbp) to their origin, thereby allowing for saturation mutagenesis of particular regions of the genome. A database of locations of transposons in the mouse genome has been established (Roberg-Perez et al., 2003). Data suggested that there are only about 10^4 transposition sites/genome that can be hit efficiently following remobilization from a single source in the germ cells of a given seed mouse (Horie

---

Figure 9.7. Integration site preferences of SB transposase. (A) Integration preferences are shown for SB transposase (red double-sided arrows and dotted lines), retroviruses (green arrowheads and lines), and lentiviruses (blue arrowheads and lines). The preferences of the three classes of vectors are shown by the horizontal lines (dotted for SB transposons) with clustering indicated by the symbols. Separating transposons and viruses is a schematic of chromosomal DNA with transcriptional motifs shown as green chevrons and transcriptional units shown by the dark blue (exons) and light blue (introns) boxes. (B) Percentages of bases over 50bp on each side of TA insertion sites in mouse genomes; the TA insertion sites in the center of the chart are invariant (G. Liu and Y. Horie, unpublished result). (C) Preferential TA insertion sites in the pFV/luc plasmid. Integration sites with two or more hits vary from an expected Poisson distribution; 29% of the total hits were between base pairs 6815 and 6854, which comprises less than 0.5% of the plasmid (Liu et al., 2005). (See Color Insert.)

et al., 2003), a number that is far lower than the number of potential TA sites, estimated to be about 10^6, as discussed earlier. This suggests that about 100 founder mice would suffice for saturation insertional mutagenesis of the mouse genome using the strategy of transposon remobilization.

From the perspective of gene therapy, there have been two quite significant findings. First, no offspring with dominant mutations have been discovered following germ line remobilizations, unless the new mutations were backcrossed to homozygosity (Carlson *et al.*, 2003; Horie *et al.*, 2003). However, the number of total published and characterized insertional events is very low—less than 200 at the time of this review. Second, there are no reports of abnormalities in the mice that express the SB transposase, either ubiquitously or from a germ cell-specific promoter, even if these mice also harbor transposon transgenes. It is unclear if a similar degree of transposon insertion site complexity exists in nongerm line tissues of doubly transgenic transposase plus transposon mice. However, if a similar complexity does exist, then it suggests that of the approximately 2×10^4 different *Tc1/mariner*-type transposon insertions that might be present in any one tissue, no dominant mutations that would result in neoplasia or other observable phenotypes have been recovered. In fact, so far, no dominant germ line mutations have been recovered by germ line mutagenesis using SB vectors either. Nevertheless, work from the Largaespada laboratory shows that appropriately designed SB transposon vectors can accelerate cancer in genetically susceptible mice via an insertional mutagenesis mechanism in which gain-of-function, dominant mutations are induced (Collier *et al.*, unpublished observations).

Because SB transposase activity is often provided from its gene, as discussed in the following sections, there is a chance that some SB genes will integrate into genomes of targeted cells, resulting in expression at a low rate. Early studies on the SB system indicated that there is a high degree of binding specificity by SB transposase for the DRs of SB transposons but not for related transposons, even those of other species of fish (Ivics *et al.*, 1997). This specificity, and the requirement for a specific stoichiometry (Geurts *et al.*, 2003), presumably four transposase polypeptides per transposon, would account for the lack of detected remobilization of endogenous transposons in cells expressing SB transposase. There is no evidence to suggest that SB transposition will be accompanied by recombination or deletion events at an integration site (Izsvak and Ivics, 2004).

In contrast to SB-type DNA transposons, there are millions of retrotransposons in human and other mammalian genomes. In humans, about 100 of these retrotransposons are active due to active reverse transcriptase/integrase genes. They apparently can direct new insertions in about 12% of individuals (Han *et al.*, 2004; Kazazian, 1999, 2004). In contrast to retrotransposons, all known DNA transposons in mammalian genomes have been silent for the past

50 million years. Thus, *retro*transposition is a natural phenomenon in humans that has been associated with induction of genetic disease, e.g., hemophilia (Kazazian *et al.*, 1988), and thereby can be considered a base line against which to estimate potential adverse effects from the relatively few cells that will take up one or two copies of an SB transposon following gene therapy.

III. APPLICATIONS OF *SLEEPING BEAUTY* TRANSPOSONS FOR GENE THERAPY

DNA-mediated gene delivery holds great promise in its potential for therapeutic application, as attested by the different chapters in this volume. In some cases, a beneficial therapeutic outcome may be anticipated after a transient burst of expression from newly introduced gene sequences. Perhaps the best example of the utility of such short-term gene transfer and expression is in the development of DNA vaccines (Liu *et al.*, 2004c), in which case such short-term expression can elicit an effective immune response against a DNA-encoded antigen. However, for many other applications, a more extended period of expression following introduction of new sequences will be required in order to achieve a therapeutic benefit. The most demanding circumstance would be in the treatment of a genetic deficiency disease, in which case indefinite expression of the newly introduced gene is sought. Such extended or indefinite expression of newly introduced DNA could be brought about by one of three different approaches: (i) Maintenance of the newly introduced DNA in an extrachromosomal form. This requires stability of the plasmid in the cellular setting, lack of cell division (if copy number per cell is to remain constant), or the ability of the vector element to replicate. (ii) Tethering of the DNA element to the endogenous chromosome, thus promoting its replication and maintenance. (iii) Integration of the DNA into the host cell chromosome, thus allowing its stable maintenance and, perhaps, expression by relying on the cellular machinery for this purpose. As described earlier, *Sleeping Beauty* mediates chromosomal integration of non-viral DNA and therefore provides the potential for extended and even indefinite expression in a gene therapy setting.

The utility of the SB system for achieving long-term expression in animals was supported by experiments in zebrafish and in transgenic mice, as described earlier in Section I.B. However, more pertinent to the potential application of SB for gene therapy is its activity when introduced into somatic tissues. As the transposon component of the SB system consists of DNA, and the transposase component thus far has been provided as a transposase-encoding DNA, delivery of the SB system to somatic tissues is subject to the same constraints as any of the non-viral DNA delivery methodologies thus far reported. In addition, there may be additional constraints that are specific to the SB transposon system, such as the requirement for an optimal molecular ratio of SB transposase to transposon as detailed earlier. For these reasons, testing of the SB transposon system in somatic tissues has thus far been carried out using the most effective means of non-viral DNA delivery to these tissues. This section reviews the current status of applying the SB transposon system to achieve stable gene transfer and expression in somatic tissues of the mouse as a model for gene therapy, emphasizing key concepts in the study of transposition *in vivo* that have been addressed in this work. We also present a consideration of safety concerns in the potential clinical application of the SB transposon system.

A. SB-mediated gene expression in the liver

The ability of SB to mediate stable, long-term expression in mouse tissues was initially reported in the seminal work by Yant *et al.* (2000), in which extended expression of α_1-antitrypsin as a reporter in normal C57BL/6 mice and of human clotting factor IX as a therapeutic gene product in factor IX-deficient mice was demonstrated. These studies took advantage of the newly discovered "hydrodynamics"-based method for delivery of DNA to the mouse liver (Liu *et al.*, 1999a; Zhang *et al.*, 1999), a technique that has since become widely used for the study of gene transfer and expression in the liver, even though the mechanism of uptake is poorly understood (Andrianaivo *et al.*, 2004; Kobayashi *et al.*, 2004). In addition to demonstrating transposase-dependent long-term expression of gene products secreted into the bloodstream, Yant *et al.* (2000) also used a plasmid rescue technique to recover *NEO* transposon sequences from the liver after codelivery with a transposase-encoding plasmid, thus providing molecular demonstration of the ability of SB to mediate transposition in the liver.

As mentioned previously, it is anticipated that the overall level of long-term gene transfer achieved using non-viral delivery approaches is likely to be much lower than that observed using viral delivery systems. As shown by the work of Yant *et al.* (2000), factor IX deficiency has been an excellent model system for gene therapy studies because a therapeutic benefit can be achieved if only a small fraction (1%) of the normal level of gene product is observed in the bloodstream of the treated animal. Another approach that can be taken in gene therapy studies where a relatively low level of gene transfer is anticipated is to employ a strategy whereby the selective outgrowth of genetically corrected cells is anticipated. Montini *et al.* (2002) used the FAH deficiency model of inherited tyrosinemia to demonstrate selective outgrowth of hepatocytes that had been corrected genetically by hydrodynamics-based delivery of a FAH-encoding SB transposon. A 20-fold increase in the frequency of FAH-corrected hepatocytes was observed when an SB transposase-encoding plasmid was codelivered along with a FAH-encoding transposon. Because this system results in the clonal outgrowth of FAH-positive cells in the liver of FAH-deficient mice, they were able to genetically characterize transposition events not only by the recovery of transposon junction sites by linker-mediated PCR, but also by Southern blot hybridization, thus providing a quantitative molecular assessment of SB-mediated transposition in the mouse liver.

The ability of the SB system to accommodate long sequences of DNA (Geurts *et al.*, 2003) allows it to be used for genes or their derivatives, such as the B domain-deleted version of FVIII (Chuah *et al.*, 2001; Miao *et al.*, 2004; Saenko *et al.*, 2003) for gene delivery to the liver. Ohlfest *et al.* (2005) showed that a B domain-deleted version of the FVIII gene could be delivered to livers of FVIII-deficient mice for correction of hemophilia A and expressed for several months at a constant level. As in other previous studies employing viral vectors (Follenzi *et al.*, 2004; Kootstra *et al.*, 2003; Ragni, 2002), a major problem with immune responses to secreted FVIII was encountered that could only be avoided by either immunosuppression of the mutant mice or pretolerization with recombinant FVIII within 24 h of birth.

Although this chapter is devoted primarily to the application of SB to non-viral gene transfer, SB also provides the potential to contribute an integrating function to viral vectors that do not integrate as a part of their normal replicative cycle. The mouse liver is highly transducible by vectors based on human adenovirus type 5. Yant *et al.* (2002) found that SB transposons could be transduced using helper-dependent adenoviruses, but reported that effective transposition required excision of the transposon substrate from the adenovirus vector into a circular form, mediated in this case by Cre recombinase. The constructs used for this study required circularization in order to form a complete and expressible transgene, but this approach prevented control experiments in which the circularization step was omitted. Nonetheless, these results demonstrated the effectiveness of this strategy in combining the efficiency of adenovirus-mediated gene delivery with the integrating function of the SB transposon system to achieve integration and long-term expression in the liver.

Applying the SB system for gene transfer and integration into somatic tissues *in vivo* faces the challenge of delivering both transposon and transposase components to the same cells in a setting where the frequency of transfected cells may be limited. Under these circumstances, what makes the most sense conceptually is to deliver both the transposon and the transposase-encoding sequence to the cells of the liver on the same plasmid (i.e., in *cis*). We found that this approach was, in fact, required in order to achieve efficient gene transfer into cultured HuH7 human hepatoma cells (Converse *et al.*, 2004; Kren *et al.*, 2003). However, early *in vivo* studies testing such plasmids in which the SB transposase was regulated by the relatively strong CMV early promoter/enhancer yielded lower levels of long-term expression than when the transposon and transposase-encoding sequences were delivered on separate plasmids (P. Score *et al.*, manuscript in preparation). Mikkelsen *et al.* (2003) screened a series of promoters exhibiting a range of activities after transient delivery into the liver and found that elements such as the human phosphoglycerate kinase promoter, providing a low to moderate level of activity, were most effective in providing expression of the transposase component in mediating long-term expression in recipient animals. Similarly, we have found that the human ubiquitin C promoter, which is much less active in the liver than CMV in our hands, is much more effective in providing expression of the transposase component for SB-mediated transposition and long-term expression in the liver (P. Score *et al.*, manuscript in preparation). These results are consistent with previous observations demonstrating "overexpression inhibition" for *Tc1/mariner*-type transposons (Hartl *et al.*, 1997a,b), including SB (Geurts *et al.*, 2003), and serve to define conditions where gene delivery, integration, and expression can be achieved more efficiently using the SB system.

While accumulation of the evidence described earlier has supported the effectiveness of the SB system for long-term expression in the liver, numerous studies have emerged reporting high-level and long-term gene expression in the liver after the delivery of plasmids, i.e., without the benefit of a transposon (Miao *et al.*, 2001, 2003; Zhang *et al.*, 2000a). In our experiments, we observed that gene expression became extinguished in most animals injected with a reporter transposon in the absence of a transposase-encoding plasmid; in some cases, long-term expression was observed in these control animals. These results made us wonder: in animals injected with both a transposon plasmid and a transposase-encoding plasmid, how much of the observed expression is due to transposition and how much is due to maintenance of the plasmid as an extrachromosomal element or potentially due to random recombination? We therefore devised a strategy whereby induced expression of Cre recombinase in the liver mediates excision of the promoter from a reporter transposon *unless* the reporter transposon has been excised from the plasmid by SB, segregating it from one of the LoxP sites and rendering it nonfunctional for Cre-mediated recombination (Score *et al.*, 2003). We found that induction of Cre recombinase resulted in a two-log-fold decrease in gene expression unless the SB-encoding plasmid was codelivered along with the reporter transposon. These results indicated that most of the expression observed in these experiments is associated with SB-mediated transposition.

The results just described provide considerable support for activity of the SB transposon system for mediating gene insertion and long-term expression in the liver. There are many challenges to be faced. Currently, the most effective way of delivering transposon and transposase-encoding DNAs is by rapid, high-volume, tail vein injection in mice (Liu *et al.*, 1999a; Zhang *et al.*, 1999). While it may be argued whether DNA can be delivered directly into the hepatic circulation under increased pressure for therapeutic purposes in a large mammal, including humans (Zhang *et al.*, 2002), it is clear that the hydrodynamics-based DNA delivery technique provides an accurate technique to evaluate the effects of genes that are delivered to the liver. Nevertheless, evaluation of the effectiveness of the SB transposon system after delivery to the liver under more clinically applicable conditions is necessary. Additionally, while the hydrodynamics-based gene transfer approach may have provided a sufficient level of gene transfer and expression for preclinical treatment of hemophilia B (Yant *et al.*, 2000), hereditary tyrosinemia (Montini *et al.*, 2002), and hemophilia A (Ohlfest *et al.*, 2004), a higher level of gene transfer may be necessary for the treatment of other diseases, such as metabolic disorders in which the effectiveness of transposon-mediated gene expression is likely to be cell autonomous. Reports of using retroductal delivery of plasmids in a mixture of various agents that prolong the lifetime of the circular DNA molecules before uptake have been reported (Niedzinski et al., 2003a,b). Tests are ongoing in our laboratories for the efficacy of ligand-conjugated DNA-condensing agents (Kren *et al.*, 2003) and nanoparticles/nanocapsules for the delivery of transposons to liver.

In addition to increased delivery of transposon and transposase-encoding DNAs to the liver, the effectiveness of the transposon system may be increased by providing improvements in activity of the SB transposase and in the ability of SB transposons to serve as substrates for transposition, as described earlier in Section II.B. Tests of improved SB transposon systems are underway in several laboratories that seek to treat liver diseases with non-viral vectors.

B. SB-mediated gene expression in the lung

One of the most effective methods of non-viral gene delivery to the lung is by intravenous injection of DNA complexed with linear polyethyleneimine (PEI) (Lawton *et al.*, 2002). Belur *et al.* (2003) used this technique to demonstrate long-term expression of luciferase transposons in mouse lung when provided with a source of SB transposase. Transposase, in this case, was provided either by coinjection of an SB transposase-encoding plasmid or by injection of luciferase transposon into transgenic mice expressing the SB transposase. At least a 100-fold increase in expression was observed in comparison with control animals injected with transposon alone, and immunohistochemistry studies indicated that a high proportion of the transduced cells were type 2 alveolar pneumocytes. Liu *et al.* (2004b) further demonstrated that expression could be directed to endothelial cells of the lung using the endothelian-1 promoter to regulate transgene expression in the context of an SB transposon delivered as a PEI complex. Both of the studies described earlier delivered DNA to the lung after intravascular injection. The potential for gene delivery through the airway remains an alternative for SB transposons. Airway delivery of SB, which has been reported previously (Lawton *et al.*, 2002), could thus provide the means for a longer lasting, stable gene therapy for cystic fibrosis. Other complexes could be tested for gene transfer in other specific cell types of the lung for potential treatments by SB-mediated gene transfer.

C. SB-mediated gene expression in hematopoietic cells

The integrating capacity of the SB transposon system is its key asset for treatments that involve extensively diving hematopoietic and lymphoid cell populations. Non-viral gene transfer techniques have been developed for cultured hematopoietic cells, and preliminary results from transposition studies in established lymphoid cultures (J. Essner *et al.*, unpublished result) have been encouraging. Non-viral, DNA-mediated gene transfer in primary hematopoietic cells has been limited to myeloid–erythroid colony-forming cells following electroporation (Wu *et al.*, 2001a,b). There has been one preliminary report of SB-mediated gene transfer into primitive hematopoietic stem cells in a mouse model of Fanconi anemia complementation group C (Noll *et al.*, 2003). This work took advantage of the selective outgrowth of stem cells genetically corrected by transposition with a FANC-C encoding transposon. Gene transfer into stem cells was demonstrated by transplantation into irradiated secondary transplant recipients with subsequent genetic analysis for presence of the FANC-C transposon. Because such transposition events may have been rare in this case, future studies must focus on more effective means of DNA delivery in order to effectively apply the SB transposon system to mediated gene transfer into hematopoietic stem cells.

D. SB-mediated gene expression in tumors

Virus-mediated gene transfer has been used to treat tumors in murine models and in some clinical trials (Gottesman, 2003). Retroviral and adenoviral vectors have been effective gene transfer vehicles for this purpose; however, there are some disadvantages associated with their use (Scanlon, 2004). Many viral vectors are highly immunogenic, which diminishes the efficacy of repeated administration and raises biosafety concerns (Hernandez *et al.*, 1999; Lam and Breakefield, 2001; Yang *et al.*, 1995). In addition, although most viral vectors are designed to be nonreplicating, recombination events have been reported yielding undesired effects (Zhu *et al.*, 1999). Traditionally, viral vectors have been seen as much better than plasmid-based vectors due to their ability to support long-term expression of the transgene and an overall higher gene transfer efficiency (Nishikawa and Huang, 2001).

However, the SB transposon system was designed for providing prolonged expression without using viral vectors and consequently may be useful for cancer gene therapy. The use of genes for cancer therapy requires special consideration, as depending on the vector system used, one seeks to achieve long-term expression for an acute disease. However, there are several situations in which long-term expression would be desirable for cancer gene therapy. First of all, it might be possible to achieve a lifelong cure that would depend on lifelong expression or, if not lifelong, an effective therapy might require prolonged expression over many months. Prolonged expression is desirable if the gene therapy is used to prevent tumor recurrence, which might occur years after initial therapy. Delivery of appropriate genes to a tumor and/or tumor-associated stromal or endothelial cells could theoretically affect a long-term cure and prevent local recurrence. Systemic therapy for the suppression of metastases is another potential application of cancer gene therapy, but may be more difficult to achieve, as the vector, or its product, must be delivered throughout the body. In either situation, some genes will be more effective if stable long-term expression is achieved.

Three main approaches for cancer gene therapy have been proposed (Gottesman, 2003; Scanlon, 2004): antiangiogenesis genes, e.g., endostatin and angiostatin (Zhang *et al.*, 2000b); cell suicide induction using enzymes that activate a prodrug, e.g., herpes simplex virus thymidine kinase (HSV-TK) (Fillat *et al.*, 2003; Rainov, 2000), or that activate intrinsic apoptotic pathways (e.g., p53 or Bax); and delivery of genes that promote an immune response to tumor cells (e.g., GM-CSF, IL-2). Several studies have shown that constant low-level delivery of an antiangiogenesis inhibitor is more effective than cycling or bolus dosing (Kerbel and Folkman, 2002; Morris, 2003; Pawliuk *et al.*, 2002). This may be due to the fact that tumors continuously make proangiogenic factors that must be counteracted by the continuous presence of antiangiogenic factors. Some gene therapies for tumor cell killing, such as HSV-TK activation of the prodrug ganciclovir, act only on dividing cells (Ram *et al.*, 1993). Because not all cells within a tumor are dividing at any one time, particularly migratory/metastatic cells, such therapies must be delivered long term to be effective. Ultimately, the choice of therapeutic gene used depends on the type and anatomic location of the tumor being targeted, the percentage of cells that can be successfully transduced or transfected with the vector being used, and the clinical situation being addressed (e.g., front line or consolidation therapy). One paper addressed the potential of SB for cancer gene therapy in terms of the percentage of cells that can be transfected (Ohlfest *et al.*, 2004).

SB was tested for gene transfer and long-term expression in xenografted human glioblastoma cells growing in nude mice (Ohlfest *et al.*, 2004). In these experiments, transposon vectors expressing the Neo, luciferase, or GFP genes were used with SB10 or the catalytically inactive transposase DDE (Fig. 9.2). The plasmid DNAs were complexed in PEI and injected directly into xenografted tumors growing subcutaneously in nude mice. Two weeks after injection, all tumors that did not receive active SB transposase had lost detectable luciferase expression. At 4 weeks, a very noticeable increase in luciferase expression was observed in tumors that received CMV-driven SB at a 1:20 ratio of SB to luciferase transposon plasmid. One explanation for increasing expression over time would be clonal expansion of cells that harbor transposon insertions. Indeed, when the explanted-treated cells were plated in G418, about 8% of the clonable U87 cells were resistant to G418, whereas none of the tumor cells injected with the catalytically inactive transposase yielded G418-resistant colonies. Ten of 10 cloned insertions from these G418-resistant clones demonstrated that the insertions were the result of transposition events into TA dinucleotide sites in the human genome. These data establish that tumors growing *in vivo* can be targets for long-term gene transfer and expression using SB; however, only about 10% of the tumor mass might be successfully transfected long term. The potential advantages that SB has in terms of reduced immunogenicity and better scalabilty should also apply to the treatment of cancer. However, given the lethal outcome of the cancers being considered for treatment by SB, a different risk/benefit analysis is certainly in order.

E. Safety issues for transposon-mediated gene therapy

SB shares safety issues in common with many other gene therapy vectors, including unintentional induction of innate or adaptive immune responses and insertional mutagenesis. The published literature on safety testing for SB is scant. Clearly, this is an important area for future research.

As with other non-viral systems for gene therapy, delivery of naked DNA might provoke the innate immune response (Krieg and Davis, 2001), resulting in an adjuvant-like effect and making an immune response against the encoded transgene more likely (Hodges *et al.*, 2004). Indeed, a serious safety concern is that the delivery of genes via the SB system will provoke an immune response and formation of inhibitory antibodies to a gene product that is benefiting the patient via enzyme replacement therapy. Thus, the form of the plasmid that is delivered, the quantity of DNA delivered, and complexing agents used must be studied carefully. Hypomethylated CpG dinucleotides are recognized by the Toll-like receptor 9 (TKR9) and are potent adjuvants (Krieg and Davis, 2001). Methylation of the transposon and transposase-encoding plasmids may help prevent activation of the innate immune system and subsequent adaptive immunity to the transgene-encoded protein. This has been demonstrated in gene transfer experiments using naked DNA (Reyes-Sandoval and Ertl, 2004). The effects of methylation of plasmids on SB transposition (Yusa *et al.*, 2004) were addressed earlier in Section II.C. The effect that the SB transposase protein itself has on the immune response is currently unknown. Presumably, transposase will be recognized as a foreign protein and, depending on the cell types that express the SB protein, an immune response may be generated. Not only might such a response prevent repeated administration of SB vectors, there might be concern that an autoimmune response against endogenously expressed *Tc1/mariner*-type transposase proteins or peptides (if they exist) might be provoked. Current efforts are underway to determine immunological responses to SB transposase in mice that constitutively express the protein. Nevertheless, this issue requires more study.

In common with integrating vectors such as lentiviruses, MLV, and AAV, SB causes chromosomal integration of vector DNA. As described earlier, SB does not prefer to integrate near promoters or within genes, as do the retroviral vectors. Instead, SB integration at TA dinucleotides is similar to what one would expect by chance (Carlson *et al.*, 2003; Horie *et al.*, 2003; Vigdal *et al.*, 2002). Nevertheless, given a large number of genomic integration events, many will land near or within genes, including those genes that might trigger cancer if mutated appropriately. Studies are underway to determine if lifelong somatic transposition of chromosomally resident transposon vectors, deliberately designed to activate or inactivate genes, will cause cancer in mice. It will be important to determine whether promoters used to drive transgene expression can activate endogenous genes as effectively as MLV long terminal repeats. Of course, a Moloney-based vector used to treat X-linked severe combined immunodeficiency disease (SCID) did cause leukemia in patients via an insertional mutagenesis mechanism (Hacein-Bey-Abina *et al.*, 2003). The use of insulator elements (Recillas-Targa *et al.*, 1999), which could prevent enhancement of endogenous promoters, might ameliorate risk due to insertional mutagenesis. However, another form of mutagenesis risk would be from gene truncation due to promoter or splice acceptor insertion into cancer genes. Another important issue is the likely number of unique insertion events that will occur over time. If the SB transposase plasmid is delivered as a gene on a plasmid, then there is some potential for achieving the unintended long-term transposase expression. The result of long-term expression of SB transposase in a cell that harbors one or more transposon vectors could be ongoing transposition and many different insertion events. Research into the relative risk due to this effect could be accomplished using SB transposase transgenic mice (Dupuy *et al.*, 2001). Alternate methods for delivering transposase are highly desirable. Clearly, it is possible to deliver transposase as *in vitro*-transcribed mRNA in microinjected embryos (Dupuy *et al.*, 2002). Whether SB transposase mRNA can be practically delivered in a gene therapy setting is unclear. Another possibility would be to deliver transposon DNA together with the purified SB transposase protein. However, active purified SB transposase has not been reported. Theoretically, it is possible that an ideal stoichiometry among transposon, transposase, and required cofactors could be preassembled and delivered to cells as one unit, the "transposasome," which would maximize efficiency while maintaining safety.

For both sources of risk described earlier, it behooves the gene therapist to determine the lowest possible dose of vector that will confer a beneficial clinical outcome. Use of a less vector plasmid should reduce the risk of an immune response and possibly, by limiting the number of total insertion events, may reduce the risk of mutagenesis. Thus, improvements in the activity of the SB system are an important aspect of SB safety research.

IV. FUTURE DIRECTIONS

A. Delivery methods for the SB transposon system

Transposon-mediated gene delivery is in its infancy compared to the use of viral methods. The challenges for using transposons to correct genetic diseases are identical to those for all of the other methods of non-viral gene delivery—efficacious transfer with minimal undesirable side effects. The challenge of delivery to specific organs and tissues has commanded considerable interest, as discussed here and elsewhere in this volume. Use of a variety of DNA "coatings" that are decorated with cell-specific ligands has and will continue to allow improvement of targeting of all types of non-viral vectors, including transposons (Schmidt-Wolf and Schmidt-Wolf, 2003). For instance, PEI conjugated with galactose has been used to target SB transposons to mouse liver following injection into the tail vein (Kren et al., 2003). Other methods, including nanoparticles and nanocapsules, are becoming available for gene therapy (Allen and Cullis, 2004; Reynolds et al., 2003). The greater use of tissue-specific and cell type-specific promoters that can be regulated will compensate for misdirected vectors. Improvements on both fronts are ongoing in many laboratories and they can be incorporated into protocols that employ SB transposons.

B. Efficiency and evolution of the SB transposon system

A second area of activity is improvement of the efficiency of SB transposition based on improvement of the transposase enzyme. As discussed earlier, the combination of target site recognition motifs embedded in the cut-and-paste catalytic domain is a major impediment to rational design of a site-specific SB transposase. Our experience is that the simple addition of amino acid sequences to termini of SB transposase generally has inactivated the enzyme completely (unpublished result) so that the addition of targeting sequences is unlikely to be successful. Until a clear, three-dimensional structure of the SB transposase is developed, all improvements will come from random mutagenesis with either a genetic selection screen or laborious testing of each derivative. The amino terminus of the Tc3 transposase of *Caenorhabditis elegans* is the only crystal structure of a *Tc1/mariner*-type transposon that has been published and this structure includes only the transposon-binding sites and not the catalytic domain.

C. Safety issues for the SB transposon system

The most formidable challenge is the issue of random integration (Check, 2003; Linden, 2002; Williams and Baum, 2003) based on the results from the French X-SCID clinical trial in which two cases of a leukemia-like syndrome apparently resulting from insertional activation of the LMO2 oncogene were reported (Hacein-Bey-Abina *et al.*, 2003). There are two ways to avoid the *observed* problems of random or semirandom integration: the use of site-specific integrases and blocking enhancer activity within the therapeutic vector from the activation of endogenous genes. The Streptomyces ϕC31 phage integrase appears to direct integration of the transgenic construct into a relatively small number of sites in human chromosomes that resemble its normal recognition site (Groth and Calos, 2004; Groth *et al.*, 2000; Olivares *et al.*, 2002). However, potential problems associated with this integrase, including chromosomal deletions mediated by the integrase, have not been investigated thoroughly. The second strategy, to use border/insulator elements to block enhancers in vectors from activating chromosomal genes, is discussed next.

For many gene therapy applications, long-term expression of genes is essential. Transcriptional silencing of retroviruses poses a major obstacle to their use as gene therapy vectors, and border elements have been proposed as a solution (Pannell and Ellis, 2001). Consequently, a number of investigators have incorporated border elements into their constructs to block methylation of the therapeutic gene (Ellis and Pannell, 2001; Rivella *et al.*, 2000; Steinwaerder and Lieber, 2000). However, none of these studies has addressed the blocking of enhancer effects of the transgenic constructs on endogenous chromosomal genes.

How might border/insulator elements be used? Although the DNA in nuclei is often envisaged as a tangled mess of chromosomal fibers, domains of activated genes appear to be structurally divided by sequences called matrix attachment regions (MARs) or scaffold attachment regions (Schedl and Grosveld, 1995; Wolffe, 1994). The sequences of MARs proximal to different genes are not identical. MARs appear to be able to ensure the proper expression of transgenes in terms of tissue specificity and timing but do not seem to protect genes against repression when integrated into some regions of chromosomes (McKnight *et al.*, 1992; Thompson *et al.*, 1994). Insulators are a second type of DNA sequence that appear to function by a different mechanism than MARs to alleviate position effects in transgenic animals (Bell *et al.*, 2001; Roseman *et al.*, 1993; West *et al.*, 2002). The 5′ constitutive DNase I hypersensitive site from the chicken β-globin locus control region is the most studied insulator element (Bell *et al.*, 1999; Farrell *et al.*, 2002; Mutskov *et al.*, 2002; Prioleau *et al.*, 1999; Walters *et al.*, 1999). This insulator has several motifs that confer either of two properties. The first property is to act as a barrier to block heterochromatization by the spread of methylation that can permanently shut down the expression of transgenes. The second property is an enhancer-blocking activity mediated by a site that binds a protein known as CTCF. CTCF binding to DNA blocks enhancers neighboring a transgenic construct from influencing the activity of the transgene (Yao *et al.*, 2003; Yusufzai and Felsenfeld, 2004).

   All of the evidence described here serves to emphasize the following common features of border elements. First, they are active only when part of chromatin. Second, none of the border elements alters the expression levels of genes in transient assays. Third, neither MARs nor insulators alter tissue specificity of transgene expression. Essentially MARs appear to block enhancer activity, but they do not exhibit silencing activity. Many insulators have the ability to block both silencing and enhancing activities. However, boundary/insulator elements have a role in establishing domains of open chromatin characterized by global changes in histone modifications. As a result, the effects of incorporating boundary/insulator elements randomly in human chromosomes are unknown. For instance, random integration of an insulator-flanked vector could lead to repression of a tumor suppressor gene if the vector integrated between a critical enhancer and the transcriptional unit of the gene. Future studies must therefore consider the implications of widespread insertion of such elements as a safety precaution.

D. Conclusion

Since its creation in 1997, the SB transposon system has been on a rapid course of development for employment as a vector for non-viral gene therapy. Although problems and questions remain, within the first 7 years of its appearance, there have been 22 papers on its basic properties, 21 papers on its activities in mice, and about half a dozen papers on its use in other vertebrates. When combined with developments for the delivery of various forms of non-viral constructs, this record suggests that the SB transposon system is on a relatively rapid adaptation pathway for use in gene therapy as a vector that combines the advantages of viral vectors—high rates of chromosomal integration and integration of single copies of a therapeutic gene—with the advantages of non-viral vectors—the absence of protein factors that elicit adverse responses.

Acknowledgments

We thank the Arnold and Mabel Beckman Foundation for support of our work and all members of the Beckman Center for Transposon Research for a long history of contributions of ideas and results. We are especially grateful to Dr. Elena Aronovich and Kirk Wangensteen for careful proofreading of the manuscript. The authors were also supported by NIH Grants 1PO1 HD32652-07 (PBH and RSM), R43 HL076908-01 (PBH and RSM) DA014764 (DAL), and 1RO1-DA14546-01 (SCE).

References

Acsadi, G., Jiao, S. S., Jani, A., Duke, D., Williams, P., Chong, W., and Wolff, J. A. (1991). Direct gene transfer and expression into rat heart in vivo. New Biol. **3**, 71–81.

Allen, T. M., and Cullis, P. R. (2004). Drug delivery systems: Entering the mainstream. Science **303**, 1818–1822.

Andrianaivo, F., Lecocq, M., Wattiaux-De Coninck, S., and Jadot, M. (2004). Hydrodynamics-based transfection of the liver: Entrance into hepatocytes of DNA that causes expression takes place very early after injection. J. Gene Med. **6**, 877–883.

Balciunas, D., Davidson, A. E., Sivasubba, S., Hermanson, S. B., Welle, Z., and Ekker, S. C. (2004). Enhancer detection in zebrafish using the Sleeping Beauty transposon. BCN Genom. **5**, 62.

Baum, C., von Kalle, C., Staal, F. J., Li, Z., Fehse, B., Schmidt, M., Weerkamp, F., Karlsson, S., Wagemaker, G., and Williams, D. A. (2004). Chance or necessity? Insertional mutagenesis in gene therapy and its consequences. Mol. Ther. **9**, 5–13.

Bell, A. C., West, A. G., and Felsenfeld, G. (1999). The protein CTCF is required for the enhancer blocking activity of vertebrate insulators. Cell **98**, 387–396.

Bell, A. C., West, A. G., and Felsenfeld, G. (2001). Insulators and boundaries: Versatile regulatory elements in the eukaryotic genome. Science **291**, 447–450.

Belur, L. R., Frandsen, J. L., Dupuy, A. J., Largaespada, D. A., Hackett, P. B., and McIvor, R. S. (2003). Gene insertion and long-term expression in lung mediated by the Sleeping Beauty transposon system. Mol. Ther. **8**, 501–507.

Bird, A. (2002). DNA methylation patterns and epigenetic memory. Genes Dev. **16**, 6–21.

Carlson, C. M., Dupuy, A. J., Fritz, S., Roberg-Perez, K. J., Fletcher, C. F., and Largaespada, D. A. (2003). Transposon mutagenesis of the mouse germline. Genetics **165**, 243–256.

Check, E. (2003). Harmful potential of viral vectors fuels doubts over gene therapy. Nature **423**, 573–574.

Chevalier-Mariette, C., Henry, I., Montfort, L., Capgras, S., Forlani, S., Muschler, J., and Nicolas, J. F. (2003). CpG content affects gene silencing in mice: Evidence from novel transgenes. Genome Biol. **4**, R53.

Chuah, M. K., Collen, D., and VandenDriessche, T. (2001). Gene therapy for hemophilia. J. Gene Med. **3**, 3–20.

Clark, K., Geurts, A. M., Bell, J., and Hackett, P. B. (2004). Transposon vectors for gene-trap insertional mutagenesis in vertebrates. Genesis **29**, 225–233.

Converse, A., Belur, L., Gori, J., Liu, G., Amaya, F., Aguilar-Cordova, E., Hackett, P. B., and McIvor, R. S. (2005). Counterselection and co-delivery of transposon and transposase functions for Sleeping Beauty-mediated transposition in cultured mammalian cells. Som. Cell Mol. Genet. in press.

Craig, N. L. (1995). Unity in transposition reactions. Science **270**, 253–254.

Craig, N. L. (1997). Target site selection in transposition. Annu. Rev. Biochem. **66**, 437–474.

Cui, Z., Guerts, A. M., Liu, G., Kaufman, C. D., and Hackett, P. B. (2002). Structure-function analysis of the inverted terminal repeats of the *Sleeping Beauty* transposon. *J. Mol. Biol.* **318,** 1221–1235.

Dave', U. P., Jenkins, N. A., and Copeland, N. G. (2004). Gene therapy insertional mutagenesis insights. *Science* **303,** 33.

Davidson, A. E., Balciunas, D., Mohn, D., Shaffer, J., Hermanson, S., Sivasubbu, S., Cliff, M. P., Hackett, P., and Ekker, S. (2003). Efficient gene delivery and gene expression in zebrafish using Sleeping Beauty transposons. *Dev. Biol.* **263,** 191–202.

Davies, D. R., Goryshin, I. Y., Reznikoff, W. S., and Rayment, I. (2000). Three-dimensional structure of the Tn5 synaptic complex transposition intermediate. *Science* **289,** 77–85.

Dupuy, A. J., Clark, K., Carlson, C. M., Fritz, S., Davidson, A. E., Markley, K. M., Finley, K., Fletcher, C. F., Ekker, S. C., Hackett, P. B., Horn, S., and Largaespada, D. A. (2002). Mammalian germ-line transgenesis by transposition. *Proc. Natl. Acad. Sci. USA* **99,** 4495–4499.

Dupuy, A. J., Fritz, S., and Largaespada, D. A. (2001). Transposition and gene disruption using a mutagenic transposon vector in the male germline of the mouse. *Genesis* **30,** 82–88.

Ellis, J., and Pannell, D. (2001). The beta-globin locus control region versus gene therapy vectors: A struggle for expression. *Clin. Genet.* **59,** 17–24.

Fadool, J. M., Hartl, D. L., and Dowling, J. E. (1998). Transposition of the mariner element from Drosophila mauritiana in zebrafish. *Proc. Natl. Acad. Sci. USA* **95,** 5182–5186.

Farley, A. H., Luning-Prak, E. T., and Kazazian, H. H. J. (2004). More active human L1 retro-transposons produce longer insertions. *Nucleic Acids Res.* **32,** 502–510.

Farrell, C. M., West, A. G., and Felsenfeld, G. (2002). Conserved CTCF insulator elements flank the mouse and human beta-globin loci. *Mol. Cell. Biol.* **22,** 3820–3831.

Ferkol, T., Perales, J. C., Eckman, E., Kaetzel, C. S., Hanson, R. W., and Davis, P. B. (1995). Gene transfer into the airway epithelium of animals by targeting the polymeric immunoglobulin receptor. *J. Clin. Invest.* **95,** 493–502.

Ferrari, S., Moro, E., Pettenazzo, A., Behr, J. P., Zacchello, F., and Scarpa, M. (1997). ExGen 500 is an efficient vector for gene delivery to lung epithelial cells *in vitro* and *in vivo*. *Gene Ther.* **4,** 1100–1106.

Fillat, C., Carrio, M. A. C., and Sangro, B. (2003). Suicide gene therapy mediated by the herpes simplex virus thymidine kinase gene/ganciclovir system: Fifteen years of application. *Curr. Gene Ther.* **3,** 13–26.

Fischer, S. E., Wienholds, E., and Plasterk, R. H. (2001). Regulated transposition of a fish transposon in the mouse germ line. *Proc. Natl. Acad. Sci. USA* **98,** 6759–6764.

Follenzi, A., Battaglia, M., Lombardo, A., Annoni, A., Roncarolo, M. G., and Naldini, L. (2004). Targeting lentiviral vector expression to hepatocytes limits transgene-specific immune response and establishes long-term expression of human antihemophilic factor IX in mice. *Blood* **103,** 3700–3709.

Geurts, A. M., Yang, Y., Clark, K. J., Cui, Z., Dupuy, A. J., Largaespada, D. A., and Hackett, P. B. (2003). Gene transfer into genomes of human cells by the *Sleeping Beauty* transposon system. *Mol. Ther.* **8,** 108–117.

Gibbs, P. D. L., Gray, A., and Thorgard, G. (1994). Inheritance of P element and reporter gene sequences in zebrafish. *Mol. Mar. Biol. Biotech.* **3,** 317–326.

Gottesman, M. (2003). Cancer gene therapy: An awkward adolescence. *Cancer Gene Ther.* **10,** 501–508.

Grabher, C., Henrich, T., Sasado, T., Arenz, A., Wittbrodt, J., and Furutani-Seiki, M. (2003). Transposon-mediated enhancer trapping in medaka. *Gene* **322,** 57–66.

Grabher, C., and Wittbrodt, J. (2004). Efficient activation of gene expression using a heat-shock inducible Gal4/Vp16-UAS system in medaka. *BMC Biotechnol.* **4,** 26.

Groth, A. C., and Calos, M. P. (2004). Phage integrases: Biology and applications. *J. Mol. Biol.* **335,** 667–678.

Groth, A. C., Olivares, E. C., Thyagarajan, B., and Calos, M. P. (2000). A phage integrase directs efficient site-specific integration in human cells. *Proc. Natl. Acad. Sci. USA* **97**, 5995–6000.

Hacein-Bey-Abina, S., von Kalle, C., Schmidt, M., Le Deist, F., Wulffraat, N., McIntyre, E., Radford, I., Villeval, J. L., Fraser, C. C., Cavazzana-Calvo, M., and Fischer, A. (2003). A serious adverse event after successful gene therapy for X-linked severe combined immunodeficiency. *N. Engl. J. Med.* **348**, 255–256.

Hackett, P. B., Ekker, S. C., and Essner, J. J. (2004). Applications of transposable elements in fish for transgenesis and functional genomics. *Fish Dev. Biol. Genet.* **13**, 454–475.

Han, J. S., Szak, S. T., and Boeke, J. D. (2004). Transcriptional disruption by the L1 retrotransposon and implications for mammalian transcriptomes. *Nature* **429**, 268–274.

Harris, J. W., Strong, D. D., Amoui, M., Baylink, D. J., and William Lau, K. (2002). Construction of a Tc1-like transposon *Sleeping Beauty*-based gene transfer plasmid vector for generation of stable transgenic mammalian cell clones. *Anal. Biochem.* **310**, 15–26.

Hartl, D. L., Lohe, A. R., and Lozovskaya, E. R. (1997a). Regulation of the transposable element mariner. *Genetica* **100**, 177–184.

Hartl, D. L., Lozovskaya, E. R., Nurminsky, D. I., and Lohe, A. R. (1997b). What restricts the activity of mariner-like transposable elements. *Trends Genet.* **13**, 197–201.

He, C. X., Shi, D., Wu, W. J., Ding, Y. F., Feng, D. M., Lu, B., Chen, H. M., Yao, J. H., Shen, Q., Lu, D. R., and Xue, J. L. (2004). Insulin expression in livers of diabetic mice mediated by hydrodynamics-based administration. *World J. Gastroenterol.* **10**, 567–572.

Hernandez, Y. J., Wang, J., Kearns, W. G., Loiler, S., Poirier, A., and Flotte, T. R. (1999). Latent adeno-associated virus infection elicits humoral but not cell-mediated immune responses in a nonhuman primate model. *J. Virol.* **73**, 8549–8558.

Hodges, B. L., Taylor, K. M., and al., e. (2004). Long-term transgene expression from plasmid DNA gene therapy vectors is negatively affected by CpG dinucleotides. *Mol. Ther.* **10**, 269–278.

Horie, K., Kuroiwa, A., Ikawa, M., Okabe, M., Kondoh, G., Matsuda, Y., and Takeda, J. (2001). Efficient chromosomal transposition of a Tc1/mariner-like transposon Sleeping Beauty in mice. *Proc. Natl. Acad. Sci. USA* **98**, 9191–9196.

Horie, K., Yusa, K., Yae, K., Odajima, J., Fischer, S. E., Keng, V. W., Hayakawa, T., Mizuno, S., Kondoh, G., Ijiri, T., Matsuda, Y., Plasterk, R. H., and Takeda, J. (2003). Characterization of Sleeping Beauty transposition and its application to genetic screening in mice. *Mol. Cell. Biol.* **23**, 9189–9207.

Ivics, Z., Hackett, P. B., Plasterk, R. H., and Izsvak, Z. (1997). Molecular reconstruction of *Sleeping Beauty*, a Tc1-like transposon from fish, and its transposition in human cells. *Cell* **91**, 501–510.

Ivics, Z., and Izsvak, Z. (2004). Transposable elements for transgenesis and insertional mutagenesis in vertebrates: A contemporary review of experimental strategies. *Methods Mol. Biol.* **260**, 255–276.

Ivics, Z., Izsvák, Z., Minter, A., and Hackett, P. B. (1996). Identification of functional domains and evolution of Tc1-like transposable elements. *Proc. Natl. Acad. Sci. USA* **93**, 5008–5013.

Ivics, Z., Kaufman, C. D., Zayed, H., Miskey, C., Walisko, O., and Izsvak, Z. (2004). The Sleeping Beauty transposable element: Evolution, regulation and genetic applications. *Curr. Issues Mol. Biol.* **6**, 43–55.

Izsvak, Z., and Ivics, Z. (2004). Sleeping Beauty transposition: Biology and applications for molecular therapy. *Mol. Ther.* **9**, 147–156.

Izsvak, Z., Ivics, Z., and Hackett, P. B. (1995). Characterization of a Tc1-like transposable element in zebrafish (*Danio rerio*). *Mol. Gen. Genet.* **247**, 312–322.

Izsvak, Z., Ivics, Z., and Hackett, P. B. (1997). Repetitive elements and their genetic applications in zebrafish. *Biochem. Cell Biol.* **75**, 507–523.

Izsvak, Z., Ivics, Z., and Plasterk, R. H. (2000). Sleeping Beauty, a wide host-range transposon vector for genetic transformation in vertebrates. *J. Mol. Biol.* **302**, 93–102.

Izsvak, Z., Kahare, D., Behlke, J., Heiinemann, U., Plasterk, R. H., and Ivics, Z. (2002). Involvement of a bifunctional, paired-lilke DNA-binding domain and a transpositional enhancer in Sleeping Beauty transposition. *J. Biol. Chem.* **277,** 34581–34588.

Izsvak, Z., Stuwe, E. E., Fiedler, D., Katzer, A., Jeggo, P. A., and Ivics, Z. (2004). Healing the wounds inflicted by Sleeping Beauty transposon by double-strand break repair in mammalian somatic cells. *Mol. Cell* **9,** 279–290.

Jones, P. A., and Talai, D. (2001). The role of DNA methylation in mammalian epigenetics. *Science* **293,** 1068–1070.

Kalodimos, C. G., Biris, N., Bonvin, A. M., Levandoski, M. M., Guennuegues, M., Boelens, R., and Kaptein, R. (2004). Structure and flexibility adaptation in nonspecific and specific protein-DNA complexes. *Science* **305,** 386–389.

Kaminski, J. M., Huber, M. R., Summers, J. B., and Ward, M. B. (2002). Design of a non-viral vector for site-selective, efficient integration into the human genome. *FASEB J.* **16,** 1242–1247.

Karsi, A., Moav, B., Hackett, P., and Liu, Z. J. (2001). Effects of insert size on transposition efficiency of the Sleeping Beauty transposon in mouse cells. *Mar. Biotech.* **3,** 241–245.

Kawakami, K., and Shima, A. (1999). Identification of the Tol2 transposase of the medaka fish *Oryzias latipes* that catalyzes excision of a nonautonomous Tol2 element in zebrafish. *Danio rerio. Gene* **240,** 239–244.

Kay, M. A., Glorioso, J. C., and Naldini, L. (2001). Viral vectors for gene therapy: The art of turning infectious agents into vehicles of therapeutics. *Nature Med.* **7,** 33–40.

Kazazian, H. H. (1999). An estimated frequency of endogenous insertional mutations in humans. *Nature Genet.* **22,** 130.

Kazazian, H. H. (2004). Mobile elements: Drivers of genome evolution. *Science* **303,** 1626–1632.

Kazazian, H. H., and Goodier, J. L. (2002). LINE drive: Retrotransposition and genome instability. *Cell* **110,** 277–280.

Kazazian, H. H., Wong, C., Youssoufian, H., Scott, A. F., Phillips, D. G., and Antonarakis, S. E. (1988). Haemophilia A resulting from de novo insertion of L1 sequences represents a novel mechanism for mutation in man. *Nature* **332,** 164–166.

Kerbel, R., and Folkman, J. (2002). Clinical translation of angiogenesis inhibitors. *Nature Rev. Cancer* **2,** 727–739.

Ketting, R. F., Fischer, S. E., and Plasterk, R. H. A. (1997). Target choice determinants of the Tc1 transposon of. *Caenorhabditis elegans. Nucleic Acids Res.* **25,** 4041–4047.

Kleiman, N. S., Patel, N. C., Allen, K. B., Simons, M., S., Y.-H., E., G., and Dzau, V. J. (2003). Evolving revascularization approaches for myocardial ischemia. *Am. J. Cardiol.* **92,** 9N–17N.

Klinakis, A. G., Zagoraiou, I., Vassilatis, D. K., and Savakis, C. (2000). Genome-wide insertional mutagenesis in human cells by the Drosophila mobile element Minos. *EMBO Rep.* **1,** 416–421.

Kobayashi, N., Nishikawa, M., Hirata, K., and Takakura, Y. (2004). Hydrodynamics-based procedure involves transient hyperpermeability in the hepatic cellular membrane: Implication of a nonspecific process in efficient intracellular gene delivery. *J. Gene Med.* **6,** 584–592.

Koga, A., Iida, A., Kamiya, M., Hayashi, R., Hori, H., Ishikawa, Y., and Tachibana, A. (2003). The medaka fish Tol2 transposable element can undergo excision in human and mouse cells. *J. Hum. Genet.* **48,** 231–235.

Kootstra, N. A., Matsumura, R., and Verma, I. M. (2003). Efficient production of human FVIII in hemophilic mice using lentiviral vectors. *Mol. Ther.* **7,** 623–631.

Kren, B. T., Gosh, S. S., Linehan, C. L., Roy Chowdhry, N., Hackett, P. B., Roy Chowdhury, J., and Steer, C. J. (2003). Hepatocyte-targeted delivery of Sleeping Beauty mediates efficient transposition *in vivo. Gene Ther. Mol. Biol.* **7,** 229–238.

Krieg, A. M., and Davis, H. L. (2001). Enhancing vaccines with immune stimulatory CpG DNA. *Curr. Opin. Mol. Ther.* **3,** 15–24.

Lam, P. Y., and Breakefield, X. O. (2001). Potential of gene therapy for brain tumors. *Hum. Mol. Genet.* **10**, 777–787.

Lampe, D. J., Churchill, M. E., and Robertson, H. M. (1996). A purified mariner transposase is sufficient to mediate transposition *in vitro*. *EMBO J.* **15**, 5470–5479.

Lander, E. S., et al. (2001). Initial sequencing and analysis of the human genome. *Nature* **409**, 860–921.

Lawton, A. E., Pringle, I. A., Davies, L. A., Hyde, S. C., and Gill, D. R. (2002). Non-viral integrating vectors for the lung: Sleeping Beauty transposons. *Mol. Ther.* **5**, S323.

Lee, S. Y., Aihara, H., Ellenberger, T., and Landy, A. (2004). Two structural features of integrase that are critical for DNA cleavage by multimers but not by monomers. *Proc. Natl. Acad Sci. USA* **101**, 2770–2775.

Lehrman, S. (1999). Virus treatment questioned after gene therapy death. *Nature* **401**, 517–518.

Li, S., and Huang, L. (1997). *In vivo* gene transfer via intravenous administration of cationic lipid-protamine-DNA (LPD) complexes. *Gene Ther.* **4**, 891–900.

Linden, R. M. (2002). Gene therapy gets the *Beauty* treatment. *Nature Biotech.* **20**, 987–988.

Lippman, Z., Gendrel, A.-V., Black, M., Vaughn, M. W., Dedhia, N., McCombie, W. R., and Martienssen, R. A. (2004). Role of transposable elements in heterochromatin and epigenetic control. *Nature* **430**, 471–476.

Liu, F., Song, Y., and Liu, D. (1999a). Hydrodynamics-based transfection in animals by systemic administration of plasmid DNA. *Gene Ther.* **6**, 1258–1266.

Liu, G., Aronovich, E. L., Cui, Z., Whitley, C. B., and Hackett, P. B. (2004a). Excision of Sleeping Beauty transposons: Parameters and applications to gene therapy. *J. Gene Med.* **6**, 574–583.

Liu, G., Geurts, A. M., Yae, K., Srinivassan, A. R., Fahrenkrug, S. C., Largaespada, D. A., Takeda, J., Horie, K., Olson, W. K., and Hackett, P. B. (2005). Target-site preference for Sleeping Beauty transposons. *J. Mol. Biol.* **346**, 161–173.

Liu, L., Sanz, S., Heggestad, A. D., Antharam, V., Notterpek, L., and Fletcher, B. S. (2004b). Endothelial targeting of the Sleeping Beauty transposon within lung. *Mol. Ther.* **10**, 97–105.

Liu, M., Acre, s.B., Balloul, J. M., Bizouarne, N., Paul, S., Slos, P., and Squiban, P. (2004c). Gene-based vaccines and immunotherapeutics. *Proc. Natl. Acad. Sci. USA* **101**, 14567–14571.

Liu, Y., Thor, A., Shtivelman, E., Cao, Y., Tu, G., Heath, T. D., and Debs, R. J. (1999b). Systemic gene delivery expands the repertoire of effective antiangiogenic agents. *J. Biol. Chem.* **274**, 13338–13344.

Lohe, A. R., De Aguiar, D., and Hartl, D. L. (1997). Mutations in the mariner transposase: The D,D (35)E consensus sequence is nonfunctional. *Proc. Natl. Acad. Sci. USA* **94**, 1293–1297.

Lozier, J. N., Yankaskas, J. R., Ramsey, W. J., Chen, L., Berschneider, H., and Morgan, R. A. (1997). Gut epithelial cells as targets for gene therapy of hemophilia. *Hum. Gene Ther.* **8**, 1481–1490.

Luning Prak, E. T., and Kazazian, H. H. J. (2000). Mobile elements and the human genome. *Nature Rev. Genet.* **1**, 134–144.

Luo, G., Ivics, Z., Izsvak, Z., and Bradley, A. (1998). Chromosomal transposition of a Tc1/mariner-like element in mouse embryonic stem cells. *Proc. Natl. Acad. Sci. USA* **95**, 10769–10773.

Lupas, A., Van Dyke, M., and Stock, J. (1991). Predicting coiled coil from protein sequences. *Science* **252**, 1162–1164.

Mannucci, P. M., and Tuddenham, E. G. (2001). The hemophilias: From royal genes to gene therapy. *N. Eng. J. Med.* **344**, 1773–1779.

Martienssen, R. A. (1998). Transposons, DNA methylation and gene control. *Trends Genet.* **14**, 263–264.

McKnight, R. A., Shamay, A., Sankaran, L., Wall, R. J., and Henninghausen, L. (1992). Matrix-attachment regions can impart position-independent regulation of a tissue-specific gene in transgenic mice. *Proc. Natl. Acad. Sci. USA* **89**, 6943–6947.

Miao, C. H., Ohashi, K., Patijn, G. A., Meuse, L., Ye, X., Thompson, A. R., and Kay, M. A. (2000). Inclusion of the hepatic locus control region, an intron, and untranslated region increases and stabilizes hepatic factor IX gene expression *in vivo* but not *in vitro*. *Mol. Ther.* **1,** 522–532.

Miao, C. H., Thompson, A. R., Loeb, K., and Ye, X. (2001). Long-term and therapeutic-level hepatic gene expression of human factor IX after naked plasmid transfer *in vivo*. *Mol. Ther.* **3,** 947–957.

Miao, C. H., Ye, X., and Thompson, A. R. (2003). High-level factor VIII gene expression *in vivo* achieved by non-viral liver-specific gene therapy vectors. *Hum. Gene Ther.* **14,** 1297–1305.

Miao, H. Z., Sirachainan, N., L., P., Kucab, P., Cunningham, M. A., Kaufman, R. J., and Pipe, S. W. (2004). Bioengineering of coagulation factor VIII for improved secretion. *Blood* **103,** 3412–3419.

Mikkelsen, J. G., Yant, S., Meuse, L., Huang, Z., Xu, H., and Kay, M. A. (2003). Helper-independent *Sleeping Beauty* transposon-transposase vectors for efficient non-viral gene delivery and persistent gene expression *in vivo*. *Mol. Ther.* **8,** 654–665.

Miskey, C., Izsvak, Z., Plasterk, R. H. A., and Ivics, Z. (2003). The Frog Prince: A reconstructed transposon from *Rana pipiens* with high transpositional activity in vertebrate cells. *Nucleic Acids Res.* **31,** 6873–6881.

Misteli, T. (2004). Spatial positioning: A new dimension in genome function. *Cell* **119,** 153–156.

Mitchell, R. S., Beitzel, B. F., Schroder, A. R., Shinn, P., Chen, H., Berry, C. C., Ecker, J. R., and Bushman, F. D. (2004). Retroviral DNA integration: ASLV, HIV, and MLV show distinct target site preferences. *PLOS* **2,** 1127–1136.

Miura, A., Yonebayashi, S., Watanabe, K., Toyama, T., Shimada, H., and Kakutani, T. (2001). Mobilization of tranposons by a mutation abolishing full DNA methylation in. *Arabidopsis. Nature* **411,** 212–214.

Montgomery, M. K. (2004). RNA interference: Historical overview and significance. *Methods Mol. Biol.* **265,** 3–21.

Montini, E. P., Held, P. K., Noll, M., Morcinek, N., Al-Dhalimy, M., Finegold, M., Yant, S. R., Kay, M. A., and Grompe, M. (2002). *In vivo* correction of murine tyrosinemia type I by DNA-mediated transposition. *Mol. Ther.* **6,** 759–769.

Morishita, R. (2002). Recent progress in gene therapy for cardiovascular disease. *Circ. J.* **66,** 1077–1086.

Morris, J. C. (2003). Cancer gene therapy: Lessons learned from experiences with chemotherapy. *Mol. Ther.* **7,** 717–719.

Mutskov, V. J., Farrell, C. M., Wade, P. A., Wolffe, A. P., and Felsenfeld, G. (2002). The barrier function of an insulator couples high histone acetylation levels with specific protection of promoter DNA from methylation. *Genes Dev.* **16,** 1540–1554.

Nabel, E. G., Plautz, G., Boyce, F. M., Stanley, J. C., and Nabel, G. J. (1989). Recombinant gene expression *in vivo* within endothelial cells of the arterial wall. *Science* **244,** 1342–1344.

Nakai, H., Montini, E., Fuess, S., Storm, T. A., Grompe, M., and Kay, M. A. (2003). AAV serotype 2 vectors preferentially integrate into active genes in mice. *Nature Genet.* **34,** 297–302.

Niedzinski, E. J., Chen, Y. J., Olson, D. C., Parker, E. A., Park, H., Udove, J. A., Scollay, R., McMahon, B. M., and Bennett, M. J. (2003a). Enhanced systemic transgene expression after non-viral salivary gland transfection using a novel endonuclease inhibitor/DNA formulation. *Gene Ther.* **10,** 2133–2138.

Niedzinski, E. J., Olson, D. C., Chen, Y. J., Udove, J. A., Nantz, M. H., Tseng, H. C., Bolaffi, J. L., and Bennett, M. J. (2003b). Zinc enhancement of non-viral salivary gland transfection. *Mol. Ther.* **7,** 396–400.

Niidome, T., and Huang, L. (2002). Gene therapy progress and prospects: Non-viral vectors. *Gene Ther.* **9,** 1647–1652.

Nishikawa, M., and Huang, L. (2001). Non-viral vectors in the new millennium: Delivery barriers in gene transfer. *Hum. Gene Ther.* **12,** 861–870.

Noll, M., Yant, S. R., Mikkelsen, J. G., Bennett, R. E., Edington, J., Chen, C., Kay, M. A., and Grompe, M. (2003). *In vivo* correction of FANCC mutant hematopoietic stem cells by DNA-mediated transposition. *Mol. Ther.* **7**, S154.

Oehler, S., Amouyal, M., Kolkhof, P., von Wilcken-Bergmann, B., and Muller-Hill, B. (1994). Quality and position of the three lac operators of *E. coli* define efficiency of repression. *EMBO J.* **13**, 3348–3355.

Ohlfest, J. R., Frandsen, J. L., Fritz, S., Lobitz, P. D., Perkinson, S. G., Clark, K. J., Key, N. S., McIvor, R. S., Hackett, P. B., and Largaespada, D. A. (2005). Phenotypic correction and long-term factor VIII expression in hemophilia-A mice by immunotolerization and non-viral gene transfer using the *Sleeping Beauty* transposon system. *Blood* **105**, 2691–2698.

Ohlfest, J. R., Lobitz, P. D., Perkinson, S. G., and Largaespada, D. A. (2004). Integration and long-term expression in xenografted human glioblastoma cells using a plasmid-based transposon system. *Mol. Ther.* **10**, 260–268.

Olivares, E. C., Hollis, R. P., Chalberg, T. W., Meuse, L., Kay, M. A., and Calos, M. P. (2002). Site-specific genomic integration produces therapeutic factor IX levels in mice. *Nature Biotech.* **20**, 1124–1128.

Ortiz, S., Lin, Q., Yant, S. R., Keene, D., Kay, M. A., and Khavari, P. A. (2003). Sustainable correction of junctional epidermollysis bullosa via transposon-mediated non-viral gene transfer. *Gene Ther.* **10**, 1099–1104.

Pannell, D., and Ellis, J. (2001). Silencing of gene expression: Implications for design of retrovirus vectors. *Rev. Med. Virol.* **11**, 205–217.

Pawliuk, R., Bachelot, T., Zurkiya, O., Eriksson, A., Cao, Y., and Leboulch, P. (2002). Continuous intravascular secretion of endostatin in mice from transduced hematopoietic stem cells. *Mol. Ther.* **5**, 345–351.

Plasterk, R. H. (1993). Molecular mechanisms of transposition and its control. *Cell* **74**, 781–786.

Plasterk, R. H. A., Izsvák, Z., and Ivics, Z. (1999). Resident aliens: The Tc1/mariner superfamily of transposable elements. *Trends Genet.* **15**, 326–332.

Prak, E. T. L., and Kazazian, H. H. (2000). Mobile elements and the human genome. *Nature Rev. Genet.* **1**, 134–144.

Prioleau, M. N., Nony, P., Simpson, M., and Felsenfeld, G. (1999). An insulator element and condensed chromatin region separate the chicken beta-globin locus from an independently regulated erythroid-specific folate receptor gene. *EMBO J.* **18**, 4035–4048.

Przybylska, M., Wu, I. H., Zhao, H., Ziegler, R. J., Tousignant, J. D., Desnick, R. J., Scheule, R. K., Cheng, S. H., and Yew, N. S. (2004). Partial correction of the alpha-galactosidase A deficiency and reduction of glycolipid storage in Fabry mice using synthetic vectors. *J. Gene Med.* **6**, 85–92.

Radice, A. D., Bugaj, B., Fitch, D. H., and Emmons, S. W. (1994). Widespread occurrence of the Tc1 transposon family: Tc1-like transposons from teleost fish. *Mol. Gen. Genet.* **244**, 606–612.

Ragni, M. V. (2002). Safe Passage: A plea for safety in hemophilia gene therapy. *Mol. Ther.* **6**, 436–440.

Rainov, N. G. (2000). A phase III clinical evaluation of herpes simplex virus type 1 thymidine kinase and ganciclovir gene therapy as an adjuvant to surgical resection and radiation in adults with previously untreated glioblastoma multiforme. *Hum. Gene Ther.* **11**, 2389–2401.

Ram, Z., Culver, K. W., Walbridge, S., Blaese, R. M., and Oldfield, E. H. (1993). *In situ* retroviral-mediated gene transfer for the treatment of brain tumors in rats. *Cancer Res.* **53**, 83–88.

Raz, E., van Luenen, H. G., Schaerringer, B., Plasterk, R. H., and Driever, W. (1998). Transposition of the nematode *Caenorhabditis elegans* Tc3 element in the zebrafish. *Danio rerio. Curr. Biol.* **8**, 82–88.

Recillas-Targa, F., Bell, A. C., and Felsenfeld, G. (1999). Positional enhancer-blocking activity of the chicken beta-globin insulator in transiently transfected cells. *Proc. Natl. Acad. Sci. USA* **96**, 14354–14359.

Reyes-Sandoval, A., and Ertl, H. C. (2004). CpG methylation of a plasmid vector results in extended transgene product expression by circumventing induction of immune responses. *Mol. Ther.* **9,** 249–261.

Reynolds, A. R., Moghimi, S. M., and Hodivala-Dilke, K. (2003). Nanoparticle-mediated gene delivery to tumoour neovasculature. *Trends Mol. Med.* **9,** 2–4.

Richards, E. J., and Elgin, S. C. R. (2002). Epigenetic codes for heterochromatin formation and silencing: Rounding up the usual suspects. *Cell* **108,** 489–500.

Rivella, S., Callegari, J. A., May, C., Tan, C. W., and Sadelain, M. (2000). The cHS4 insulator increases the probability of retroviral expression at random chromosomal integration sites. *J. Virol.* **74,** 4679–4687.

Roberg-Perez, K., Carlson, C. M., and Largaespada, D. A. (2003). MTID: A database of Sleeping Beauty transposon insertions in mice. *Nucleic Acids Res.* **31,** 78–81.

Roseman, R. R., Pirrotta, V., and Geyer, P. K. (1993). The su(Hw) protein insulates expression of the *Drosophila melanogaster* white gene from chromosomal position effects. *EMBO J.* **12,** 435–442.

Saenko, E., Ananyeva, N. M., Moayeri, M., Ramezani, A., and Hawley, R. G. (2003). Development of improved factor VIII molecules and new gene transfer approaches for hemophilia A. *Curr. Gene Ther.* **3,** 27–41.

Sakai, J., and Kleckner, N. (1997). The Tn10 synaptic complex can capture a target DNA only after transposon excision. *Cell* **89,** 205–214.

Scanlon, K. J. (2004). Cancer gene therapy: Challenges and opportunities. *Anticancer Res.* **24**(2A), 501–504.

Schedl, P., and Grosveld, F. (1995). Domains and boundaries. *In* "Chromatin Structure and Gene Expression" (S. C. R. Elgin, ed.), pp. 172–196. Oxford Univ. Press, Oxford.

Schmidt-Wolf, G. D., and Schmidt-Wolf, I. H. G. (2003). Non-viral and hybrid vectors in human gene therapy: An update. *Trends Mol. Med.* **9,** 67–71.

Schouten, G. J., van Luenen, H. G., Verra, N. C., Valerio, D., and Plasterk, R. H. (1998). Transposon Tc1 of the nematode *Caenorhabditis elegans* jumps in human cells. *Nucleic Acids Res.* **26,** 3013–3017.

Schroder, A. R. W., Shinn, P., Chen, H., Berry, C., Ecker, J. R., and Bushman, F. (2002). HIV-1 integration in the human genome favors active genes and local hotspots. *Cell* **110,** 521–529.

Score, P. R., Belur, L., Frandsen, J. L., Geurts, J., Hackett, P. B., Largaespada, D. A., and McIvor, R. S. (2003). Molecular evidence for *Sleeping Beauty*-mediated transposition and long-term expression *in vivo*. *Mol Ther.* **7,** S9.

Steinwaerder, D. S., and Lieber, A. (2000). Insulation from viral transcriptional regulatory elements improves inducible transgene expression from adenovirus vectors *in vitro* and *in vivo*. *Gene Ther.* **7,** 556–567.

Svahn, M. G., Lundin, K. E., Ge, R., Törnquist, E., Simonson, E. O., Oscarsson, S., Leijon, M., Brandén, L. J., and Smith, C. I. E. (2004). Adding functional entities to plasmids. *J. Gene Med.* **6,** S36–S44.

Thomas, C. E., Ehrhardt, A., and Kay, M. A. (2003). Progress and problems with the use of viral vectors for gene therapy. *Nature Rev. Genet.* **4,** 346–357.

Thompson, E. M., Christians, E., Stinnakre, M. G., and Renard, J. P. (1994). Scaffold attachment regions stimulate HSP70.1 expression in mouse preimplantation embryos but not in differentiated tissues. *Mol. Cell. Biol.* **14,** 694–703.

van Pouderoyen, G., Ketting, R. F., Perrakis, A., Plasterk, R. H., and Sixma, T. K. (1997). Crystal structure of the specific DNA-binding domain of Tc3 transposase of *C. elegans* in complex with transposon DNA. *EMBO J.* **16,** 6044–6054.

Vastenhous, N. L., and Plasterk, R. H. A. (2004). RNAi protects the *Caenorhabditis elegans* germline against transposition. *Trends Genet.* **20,** 1630–1642.

Venter, J. C. (2001). The sequence of the human genome. *Science* **291**, 1304–1351.

Vigdal, T. J., Kaufman, C. D., Izsvak, Z., Voytas, D. F., and Ivics, Z. (2002). Common physical properties of DNA affecting target site selection of Sleeping Beauty and other Tc1/mariner transposable elements. *J. Mol. Biol.* **323**, 411–452.

von Hippel, P. H. (2004). Completing the view of transcriptional regulation. *Science* **305**, 350–352.

Vos, J. C., DeBaere, I., and Plasterk, R. H. A. (1996). Transposase is the only nematode protein required for *in vitro* transposition of Tc1. *Genes Dev.* **10**, 755–761.

Wadman, S. A., Clark, K. J., and Hackett, P. B. (2005). Fishing for answers with transposons. *Mar. Biotech.* **7**. in press.

Walters, M. C., Fiering, S., Bouhassira, E. E., Scalzo, D., Goeke, S., Magis, W., Garrick, D. E., W., and Martin, D. I. (1999). The chicken beta-globin 5′HS4 boundary element blocks enhancer-mediated suppression of silencing. *Mol. Cell. Biol.* **17**, 5656–5666.

West, A. G., Gaszner, M., and Felsenfeld, G. (2002). Insulators: Many functions, many mechanisms. *Genes Dev.* **16**, 271–278.

Williams, D. A., and Baum, C. (2003). Gene therapy: New challenges ahead. *Science* 400–401.

Williams, T. L., Jackson, E. L., Carritte, A., and Baker, T. A. (1999). Organization and dynamics opf the Mu transposome: Recombination by communication between two active sites. *Genes Dev.* **13**, 2725–2737.

Wolff, J. A., Ludtke, J. J., Acsadi, G., Williams, P., and Jani, A. (1992). Long-term persistence of plasmid DNA and foreign gene expression in mouse muscle. *Hum. Mol. Genet.* **1**, 363–369.

Wolff, J. A., Malone, R. W., Williams, P., Chong, W., Acsadi, G., Jani, A., and Felgner, P. L. (1990). Direct gene transfer into mouse muscle *in vivo*. *Science* **247**, 1465–1468.

Wolffe, A. P. (1994). Insulating chromatin. *Curr. Biol.* **4**, 85–87.

Wu, M. H., Liebowitz, D. N., Smith, S. L., Williams, S. F., and Dolan, M. E. (2001a). Efficient expression of foreign genes in human CD34+ hematopoietic precursor cells using electroporation. *Gene Ther.* **8**, 384–390.

Wu, M. H., Smith, S. L., Danet, G. H., Lin, A. M., Williams, S. F., Liebowitz, D. N., and Dolan, M. E. (2001b). Optimization of culture conditions to enhance transfection of human CD34+ cells by electroporation. *Bone Marrow Transplant.* **27**, 1201–1209.

Wu, X., Li, Y., Crise, B., and Burgess, S. M. (2003). Transcription start regions in human genome are favored targets for MLV integration. *Science* **300**, 1749–1751.

Yang, Y., Li, Q., Ertl, H. C., and Wilson, J. M. (1995). Cellular and humoral immune responses to viral antigens create barriers to lung-directed gene therapy with recombinant adenoviruses. *J. Virol.* **69**, 2004–2015.

Yant, S. R., Ehrhardt, A., Mikkelsen, J. G., Meuse, L., Pham, T., and Kay, M. A. (2002). Transposition from a gutless adeno-transposon vector stabilizes transgene expression *in vivo*. *Nature Biotech.* **20**, 999–1005.

Yant, S. R., and Kay, M. A. (2003). Nonhomologous-end-joining factors regulate DNA repair fidelity during Sleeping Beauty element transposition in mammalian cells. *Mol. Cell. Biol.* **23**, 8505–8508.

Yant, S. R., Meuse, L., Chiu, W., Ivics, Z., Izsvak, Z., and Kay, M. A. (2000). Somatic integration and long-term transgene expression in normal and haemophilic mice using a DNA transposon system. *Nature Genet.* **25**, 35–41.

Yant, S. R., Park, J., Huang, Y., Mikkelsen, J. G., and Kay, M. A. (2004). Mutational analysis of the N-terminal DNA-binding domain of Sleeping Beauty transposase: Critical residues for DNA binding and hyperactivity in mammalian cells. *Mol. Cell. Biol.* **24**, 9239–9247.

Yant, S. R., Wu, X., Huang, Y., Garrison, B. A., Burgess, S. M., and Kay, M. A. (2005). High-resolution genome-wide mapping of transposon integration in mammals. *Mol. Cell. Biol.* **25**, 2085–2094.

Yao, S., Osborne, C. S., Bharadwaj, R. R., Pasceri, P., Sukonnik, T., Pannell, D., Recillas-Targa, F., West, A. G., and Ellis, J. (2003). Retrovirus silencer blocking by the cHS4 insulator is CTCF independent. *Nucleic Acids Res.* **31,** 5317–5323.

Yoder, J. A., Walsh, C. P., and Bestor, T. H. (1997). Cytosine methylation and the ecology of intragenomic parasites. *Trends Genet.* **13,** 335–340.

Yoshimura, K., Rosenfeld, M. A., Nakamura, H., Scherer, E. M., Pavirani, A., Lecocq, J. P., and Crystal, R. G. (1992). Expression of the human cystic fibrosis transmembrane conductance regulator gene in the mouse lung after *in vivo* intratracheal plasmid-mediated gene transfer. *Nucleic Acids Res.* **20,** 3233–3240.

Yusa, K., Takeda, J., and Horie, K. (2004). Enhancement of *Sleeping Beauty* transposition by CpG methylation: A possible role of heterochromatin formation. *Mol. Cell. Biol.* **24,** 4004–4018.

Yusufzai, T. M., and Felsenfeld, G. (2004). The 5′-HS4 chicken-globin insulator is a CTCF-dependent nuclear matrix-associated element. *Proc. Natl. Acad. Sci. USA* **101,** 8620–8624.

Zanta, M. A., Boussif, O., Adib, A., and Behr, J. P. (1997). *In vitro* gene delivery to hepatocytes with galactosylated polyethylenimine. *Bioconjug. Chem.* **8,** 839–844.

Zayed, H., Izsvak, Z., Khare, D., Heinemann, U., and Ivics, Z. (2003). The DNA-bending protein HMGB1 is a cellular cofactor of Sleeping Beauty transposition. *Nucleic Acids Res.* **31,** 2313–2322.

Zayed, H., Izsvak, Z., Walisko, O., and Ivics, Z. (2004). Development of hyperactive Sleeping Beauty transposon vectors by mutational analysis. *Mol. Ther.* **9,** 292–304.

Zhang, G., Budker, V., Williams, P., Hanson, K., and Wolff, J. A. (2002). Surgical procedures for intravascular delivery of plasmid DNA to organs. *Methods Enzymology* **346,** 125–133.

Zhang, G., Budker, V., and Wolff, J. A. (1999). High levels of foreign gene expression in hepatocytes after tail vein injections of naked plasmid DNA. *Hum. Gene Ther.* **10,** 1735–1737.

Zhang, G., Song, Y. K., and Liu, D. (2000a). Long-term expression of human alpha1-antitrypsin gene in mouse liver achieved by intravenous administration of plasmid DNA using a hydrodynamics-based procedure. *Gene Ther.* **7,** 1344–1349.

Zhang, L., Chen, Q. R., and Mixon, A. J. (2000b). Antiangiogenic gene therapy in cancer. *Curr. Genom.* **1,** 117–133.

Zhang, L., Sankar, U., Lampe, D. J., Robertson, H. M., and Graham, F. L. (1998). The Himar1 mariner transposase cloned in a recombinant adenovirus vector is functional in mammalian cells. *Nucleic Acids Res.* **26,** 3687–3693.

Zhu, J., Grace, M., Casale, J., Chang, A. T., Musco, M. L., Bordens, R., Greenberg, R., Schaefer, E., and Indelicato, S. R. (1999). Characterization of replication-competent adenovirus isolates from large-scale production of a recombinant adenoviral vector. *Hum. Gene Ther.* **10.**

Zhu, N., Liggitt, D., Liu, Y., and Debs, R. (1993). Systemic gene expression after intravenous DNA delivery into adult mice. *Science* **261,** 209–211.

Section 3

ANIMAL MODEL AND CLINICAL STUDY

Cancer-Specific Gene Therapy

Hui-Wen Lo, Chi-Ping Day, and Mien-Chie Hung
Department of Molecular and Cellular Oncology
The University of Texas M.D. Anderson Cancer Center
Houston, Texas 77030

ABSTRACT

Cancer cells transcriptionally activate many genes that are important for uncontrolled proliferation and cell death. Deregulated transcriptional machinery
in tumor cells usually consists of increased expression/activity of transcription
factors. Ideally, cancer-specific killing can be achieved by delivering a therapeutic gene under the control of the DNA elements that can be activated by
transcription factors that are overexpressed and/or constitutively activated in
cancer cells. Additionally, tumor-specific translation of tumor-killing genes has
been also exploited in cancer gene therapy. Based on these rationales, cancer-
specific expression of a therapeutic gene has emerged as a potentially successful
approach for cancer gene therapy.

Advances in Genetics, Vol. 54
Copyright 2005, Elsevier Inc. All rights reserved.

0065-2660/05 $35.00
DOI: 10.1016/S0065-2660(05)54010-0

To achieve tumor-specific expression, cancer-specific vectors are generally composed of promoters, enhancers, and/or 5′-UTR that are responsive to tumor-specific transcription factors. A number of cancer-specific promoters have been reported, such as those of probasin, human telomerase reverse transcriptase, survivin, ceruloplasmin, HER-2, osteocalcin, and carcinoembryonic antigen. Evidences suggest that the enhancer element targeted by β-catenin can be useful to target colon cancer cells. The 5′-UTR of the basic fibroblast growth factor-2 has been reported to provide tumor specificity. Moreover, a variety of therapeutic genes demonstrated direct antitumor effects such as those encoding proapoptotic proteins p53, E1A, p202, PEA3, BAX, Bik, and prodrug metabolizing enzymes, namely thymidine kinase and cytosine deaminase.

As cancerous cells of different origins vary significantly in their genetic, transcriptional/translational, and cellular profiles, the success of a cancer gene therapy will not be promised unless it is carefully designed based on the biology of a specific tumor type. Thus, tremendous research efforts have been focused on the development of non-viral vectors that selectively target various tumors resulting in minimal toxicity in the normal tissues. Significant progresses were also made in the exploitation of various novel apoptotic, cytotoxic genes as therapeutic tools that suppress the growth of different tumors. Together, these recent advances provide rationales for future clinical testing of transcriptionally targeted non-viral vectors in cancer patients.   © 2005, Elsevier Inc.

I. INTRODUCTION

Gene therapy is an attractive approach for cancer therapeutics. DNA vectors encoding therapeutic molecules can be systemically administered and delivered to the tumor sites and, ideally, to micrometastatic tumors and cancer cells in the bloodstream. Importantly, the expression of the transgene can be carefully tailored to meet the specific needs for caner therapeutics, including (1) tumor selectivity to specifically eradicate tumor cells and reduce toxicity to normal cells, (2) robust expression of therapeutic molecules, and (3) tumor type-specific killing. Therefore, gene therapy may display therapeutic advantages over the chemotherapeutic agent and radiotherapy, which frequently result in unwanted toxicity to normal tissues.

Cancer-specific expression of a tumor-killing molecule or a nontoxic prodrug converting enzyme has emerged as a potentially important approach for cancer gene therapy. A variety of therapeutic genes demonstrated direct antitumor effects such as those encoding proapoptotic proteins p53, E1A, p202, PEA3, BAX, Bik, and prodrug metabolizing enzymes (Flint and Shenk, 1989). The so-called suicide genes encode non-mammalian enzymes, including thymidine kinase (TK) and cytosine deaminase (CD), that convert nontoxic prodrugs to toxic metabolites, leading to cell death.

Cancer-specific promoters have been tested extensively in the viral vector system and have been shown to specifically drive the expression of therapeutic genes with both preclinical and clinical success. Such selective expression can be achieved transcriptionally by placing a cancer-specific promoter (CSP) or a cancer-specific 5′-UTR at 5′ of the transgene. This approach is based on the rationale that cancer cells often express higher levels of growth-promoting transcription factors compared to normal tissues. Alternatively, selective transgene expression can occur via translational mechanisms. For example, insertion of specific 5′-UTR has been shown to facilitate tumor-specific translation of transgene mRNA.

However, major challenges remain for non-viral vectors under the regulation of CSPs as these vectors often display (1) insufficient expression of the therapeutic genes, (2) leaky expression, resulting in toxicity to normal tissues, and (3) inefficient delivery by the existing delivery systems, i.e., cationic liposomes and polycationic polymers. To date, the efficacy of these CSP-controlled non-viral vectors has not been extensively tested clinically. This chapter, therefore, focuses on the recent advances made in the development of cancer-specific non-viral vectors as well as their preclinical efficacy in cell culture systems and in animal models in order to facilitate the understanding of this area and help to develop clinical trials in the near future.

II. DEVELOPMENT OF CANCER-SPECIFIC VECTORS

A. Overviews

In addition to a cDNA encoding a therapeutic protein that directly kills cancer cells or a prodrug converting enzyme that produces metabolites toxic to cells, cancer-specific vectors are generally composed of one of the following key components to reach tumor-specific expression: cancer-specific promoter/ enhancer and cancer-specific 5′-UTR. A number of cancer-specific promoters have been reported, such as those of the probasin (Andriani et al., 2001; Kakinuma et al., 2003; Wen et al., 2003; Yu et al., 2004a), human TERT (Abdul-Ghani et al., 2000; Gu et al., 2000; Lin et al., 2002; Liu et al., 2002; Song et al., 2003), survivin (Chen et al., 2004; Yang et al., 2004), ceruloplasmin (Lee et al., 2004), HER-2 (Anderson et al., 2000; Harris et al., 1994; Pandha et al., 1999; Yu et al., 2004b), osteocalcin (Kubo et al., 2003), and carcinoembryonic antigen (Kijima et al., 1999; Okabe et al., 2003). The enhancer element targeted by β-catenin has been proven effective to target colon cancer cells (Chen and McCormick, 2001; Kwong et al., 2002; Lipinski et al., 2004). Several cancer-specific 5′-UTRs have also been reported, such as that of the basic fibroblast growth factor-2 (DeFatta *et al.*, 2002a,b; Kevil *et al.*, 1995; Rosenwald, 1996).

B. Therapeutic genes

A number of genes have been tested for their use in cancer gene therapy. Most of these genes encode cellular proteins that are involved in apoptosis and/or anti-proliferation, such as p53, Fas, p202, E1A, BAX, Bik, and PEA3. However, some of these transgenes encode for the antisense transcripts of growth-promoting, anti-apoptosis, or angiogenesis genes, such as EGFR, Bcl-2, and VEGF, respectively. In suicide gene therapy, nontoxic nonmammalian TK and CD enzymes are expressed to convert prodrugs into toxic metabolites and kill tumor cells. Among these molecules, p202 (Van Tendeloo *et al.*, 2001), BAX (Tai *et al.*, 1999), and the Bik mutant (Chen *et al.*, 2004) have emerged as important therapeutic tools with selective tumor-suppressing efficacy when their expression was under the control of CSPs. Hence, the physiological functions of these genes are summarized as follows.

1. p53

Mutation of the p53 tumor suppressor gene is found in many inherited and sporadic cancers and approximately 50% of human tumors express mutant defective p53 genes (Hollstein *et al.*, 1991; Wang and El-Deiry, 2004). p53 exerts its tumor-suppressing function by inducing cell cycle arrest or apoptosis in response to genome stress, such as DNA damage (Shimada *et al.*, 2002; Vogelstein *et al.*, 2000; Wang and El-Deiry, 2004). Conversely, loss of wild-type p53 results in genome instability as well as resistance to chemotherapy and radiotherapy (Bush and Li, 2002; Lowe *et al.*, 1993b). As such, restoration of p53 expression/function is considered as an attractive anticancer strategy (Campling and El-Deiry, 2003; El-Deiry, 2001; Roth *et al.*, 2001).

Many p53-based gene therapy protocols are tested in clinical trials for treating various types of cancers. For example, the adenoviral vector SCH58500 (Schering-Plough) expresses the p53 transgene under control of the CMV promoter and has been tested in phase I/II trials for treating patients with nonsmall cell lung cancer (NSCLC) (Nemunaitis *et al.*, 2000) and ovarian cancer (Buller *et al.*, 2002). SCH58500 was administered intratumorally and intraperitoneally to these patients and showed clinical safety and feasibility as a single regimen or when used in combination with chemotherapy. For ovarian cancer, promising results from both preclinical studies and phase I trials led to a large international *p53* gene therapy phase II/III trial in patients with primary stage III ovarian cancer with *p53* mutations (Zeimet and Marth,

2003). However, the study was closed after the first interim analysis due to an inadequate therapeutic effect, which prompted more research efforts in the understanding of tumors with a deregulated p53 network.

Another p53-based adenoviral vector, INGN 201/ADVEXIN (Introgen Therapeutics), also utilizes the human CMV promoter to drive p53 gene expression (Wills et al., 1994; Zhang et al., 1993). When tested in preclinical models, INGN 201/ADVEXIN was shown to be effective against several different cancer types, including those of breast (Parker et al., 2000), lung (Swisher et al., 1999), glioma (Kock et al., 1996), colorectal (Spitz et al., 1996), head and neck (Clayman et al., 1998), and ovary (Modesitt et al., 2001). It also demonstrates a synergistic effect when administered in combination with chemotherapeutic agents (Lebedeva et al., 2001) and irradiation (Colletier et al., 2000; Lang et al., 1998; Sah et al., 2003). In subsequent phase I trials, INGN 201/ADVEXIN displayed minimal toxicity in patients with NSCLC (Nemunaitis et al., 2000; Roth et al., 1998; Swisher et al., 1999), glioma (Lang et al., 2003), bladder cancer (Pagliaro et al., 2003), and head and neck cancer (Clayman et al., 1998, 1999; Edelman and Nemunaitis, 2003).

INGN 201/ADVEXIN was further tested in patients with platinum- and paclitaxel-resistant epithelial ovarian cancer (Wolf et al., 2004). Although well tolerated, the dosing schedule and the amount were not concluded from this study and thus further refinement will be required before additional trials. Moreover, the phase I trial combining intratumoral injection of INGN 201/ADVEXIN and radiotherapy was well tolerated and demonstrated evidences of tumor regression (Swisher et al., 2003). In patients with prostate cancer who underwent radical prostatectomy, INGN 201/ADVEXIN administration also demonstrated clinical safety and increased p53 expression, leading to apoptosis (Pisters et al., 2004). Together, p53-based gene therapies showed promising preclinical and clinical efficacy in several cancer types, leading to a number of advanced trials.

Furthermore, a p53-based adenoviral vector was approved by the Chinese State Federal Food and Drug Administration to treat head and neck squamous cell carcinomas in late 2003 (Pearson et al., 2004). The recombinant Ad-p53, also known as Genedicine, was developed by Shenzhen SiBiono Gene Technologies and demonstrated substantial clinical success alone and combined with chemotherapy and radiotherapy (Chen et al., 2003; Han et al., 2003; Zhang et al., 2003). This was the first approval of the commercial production of a gene therapy product.

2. E1A

The early region 1A (E1A) of human adenovirus type 5 (Ad5) encodes a viral protein that immortalizes primary rodent cells and transforms them when co-transfected with H-ras (Flint and Shenk, 1989; Graham et al., 1974; Ruley, 1983).

In human cells, E1A expression does not result in malignant transformation. In contrast, E1A elicits an antitumor effect in many cancer types and is thus considered as a tumor suppressor (Frisch, 1991; Frisch and Mymryk, 2002; Frisch et al., 1990; Hung et al., 2000; Mymryk, 1996; Pozzatti et al., 1988a,b; Yan et al., 2002; Yu et al., 1991). E1A-mediated inhibition of tumor development involves downregulating HER2/neu expression (Chang et al., 1997; Yu et al., 1990) as well as other mechanisms unrelated to HER-2 (Frisch and Dolter, 1995; Yan et al., 2002). For example, E1A negatively regulates cellular proteins that are important for gene transcription, such as p300/CBP, TATA-binding proteins, TBP-associated factors, NF-κB, ATF-4, and c-Jun (Brockmann et al., 1995; Chen and Hung, 1997; Liang and Hai, 1997; Liao and Hung, 2004; Liao et al., 2004; Sang et al., 1997; Shao et al., 1997, 1999).

Conversely, E1A increases the expression of the cyclin-dependent kinase inhibitor p21$^{WAF1/CIP1}$, which inhibits cell growth (Najafi et al., 2003). The E1A protein has been shown to induce sensitization to anticancer drug-induced apoptosis (Brader et al., 1997; Frisch and Dolter, 1995; Liao and Hung, 2004; Liao et al., 2004; Lowe and Ruley, 1993; Lowe et al., 1993a; Viniegra et al., 2002). The underlying mechanisms appear to involve the downregulation of Akt (Viniegra et al., 2002), activation of p38 mitogen-activated protein kinase (Liao and Hung, 2003), and upregulation of the catalytic subunit of protein phosphatase 2A (Liao and Hung, 2004). More recently, E1A has been shown to form a tricomplex with p53 and Mdm-4 and inhibits Mdm-2 binding to p53, resulting in p53 stabilization and decreased p53 nuclear export (Li et al., 2005).

In addition, E1A reduces tumor metastasis (Shao et al., 2000; Yu et al., 1992, 1993) and promotes apoptosis induced by serum deprivation, tumor necrosis factor α (TNF-α), irradiation, and anticancer agents such as taxol, etoposide, gemcitabine, and adriamycin (Liao et al., 2004; Shao et al., 1997, 1999, 2000; Ueno et al., 1997; Zhou et al., 2001). For these reasons, E1A-based viral and non-viral gene therapies have been exploited extensively pre-clinically and clinically (Hortobagyi et al., 1998, 2001; Hung et al., 1995, 2000; Ueno et al., 2002; Xing et al., 1998; Yoo et al., 2001; Yu et al., 1995). E1A gene therapy has recently been tested in multiple clinical trials in patients with head and neck, ovarian, and breast cancers (Hortobagyi et al., 2001; Hung et al., 2000; Madhusudan et al., 2004; Ueno et al., 2001; Villaret et al., 2002; Yoo et al., 2001).

3. p202

p202 is an IFN-inducible transcriptional repressor that interacts with several transcriptional regulators, resulting in the transcriptional repression of genes important for tumor cell growth (Choubey, 2000; Johnstone and Trapani, 1999).

Transcription factors that are negatively regulated by p202 include E2Fs, Fos/Jun, c-Myc, NF-κB, and p53BP-1 (Choubey, 2000; Choubey and Lengyel, 1995; Johnstone and Trapani, 1999). Forced expression of p202 results in growth reduction in tumors of the prostate (Wen et al., 2003; Yan et al., 1999), breast (Ding et al., 2002; Wen et al., 2000), and pancreas (Ding et al., 2002; Wen et al., 2001). In breast cancer cells, stable expression of p202 sensitized tumor cells to TNF-α-induced apoptosis (Wen et al., 2000). For these reasons, p202 is considered as an attractive therapeutic gene for cancer gene therapy. A number of adenoviral and non-viral vectors have been recently developed and tested in vitro and in vivo for cancer gene therapy (Ding et al., 2002; Wen et al., 2003).

4. BAX

BAX is a proapoptotic molecule that belongs to the Bcl-2 family of proteins (Cory et al., 2003; Crompton, 2000). It is located on the outer membrane of the mitochondria and, upon apoptotic stress, undergoes oligomerization, leading to membrane permeabilization and subsequent release of cell death-activating factors such as cytochrome C, Smac/Diablo, EndoG, and AIF from mitochondria to the cytosol (Crompton, 2000; Hardwick and Polster, 2002). As the anti-apoptotic Bcl-2 is frequently overexpressed in human cancers, forced expression of apoptotic BAX has become a well-rationalized approach to induce tumor cell death (Coultas and Strasser, 2003; Norris et al., 2001). In fact, BAX gene therapy demonstrated much success in cancers of lung (Kaliberov et al., 2002), prostate (Lowe et al., 2001; Norris et al., 2001), colon (Gu et al., 2000), ovary (Tai et al., 1999), and cervix (Gu et al., 2000; Huh et al., 2001).

5. Bik

In addition to BAX, Bik is another potent pro-apoptotic protein that plays an essential role in apoptosis (Boyd et al., 1995; Coultas and Strasser, 2003; Han et al., 1996). The Bik protein heterodimerizes with various antiapoptotic proteins, such as Bcl-2 and Bcl-XL, and thus antagonizes their anti-apoptotic function (Han et al., 1996). Similar to BAX, forced Bik expression inhibits the growth and metastasis of human breast cancer xenograft in nude mice (Zou et al., 2002) and sensitizes tumor cells to apoptosis induced by certain chemotherapeutic agents (Daniel et al., 1999; Panaretakis et al., 2002). More recently, it was found that the Bik mutant, designated BikDD, with mutations at threonine-33 and serine-35 to aspartic acid displayed an enhanced ability to associate with Bcl-2 and Bcl-XL, leading to increased apoptotic death in tumors (Chen et al., 2004; Li et al., 2003). As such, BikDD has emerged as a potent therapeutic gene suitable for cancer gene therapy.

6. Suicide genes

The use of suicide genes is one of the most promising approaches for cancer gene therapy. The two common genes for this strategy are those encoding the herpes simplex virus type-1 thymidine kinase (HSV-TK) and the _Candida albicans_ cytosine deaminase (CD). HSV-TK catalyzes the conversion of the nontoxic ganciclovir (GCV) prodrug to toxic phosphorylated metabolites, rendering cell death (Nishiyama and Rapp, 1979). Similarly, CD converts the prodrug 5-fluorocytosine (5-FC) to the cytotoxic 5'-fluorouracil (5-FU) (Mullen _et al._, 1992). 5-FU is a widely used chemotherapeutic agent for a variety of cancers (Lokich, 1998). Importantly, a number of HSV-TK- and CD-based viral vectors are currently tested in clinical trials whereas the non-viral vectors are examined preclinically.

Cancer-specific expression of suicide genes has demonstrated promising therapeutic effects in a number of tumor types. For example, a phase I trial showed the safety and efficacy of a tumor-specific CD/prodrug construct, controlled by the human erbB-2 gene promoter in patients with breast cancer (Pandha _et al._, 1999). In this trial, 12 breast cancer patients received the gene therapy intratumorally and 90% of cases showed targeted gene expression, indicating the selectivity of the approach. Moreover, another phase I trial was undertaken using the promoter of osteocalcin (OC) to drive the expression of HSV-TK to target both prostate cancer cells and their surrounding stromal cells (Kubo _et al._, 2003). OC, a noncollagenous bone matrix protein, is expressed in prostate cancer epithelial cells, adjacent fibromuscular stromal cells, and osteoblasts in locally recurrent prostate cancer and prostate cancer bone metastasis. This OC-OSV-TK phase I trial was the first clinical trial to inject therapeutic adenoviruses directly into prostate cancer lymph node and bone metastasis. Their results showed HSV-TK protein expression in clinical specimens and thus promised further development.

In preclinical studies, an adenovirus containing the HSV-TK driven by the gastrin-releasing peptide (GRP) promoter targeted most types of small cell lung carcinoma (Pandha _et al._, 1999). In nude mice, a subcutaneously inoculated tumor infected with this virus regressed completely after an intraperitoneal administration of GCV (Morimoto _et al._, 2001). The carcinoembryonic antigen (CEA) has been used to drive tumor-specific expression and killing of both HSV-TK and CD in cancers of lung (Kijima _et al._, 1999; Okabe _et al._, 2003) and colon (Kijima _et al._, 1999; Okabe _et al._, 2003) and in hepatic metastatic cancer in mice (Terazaki _et al._, 2003). Additionally, TERT promoter-driven HSV-TK vectors demonstrated an _in vitro_ killing effect in NSCLC and ovarian cancer cells (Song, 2004; Song and Kim, 2004; Song _et al._, 2003).

C. Vector design for tumor-specific expression

1. Transcriptional control by tumor-specific promoters

a. Probasin (PB) promoter in prostate cancer

Probasin is an abundant protein located in nuclei of the prostate and is a well-established prostate-specific gene (Kakinuma et al., 2003). Its expression is responsive to androgen but not to other steroid hormones; therefore, the PB promoter has been extensively characterized for androgen responsiveness. A small composite PB promoter, designated ARR_2PB, is crucial for the prostate-specific expression of PB (Zhang et al., 2000). When fused to therapeutic genes, ARR_2PB demonstrated prostate-specific expression and antitumor cytotoxicity (Andriani et al., 2001; Wen et al., 2003; Yu et al., 2004a).

For example, an adenoviral vector composed of the ARR_2PB region and the proapoptotic protein BAX induced apoptosis in prostate cancer cell lines (Andriani et al., 2001). While BAX overexpression often results in apoptosis in normal organs, delivery of the adenoviral ARR_2PB-BAX selectively killed prostate xenografts, via apoptosis, leaving normal tissues unharmed (Andriani et al., 2001).

Moreover, the ARR_2PB region was fused with the sodium iodide symporter (NIS) cDNA, which encodes an intrinsic membrane protein responsible for the thyroid gland to transport and concentrate iodide (Kakinuma et al., 2003). Clinically, NIS expression is important for radioactive iodine treatment in patients with thyroid tumors. As prostate tissues lack NIS expression, delivery of a NIS transgene that is under the control of prostate-specific promoter ARR_2PB became an attractive strategy to target prostate tumors. Indeed, an adenoviral vector carrying ARR_2PB and human NIS cDNA selectively increases the uptake of radioactive iodine in androgen-producing prostate cancer cell lines compared to those of other origins (Kakinuma et al., 2003). However, the antitumor effect of this approach remains uninvestigated.

More recently, a non-viral vector carrying the ARR_2PB fragment and the cDNA for the growth-suppressing p202 was generated and tested for its ability to target prostate tumors (Wen et al., 2003). This vector is built on the pGL3 enhancer vector (Promega, Madison, WI) in which p202 cDNA replaces the luciferase cDNA. The probasin ARR_2PB gene promoter was further inserted into the vector and was found to direct p202 expression in an androgen-dependent fashion both in vitro and in vivo (Wen et al., 2003). The ARR_2PB-p202 vector was delivered using the gene delivery system, designated SN, with a cationic liposome formulation composed of dipalmitoylethylphosphocholine, dioleoylphosphoethanolamine, dipalmitoylphospho-ethanoamine, and polyethylene-glycol (Zou et al., 2002). Systemic administration of ARR_2PB-p202/liposome to mice bearing AR-positive prostate tumors suppressed prostate tumor growth (Wen _et al._, 2003). Taken together, the probasin ARR$_2$PB promoter serves as a prostate cancer-specific promoter that displays a high potential for future use in prostate-specific cancer gene therapy.

b. Telomerase reverse transcriptase (TERT) promoter

TERT activity is frequently detected in human malignancies, but is absent in most normal cells (Shay and Bacchetti, 1997). Telomerase functions as a DNA polymerase that is responsible for replicating telomeres, chromosome ends. Telomerase consists of the RNA template, encoded by TER gene, and the catalytic subunit, encoded by TERT (Meyerson _et al._, 1997). TERT plays a crucial role in carcinogenesis as it is the key determinant for telomerase activity (Meyerson _et al._, 1997). The potential use of the TERT promoter as a cancer-specific promoter has been exploited by several studies (Abdul-Ghani _et al._, 2000; Gu _et al._, 2000; Lin _et al._, 2002; Liu _et al._, 2002; Song, 2004; Song and Kim, 2004).

It has been shown (Gu _et al._, 2000) that an adenoviral construct expressing the BAX gene that is under the transcriptional control of the human TERT (hTERT) promoter demonstrated tumor-specific expression of BAX. In mice systemically infused with this adenoviral construct, growth of the tumor xenografts was significantly suppressed, whereas minimal toxicity was detected in normal tissues. More recently, another adenoviral vector carrying the hTERT promoter and TRAIL, an apoptotic inducer, was tested for its efficacy in eliciting tumor-specific killing (Lin _et al._, 2002). Importantly, this study found that the hTERT-TRAIL vector resulted in a complete tumor regression in 50% of the breast tumor-bearing mice. Moreover, these animals remained tumor-free for over 5 months, suggesting the potential future success of using the hTERT promoter in cancer gene therapy.

An hTERT-driven non-viral vector encoding the diphtheria toxin A chain displayed specific cytotoxicity in bladder cancer cells but not in the normal counterparts (Abdul-Ghani _et al._, 2000). As human bladder tumors express high levels of hTERT compared to normal tissues (Abdul-Ghani _et al._, 2000), these tumors may thus possess the transcriptional machinery that activates the hTERT promoter. Although this vector has not been tested in an animal model, _in vitro_ results suggest the feasibility of using the hTERT promoter for targeted cancer. Similarly, intratumoral injection of the expression vector containing the hTERT promoter and a proapoptotic gene, FADD (Thorburn, 2004), reduced the growth of subcutaneous tumors in nude mice (Koga _et al._, 2001). Taken together, hTERT demonstrates a high potential for future use in cancer gene therapy to specifically drive expression of the therapeutic gene in tumor sites but not in the normal tissues.

c. Survivin promoter

Survivin is an inhibitor of the apoptosis protein that is expressed in many human cancer cells but not in most normal adult tissues (Altieri, 2003; Chen et al., 2004). Given its negative regulatory role in apoptosis, survivin is considered as a candidate target for cancer therapeutics (Altieri, 2003). The feasibility of the survivin promoter in driving cancer-specific expression of therapeutic genes has been exploited by two studies (Chen et al., 2004; Yang et al., 2004).

In a non-viral vector carrying the BikDD cDNA, the survivin promoter was found to drive a specific cytotoxic effect of BikDD in vitro and in vivo (Chen et al., 2004). The SRVN-BikDD vector is a composite vector originated from the pGL3-basic vector (Promega, Madison, WI) in which a 977-bp human survivin promoter is inserted and BikDD cDNA replaces the luciferase cDNA. The delivery system used was the SN liposome as described earlier for the ARR2PB-p202 vector (Wen et al., 2003; Zou et al., 2002). The SRVN-BikDD vector elicited a significantly higher killing activity in various cancer cells but not in immortalized normal cells (Chen et al., 2004). The ability of the survivin promoter in driving tumor-specific expression of the therapeutic gene was further demonstrated in a non-viral vector system expressing an autocatalytic reverse caspase-3 gene (Yang et al., 2004).

Moreover, the survivin promoter was found to be activated under hypoxic conditions (Yang et al., 2004). This is of particular importance, as most solid tumors contain regions that are deficient in oxygen, and an association between hypoxia and tumor metastasis has been reported (Zhong et al., 1999). Importantly, when a hypoxia-responsive element (HRE) from the vascular epithelial growth factor gene was added to the survivin promoter, an increased expression of caspase-3 was observed in hypoxic tumor cells (Yang et al., 2004). Together, these findings established a rationale for use of the survivin promoter or the chimeric survivin-HRE in targeted caner gene therapy.

d. Ceruloplasmin promoter in ovarian cancer

Ceruloplasmin, a copper-metabolizing enzyme, is preferentially expressed in ovarian tumors compared to normal ovarian tissue. Deletion/mutation studies revealed an activator protein-1 (AP-1) site in the ceruloplasmin promoter to be critical for optimal ceruloplasmin promoter activity (Lee et al., 2004). This ceruloplasmin AP-1 site was specifically recognized by c-jun both in vitro and in vivo. In nude mice carrying SKOV3.ip1 xenografts, the ceruloplasmin promoter demonstrated significantly higher activities in tumors compared to normal organs. In cell culture systems, the ceruloplasmin promoter also displayed significantly higher activities in a panel of ovarian cancer cell lines compared to normal cells (Lee et al., 2004). Although its ability to drive a therapeutic gene remains uninvestigated, these results suggest that the cerulo-plasmin promoter may be exploited as a promising cancer-specific promoter in developing new gene therapy strategies for ovarian cancer.

e. HER-2 promoter

Overexpression of HER-2, a member of the human epidermal growth factor receptor family, is a common characteristic of human cancers and is linked to more aggressive tumor behaviors (Wang and Hung, 2001; Yu and Hung, 2000). The ability of the HER-2 promoter to specifically drive the suicide genes in breast and pancreatic tumors has been reported in preclinical models (Anderson *et al.*, 2000; Harris *et al.*, 1994; Yu *et al.*, 2004b). In a clinical setting with breast cancer patients, a HER-2-CD plasmid vector was intratumorally injected combined with systemic administration of the prodrug CD (Pandha *et al.*, 1999). This protocol was shown to be safe with targeted CD gene expression in up to 90% of cases. Expression of the suicide gene was only observed in HER-2-positive tumor cells, indicating the selectivity of the HER-2 promoter. To note, this was the only targeted gene therapy that utilized a non-viral system with a cancer-specific promoter. The outcome of this study is promising for the future development of genetic activation of a therapeutic molecule that builds on the transcriptional profile of cancerous cells.

2. Translational control by 5′-untranslated region

5′-UTR involves translational control of mRNAs. Specifically, the 5′-UTR of basic fibroblast growth factor-2 (bFGF-2) is recognized by translation initiation factor eIF4E, a cap-binding protein for mRNA to become recruited to ribosomes for subsequent translation (Kevil *et al.*, 1995). The eIF4E level is elevated in many solid tumors, resulting in the translation of transcripts that are normally repressed by their 5′-UTR, including that of bFGF-2 (Kevil *et al.*, 1995). As such, the bFGF-2 5′-UTR serves as a candidate cancer-specific element for cancer gene therapy (Rosenwald, 1996). Several studies have exploited the potential use of bFGF-2 5′-UTR as a strategy for cancer-specific gene therapy (DeFatta *et al.*, 2002a,b). Specifically, a HSV-TK vector driven by bFGF-2 5′-UTR was found effective in reducing subcutaneous tumors and lung metastases following ganciclovir administration (DeFatta *et al.*, 2002a).

III. CONCLUSION

As cancerous cells of different origins vary significantly in their genetic, transcriptional/translational, and cellular profiles, the success of a cancer gene therapy will not be attained unless it is carefully designed based on the biology of a specific tumor type. To date, tremendous research efforts have been focused on the development of non-viral vectors that selectively target various tumors, resulting in minimal toxicity in the normal tissues. As a result, a number of tumor-specific promoters and elements have been successfully identified and validated as promising tools for the novel design of cancer gene therapy vectors. Significant progress has also been made in the exploitation of various novel apoptotic, cytotoxic genes as therapeutic tools that suppress the growth of different tumors. Together, these recent advances provide rationales for future clinical testing of transcriptionally targeted non-viral vectors in cancer patients.

References

Abdul-Ghani, R., Ohana, P., Matouk, I., Ayesh, S., Ayesh, B., Laster, M., Bibi, O., Giladi, H., Molnar-Kimber, K., Sughayer, M. A., de Groot, N., and Hochberg, A. (2000). Use of transcriptional regulatory sequences of telomerase (hTER and hTERT) for selective killing of cancer cells. *Mol. Ther.* **2,** 539–544.

Altieri, D. C. (2003). Validating survivin as a cancer therapeutic target. *Nature Rev. Cancer* **3,** 46–54.

Anderson, L. M., Krotz, S., Weitzman, S. A., and Thimmapaya, B. (2000). Breast cancer-specific expression of the *Candida albicans* cytosine deaminase gene using a transcriptional targeting approach. *Cancer Gene Ther.* **7,** 845–852.

Andriani, F., Nan, B., Yu, J., Li, X., Weigel, N. L., McPhaul, M. J., Kasper, S., Kagawa, S., Fang, B., Matusik, R. J., Denner, L., and Marcelli, M. (2001). Use of the probasin promoter ARR$_2$PB to express BAX in androgen receptor-positive prostate cancer cells. *J. Natl. Cancer Inst.* **93,** 1314–1324.

Boyd, J. M., Gallo, G. J., Elangovan, B., Houghton, A. B., Malstrom, S., Avery, B. J., Ebb, R. G., Subramanian, T., Chittenden, T., Lutz, R. J., and Chinnadurai, G. (1995). Bik, a novel death-inducing protein shares a distinct sequence motif with Bcl-2 family proteins and interacts with viral and cellular survival-promoting proteins. *Oncogene* **11,** 1921–1928.

Brader, K. R., Wolf, J. K., Hung, M. C., Yu, D., Crispens, M. A., van Golen, K. L., and Price, J. E. (1997). Adenovirus E1A expression enhances the sensitivity of an ovarian cancer cell line to multiple cytotoxic agents through an apoptotic mechanism. *Clin. Cancer Res.* **3,** 2017–2024.

Brockmann, D., Bury, C., Kroner, G., Kirch, H. C., and Esche, H. (1995). Repression of the c-Jun trans-activation function by the adenovirus type 12 E1A 52R protein correlates with the inhibition of phosphorylation of the c-Jun activation domain. *J. Biol. Chem.* **270,** 10754–10763.

Buller, R. E., Runnebaum, I. B., Karlan, B. Y., Horowitz, J. A., Shahin, M., Buekers, T., Petrauskas, S., Kreienberg, R., Slamon, D., and Pegram, M. (2002). A phase I/II trial of rAd/p53 (SCH 58500) gene replacement in recurrent ovarian cancer. *Cancer Gene Ther.* **9,** 553–566.

Bush, J. A., and Li, G. (2002). Cancer chemoresistance: The relationship between p53 and multidrug transporters. *Int. J. Cancer* **98,** 323–330.

Campling, B. G., and El-Deiry, W. S. (2003). Clinical implication of p53 mutation in lung cancer. *Mol. Biotechnol.* **24,** 141–156.

Chang, J. Y., Xia, W., Shao, R., Sorgi, F., Hortobagyi, G. N., Huang, L., and Hung, M. C. (1997). The tumor suppression activity of E1A in HER-2/neu-overexpressing breast cancer. *Oncogene* **14,** 561–568.

Chen, C. B., Pan, J. J., and Xu, L. Y. (2003). Recombinant adenovirus p53 agent injection combined with radiotherapy in treatment of nasopharyngeal carcinoma: A phase II clinical trial. *Zhonghua Yi Xue Za Zhi* **83**, 2033–2035.

Chen, H., and Hung, M. C. (1997). Involvement of co-activator p300 in the transcriptional regulation of the HER-2/neu gene. *J. Biol. Chem.* **272**, 6101–6104.

Chen, J. S., Liu, J. C., Shen, L., Rau, K. M., Kuo, H. P., Li, Y. M., Shi, D., Lee, Y. C., Chang, K. J., and Hung, M. C. (2004). Cancer-specific activation of the survivin promoter and its potential use in gene therapy. *Cancer Gene Ther.* **11**, 740–747.

Chen, R. H., and McCormick, F. (2001). Selective targeting to the hyperactive beta-catenin/T-cell factor pathway in colon cancer cells. *Cancer Res.* **61**, 4445–4449.

Choubey, D. (2000). P202: An interferon-inducible negative regulator of cell growth. *J. Biol. Regul. Homeost. Agents* **14**, 187–192.

Choubey, D., and Lengyel, P. (1995). Binding of an interferon-inducible protein (p202) to the retinoblastoma protein. *J. Biol. Chem.* **270**, 6134–6140.

Clayman, G. L., el-Naggar, A. K., Lippman, S. M., Henderson, Y. C., Frederick, M., Merritt, J. A., Zumstein, L. A., Timmons, T. M., Liu, T. J., Ginsberg, L., Roth, J. A., Hong, W. K., Bruso, P., and Goepfert, H. (1998). Adenovirus-mediated p53 gene transfer in patients with advanced recurrent head and neck squamous cell carcinoma. *J. Clin. Oncol.* **16**, 2221–2232.

Clayman, G. L., Frank, D. K., Bruso, P. A., and Goepfert, H. (1999). Adenovirus-mediated wild-type p53 gene transfer as a surgical adjuvant in advanced head and neck cancers. *Clin. Cancer Res.* **5**, 1715–1722.

Colletier, P. J., Ashoori, F., Cowen, D., Meyn, R. E., Tofilon, P., Meistrich, M. E., and Pollack, A. (2000). Adenoviral-mediated p53 transgene expression sensitizes both wild-type and null p53 prostate cancer cells *in vitro* to radiation. *Int. J. Radiat. Oncol. Biol. Phys.* **48**, 1507–1512.

Cory, S., Huang, D. C., and Adams, J. M. (2003). The Bcl-2 family: Roles in cell survival and oncogenesis. *Oncogene* **22**, 8590–8607.

Coultas, L., and Strasser, A. (2003). The role of the Bcl-2 protein family in cancer. *Semin. Cancer Biol.* **13**, 115–123.

Crompton, M. (2000). BAX, Bid and the permeabilization of the mitochondrial outer membrane in apoptosis. *Curr. Opin. Cell Biol.* **12**, 414–419.

Daniel, P. T., Pun, K. T., Ritschel, S., Sturm, I., Holler, J., Dorken, B., and Brown, R. (1999). Expression of the death gene Bik/Nbk promotes sensitivity to drug-induced apoptosis in corticosteroid-resistant T-cell lymphoma and prevents tumor growth in severe combined immunodeficient mice. *Blood* **94**, 1100–1107.

DeFatta, R. J., Chervenak, R. P., and De Benedetti, A. (2002a). A cancer gene therapy approach through translational control of a suicide gene. *Cancer Gene Ther.* **9**, 505–512.

DeFatta, R. J., Li, Y., and De Benedetti, A. (2002b). Selective killing of cancer cells based on translational control of a suicide gene. *Cancer Gene Ther.* **9**, 573–578.

Ding, Y., Wen, Y., Spohn, B., Wang, L., Xia, W., Kwong, K. Y., Shao, R., Li, Z., Hortobagyi, G. N., Hung, M. C., and Yan, D. H. (2002). Proapoptotic and antitumor activities of adenovirus-mediated p202 gene transfer. *Clin. Cancer Res.* **8**, 3290–3297.

Edelman, J., and Nemunaitis, J. (2003). Adenoviral p53 gene therapy in squamous cell cancer of the head and neck region. *Curr. Opin. Mol. Ther.* **5**, 611–617.

El-Deiry, W. S. (2001). Insights into cancer therapeutic design based on p53 and TRAIL receptor signaling. *Cell Death Differ.* **8**, 1066–1075.

Flint, J., and Shenk, T. (1989). Adenovirus E1A protein paradigm viral transactivator. *Annu. Rev. Genet.* **23**, 141–161.

Frisch, S. M. (1991). Antioncogenic effect of adenovirus E1A in human tumor cells. *Proc. Natl. Acad. Sci. USA* **88**, 9077–9081.

Frisch, S. M., and Dolter, K. E. (1995). Adenovirus E1a-mediated tumor suppression by a c-erbB-2/neu-independent mechanism. *Cancer Res.* **55,** 5551–5555.

Frisch, S. M., and Mymryk, J. S. (2002). Adenovirus-5 E1A: Paradox and paradigm. *Nature Rev. Mol. Cell. Biol.* **3,** 441–452.

Frisch, S. M., Reich, R., Collier, I. E., Genrich, L. T., Martin, G., and Goldberg, G. I. (1990). Adenovirus E1A represses protease gene expression and inhibits metastasis of human tumor cells. *Oncogene* **5,** 75–83.

Graham, F. L., van der Eb, A. J., and Heijneker, H. L. (1974). Size and location of the transforming region in human adenovirus type 5 DNA. *Nature* **251,** 687–691.

Gu, J., Kagawa, S., Takakura, M., Kyo, S., Inoue, M., Roth, J. A., and Fang, B. (2000). Tumor-specific transgene expression from the human telomerase reverse transcriptase promoter enables targeting of the therapeutic effects of the BAX gene to cancers. *Cancer Res.* **60,** 5359–5364.

Han, D. M., Huang, Z. G., Zhang, W., Yu, Z. K., Wang, Q., Ni, X., Chen, X. H., Pan, J. H., and Wang, H. (2003). Effectiveness of recombinant adenovirus p53 injection on laryngeal cancer: Phase I clinical trial and follow up. *Zhonghua Yi Xue Za Zhi* **83,** 2029–2032.

Han, J., Sabbatini, P., and White, E. (1996). Induction of apoptosis by human Nbk/Bik, a BH3-containing protein that interacts with E1B 19K. *Mol. Cell. Biol.* **16,** 5857–5864.

Hardwick, J. M., and Polster, B. M. (2002). BAX, along with lipid conspirators, allows cytochrome c to escape mitochondria. *Mol. Cell* **10,** 963–965.

Harris, J. D., Gutierrez, A. A., Hurst, H. C., Sikora, K., and Lemoine, N. R. (1994). Gene therapy for cancer using tumour-specific prodrug activation. *Gene Ther.* **1,** 170–175.

Hollstein, M., Sidransky, D., Vogelstein, B., and Harris, C. C. (1991). p53 mutations in human cancers. *Science* **253,** 49–53.

Hortobagyi, G. N., Hung, M. C., and Lopez-Berestein, G. (1998). A phase I multicenter study of E1A gene therapy for patients with metastatic breast cancer and epithelial ovarian cancer that overexpresses HER-2/neu or epithelial ovarian cancer. *Hum. Gene Ther.* **9,** 1775–1798.

Hortobagyi, G. N., Ueno, N. T., Xia, W., Zhang, S., Wolf, J. K., Putnam, J. B., Weiden, P. L., Willey, J. S., Carey, M., Branham, D. L., Payne, J. Y., Tucker, S. D., Bartholomeusz, C., Kilbourn, R. G., De Jager, R. L., Sneige, N., Katz, R. L., Anklesaria, P., Ibrahim, N. K., Murray, J. L., Theriault, R. L., Valero, V., Gershenson, D. M., Bevers, M. W., Huang, L., Lopez-Berestein, G., and Hung, M. C. (2001). Cationic liposome-mediated E1A gene transfer to human breast and ovarian cancer cells and its biologic effects: A phase I clinical trial. *J. Clin. Oncol.* **19,** 3422–3433.

Huh, W. K., Gomez-Navarro, J., Arafat, W. O., Xiang, J., Mahasreshti, P. J., Alvarez, R. D., Barnes, M. N., and Curiel, D. T. (2001). BAX-induced apoptosis as a novel gene therapy approach for carcinoma of the cervix. *Gynecol. Oncol.* **83,** 370–377.

Hung, M. C., Hortobagyi, G. N., and Ueno, N. T. (2000). Development of clinical trial of E1A gene therapy targeting HER-2/neu-overexpressing breast and ovarian cancer. *Adv. Exp. Med. Biol.* **465,** 171–180.

Hung, M. C., Matin, A., Zhang, Y., Xing, X., Sorgi, F., Huang, L., and Yu, D. (1995). HER-2/neu-targeting gene therapy: A review. *Gene* **159,** 65–71.

Johnstone, R. W., and Trapani, J. A. (1999). Transcription and growth regulatory functions of the HIN-200 family of proteins. *Mol. Cell. Biol.* **19,** 5833–5838.

Kakinuma, H., Bergert, E. R., Spitzweg, C., Cheville, J. C., Lieber, M. M., and Morris, J. C. (2003). Probasin promoter (ARR(2)PB)-driven, prostate-specific expression of the human sodium iodide symporter (h-NIS) for targeted radioiodine therapy of prostate cancer. *Cancer Res.* **63,** 7840–7844.

Kaliberov, S. A., Buchsbaum, D. J., Gillespie, G. Y., Curiel, D. T., Arafat, W. O., Carpenter, M., and Stackhouse, M. A. (2002). Adenovirus-mediated transfer of BAX driven by the vascular endothelial growth factor promoter induces apoptosis in lung cancer cells. *Mol. Ther.* **6,** 190–198.

Kevil, C., Carter, P., Hu, B., and DeBenedetti, A. (1995). Translational enhancement of FGF-2 by eIF-4 factors, and alternate utilization of CUG and AUG codons for translation initiation. _Oncogene_ **11,** 2339–2348.

Kijima, T., Osaki, T., Nishino, K., Kumagai, T., Funakoshi, T., Goto, H., Tachibana, I., Tanio, Y., and Kishimoto, T. (1999). Application of the Cre recombinase/loxP system further enhances antitumor effects in cell type-specific gene therapy against carcinoembryonic antigen-producing cancer. _Cancer Res._ **59,** 4906–4911.

Kock, H., Harris, M. P., Anderson, S. C., Machemer, T., Hancock, W., Sutjipto, S., Wills, K. N., Gregory, R. J., Shepard, H. M., Westphal, M., and Maneval, D. C. (1996). Adenovirus-mediated p53 gene transfer suppresses growth of human glioblastoma cells _in vitro_ and _in vivo. Int. J. Cancer_ **67,** 808–815.

Koga, S., Hirohata, S., Kondo, Y., Komata, T., Takakura, M., Inoue, M., Kyo, S., and Kondo, S. (2001). FADD gene therapy using the human telomerase catalytic subunit (hTERT) gene promoter to restrict induction of apoptosis to tumors _in vitro_ and _in vivo. Anticancer Res._ **21,** 1937–1943.

Kubo, H., Gardner, T. A., Wada, Y., Koeneman, K. S., Gotoh, A., Yang, L., Kao, C., Lim, S. D., Amin, M. B., Yang, H., Black, M. E., Matsubara, S., Nakagawa, M., Gillenwater, J. Y., Zhau, H. E., and Chung, L. W. (2003). Phase I dose escalation clinical trial of adenovirus vector carrying osteocalcin promoter-driven herpes simplex virus thymidine kinase in localized and metastatic hormone-refractory prostate cancer. _Hum. Gene Ther._ **14,** 227–241.

Kwong, K. Y., Zou, Y., Day, C. P., and Hung, M. C. (2002). The suppression of colon cancer cell growth in nude mice by targeting beta-catenin/TCF pathway. _Oncogene_ **21,** 8340–8346.

Lang, F. F., Bruner, J. M., Fuller, G. N., Aldape, K., Prados, M. D., Chang, S., Berger, M. S., McDermott, M. W., Kunwar, S. M., Junck, L. R., Chandler, W., Zwiebel, J. A., Kaplan, R. S., and Yung, W. K. (2003). Phase I trial of adenovirus-mediated p53 gene therapy for recurrent glioma: Biological and clinical results. _J. Clin. Oncol._ **21,** 2508–2518.

Lang, F. F., Yung, W. K., Raju, U., Libunao, F., Terry, N. H., and Tofilon, P. J. (1998). Enhancement of radiosensitivity of wild-type p53 human glioma cells by adenovirus-mediated delivery of the p53 gene. _J. Neurosurg._ **89,** 125–132.

Lebedeva, S., Bagdasarova, S., Tyler, T., Mu, X., Wilson, D. R., and Gjerset, R. A. (2001). Tumor suppression and therapy sensitization of localized and metastatic breast cancer by adenovirus p53. _Hum. Gene Ther._ **12,** 763–772.

Lee, C. M., Lo, H. W., Shao, R. P., Wang, S. C., Xia, W., Gershenson, D. M., and Hung, M. C. (2004). Selective activation of ceruloplasmin promoter in ovarian tumors: Potential use for gene therapy. _Cancer Res._ **64,** 1788–1793.

Li, Y. M., Wen, Y., Zhou, B. P., Kuo, H. P., Ding, Q., and Hung, M. C. (2003). Enhancement of Bik antitumor effect by Bik mutants. _Cancer Res._ **63,** 7630–7633.

Li, Z., Day, C.-P., Yang, J.-Y., Tsai, W.-B., Lozano, G., Shih, H.-M., and Hung, M.-C. (2005). Adenoviral E1A targets Mdm4 to stabilize tumor suppressor p53. _Cancer Res._ **64,** 9080–9085.

Liang, G., and Hai, T. (1997). Characterization of human activating transcription factor 4, a transcriptional activator that interacts with multiple domains of cAMP-responsive element-binding protein (CREB)-binding protein. _J. Biol. Chem._ **272,** 24088–24095.

Liao, Y., and Hung, M. C. (2003). Regulation of the activity of p38 mitogen-activated protein kinase by Akt in cancer and adenoviral protein E1A-mediated sensitization to apoptosis. _Mol. Cell. Biol._ **23,** 6836–6848.

Liao, Y., and Hung, M. C. (2004). A new role of protein phosphatase 2a in adenoviral E1A protein-mediated sensitization to anticancer drug-induced apoptosis in human breast cancer cells. _Cancer Res._ **64,** 5938–5942.

Liao, Y., Zou, Y. Y., Xia, W. Y., and Hung, M. C. (2004). Enhanced paclitaxel cytotoxicity and prolonged animal survival rate by a non-viral-mediated systemic delivery of E1A gene in orthotopic xenograft human breast cancer. *Cancer Gene Ther.* **11**, 594–602.

Lin, T., Huang, X., Gu, J., Zhang, L., Roth, J. A., Xiong, M., Curley, S. A., Yu, Y., Hunt, K. K., and Fang, B. (2002). Long-term tumor-free survival from treatment with the GFP-TRAIL fusion gene expressed from the hTERT promoter in breast cancer cells. *Oncogene* **21**, 8020–8028.

Lipinski, K. S., Djeha, H. A., Gawn, J., Cliffe, S., Maitland, N. J., Palmer, D. H., Mountain, A., Irvine, A. S., and Wrighton, C. J. (2004). Optimization of a synthetic beta-catenin-dependent promoter for tumor-specific cancer gene therapy. *Mol. Ther.* **10**, 150–161.

Liu, J., Zou, W. G., Lang, M. F., Luo, J., Sun, L. Y., Wang, X. N., Qian, Q. J., and Liu, X. Y. (2002). Cancer-specific killing by the CD suicide gene using the human telomerase reverse transcriptase promoter. *Int. J. Oncol.* **21**, 661–666.

Lokich, J. (1998). Infusional 5-FU: Historical evolution, rationale, and clinical experience. *Oncology (Huntingt)* **12**, 19–22.

Lowe, S. L., Rubinchik, S., Honda, T., McDonnell, T. J., Dong, J. Y., and Norris, J. S. (2001). Prostate-specific expression of BAX delivered by an adenoviral vector induces apoptosis in LNCaP prostate cancer cells. *Gene Ther.* **8**, 1363–1371.

Lowe, S. W., and Ruley, H. E. (1993). Stabilization of the p53 tumor suppressor is induced by adenovirus 5 E1A and accompanies apoptosis. *Genes Dev.* **7**, 535–545.

Lowe, S. W., Ruley, H. E., Jacks, T., and Housman, D. E. (1993a). p53-dependent apoptosis modulates the cytotoxicity of anticancer agents. *Cell* **74**, 957–967.

Lowe, S. W., Schmitt, E. M., Smith, S. W., Osborne, B. A., and Jacks, T. (1993b). p53 is required for radiation-induced apoptosis in mouse thymocytes. *Nature* **362**, 847–849.

Madhusudan, S., Tamir, A., Bates, N., Flanagan, E., Gore, M. E., Barton, D. P., Harper, P., Seckl, M., Thomas, H., Lemoine, N. R., Charnock, M., Habib, N. A., Lechler, R., Nicholls, J., Pignatelli, M., and Ganesan, T. S. (2004). A multicenter phase I gene therapy clinical trial involving intraperitoneal administration of E1A-lipid complex in patients with recurrent epithelial ovarian cancer overexpressing HER-2/neu oncogene. *Clin. Cancer Res.* **10**, 2986–2996.

Meyerson, M., Counter, C. M., Eaton, E. N., Ellisen, L. W., Steiner, P., Caddle, S. D., Ziaugra, L., Beijersbergen, R. L., Davidoff, M. J., Liu, Q., Bacchetti, S., Haber, D. A., and Weinberg, R. A. (1997). hEST2, the putative human telomerase catalytic subunit gene, is up-regulated in tumor cells and during immortalization. *Cell* **90**, 785–795.

Modesitt, S. C., Ramirez, P., Zu, Z., Bodurka-Bevers, D., Gershenson, D., and Wolf, J. K. (2001). *In vitro* and *in vivo* adenovirus-mediated p53 and p16 tumor suppressor therapy in ovarian cancer. *Clin. Cancer Res.* **7**, 1765–1772.

Morimoto, E., Inase, N., Mlyake, S., and Yoshizawa, Y. (2001). Adenovirus-mediated suicide gene transfer to small cell lung carcinoma using a tumor-specific promoter. *Anticancer Res.* **21**, 329–331.

Mullen, C. A., Kilstrup, M., and Blaese, R. M. (1992). Transfer of the bacterial gene for cytosine deaminase to mammalian cells confers lethal sensitivity to 5-fluorocytosine: A negative selection system. *Proc. Natl. Acad. Sci. USA* **89**, 33–37.

Mymryk, J. S. (1996). Tumour suppressive properties of the adenovirus 5 E1A oncogene. *Oncogene* **13**, 1581–1589.

Najafi, S. M., Li, Z., Makino, K., Shao, R., and Hung, M. C. (2003). The adenoviral E1A induces p21WAF1/CIP1 expression in cancer cells. *Biochem. Biophys. Res. Commun.* **305**, 1099–1104.

Nemunaitis, J., Swisher, S. G., Timmons, T., Connors, D., Mack, M., Doerksen, L., Weill, D., Wait, J., Lawrence, D. D., Kemp, B. L., Fossella, F., Glisson, B. S., Hong, W. K., Khuri, F. R., Kurie, J. M., Lee, J. J., Lee, J. S., Nguyen, D. M., Nesbitt, J. C., Perez-Soler, R., Pisters, K. M.,

Putnam, J. B., Richli, W. R., Shin, D. M., Walsh, G. L., Merritt, J., and Roth, J. (2000). Adenovirus-mediated p53 gene transfer in sequence with cisplatin to tumors of patients with non-small-cell lung cancer. *J. Clin. Oncol.* **18,** 609–622.

Nishiyama, Y., and Rapp, F. (1979). Anticellular effects of 9-(2-hydroxyethoxymethyl) guanine against herpes simplex virus-transformed cells. *J. Gen. Virol.* **45,** 227–230.

Norris, J. S., Hyer, M. L., Voelkel-Johnson, C., Lowe, S. L., Rubinchik, S., and Dong, J. Y. (2001). The use of Fas ligand, TRAIL and BAX in gene therapy of prostate cancer. *Curr. Gene Ther.* **1,** 123–136.

Okabe, S., Arai, T., Yamashita, H., and Sugihara, K. (2003). Adenovirus-mediated prodrug-enzyme therapy for CEA-producing colorectal cancer cells. *J. Cancer Res. Clin. Oncol.* **129,** 367–373.

Pagliaro, L. C., Keyhani, A., Williams, D., Woods, D., Liu, B., Perrotte, P., Slaton, J. W., Merritt, J. A., Grossman, H. B., and Dinney, C. P. (2003). Repeated intravesical instillations of an adenoviral vector in patients with locally advanced bladder cancer: A phase I study of p53 gene therapy. *J. Clin. Oncol.* **21,** 2247–2253.

Panaretakis, T., Pokrovskaja, K., Shoshan, M. C., and Grander, D. (2002). Activation of Bak, BAX, and BH3-only proteins in the apoptotic response to doxorubicin. *J. Biol. Chem.* **277,** 44317–44326.

Pandha, H. S., Martin, L. A., Rigg, A., Hurst, H. C., Stamp, G. W., Sikora, K., and Lemoine, N. R. (1999). Genetic prodrug activation therapy for breast cancer: A phase I clinical trial of erbB-2-directed suicide gene expression. *J. Clin. Oncol.* **17,** 2180–2189.

Parker, L. P., Wolf, J. K., and Price, J. E. (2000). Adenoviral-mediated gene therapy with Ad5CMVp53 and Ad5CMVp21 in combination with standard therapies in human breast cancer cell lines. *Ann. Clin. Lab. Sci.* **30,** 395–405.

Pearson, S., Jia, H., and Kandachi, K. (2004). China approves first gene therapy. *Nature Biotechnol.* **22,** 3–4.

Pisters, L. L., Pettaway, C. A., Troncoso, P., McDonnell, T. J., Stephens, L. C., Wood, C. G., Do, K. A., Brisbay, S. M., Wang, X., Hossan, E. A., Evans, R. B., Soto, C., Jacobson, M. G., Parker, K., Merritt, J. A., Steiner, M. S., and Logothetis, C. J. (2004). Evidence that transfer of functional p53 protein results in increased apoptosis in prostate cancer. *Clin. Cancer Res.* **10,** 2587–2593.

Pozzatti, R., McCormick, M., Thompson, M. A., Garbisa, S., Liotta, L., and Khoury, G. (1988a). Regulation of the metastatic phenotype by the E1A gene of adenovirus-2. *Adv. Exp. Med. Biol.* **233,** 293–301.

Pozzatti, R., McCormick, M., Thompson, M. A., and Khoury, G. (1988b). The E1a gene of adenovirus type 2 reduces the metastatic potential of ras-transformed rat embryo cells. *Mol. Cell. Biol.* **8,** 2984–2988.

Rosenwald, I. B. (1996). Upregulated expression of the genes encoding translation initiation factors eIF-4E and eIF-2alpha in transformed cells. *Cancer Lett.* **102,** 113–123.

Roth, J. A., Grammer, S. F., Swisher, S. G., Komaki, R., Nemunaitis, J., Merritt, J., and Meyn, R. E. (2001). P53 gene replacement for cancer: Interactions with DNA damaging agents. *Acta Oncol.* **40,** 739–744.

Roth, J. A., Swisher, S. G., Merritt, J. A., Lawrence, D. D., Kemp, B. L., Carrasco, C. H., El-Naggar, A. K., Fossella, F. V., Glisson, B. S., Hong, W. K., Khurl, F. R., Kurie, J. M., Nesbitt, J. C., Pisters, K., Putnam, J. B., Schrump, D. S., Shin, D. M., and Walsh, G. L. (1998). Gene therapy for non-small cell lung cancer: A preliminary report of a phase I trial of adenoviral p53 gene replacement. *Semin. Oncol.* **25,** 33–37.

Ruley, H. E. (1983). Adenovirus early region 1A enables viral and cellular transforming genes to transform primary cells in culture. *Nature* **304,** 602–606.

Sah, N. K., Munshi, A., Nishikawa, T., Mukhopadhyay, T., Roth, J. A., and Meyn, R. E. (2003). Adenovirus-mediated wild-type p53 radiosensitizes human tumor cells by suppressing DNA repair capacity. *Mol. Cancer Ther.* **2,** 1223–1231.

Sang, N., Avantaggiati, M. L., and Giordano, A. (1997). Roles of p300, pocket proteins, and hTBP in E1A-mediated transcriptional regulation and inhibition of p53 transactivation activity. *J. Cell Biochem.* **66,** 277–285.

Shao, R., Hu, M. C., Zhou, B. P., Lin, S. Y., Chiao, P. J., von Lindern, R. H., Spohn, B., and Hung, M. C. (1999). E1A sensitizes cells to tumor necrosis factor-induced apoptosis through inhibition of IkappaB kinases and nuclear factor kappaB activities. *J. Biol. Chem.* **274,** 21495–21498.

Shao, R., Karunagaran, D., Zhou, B. P., Li, K., Lo, S. S., Deng, J., Chiao, P., and Hung, M. C. (1997). Inhibition of nuclear factor-kappaB activity is involved in E1A-mediated sensitization of radiation-induced apoptosis. *J. Biol. Chem.* **272,** 32739–32742.

Shao, R., Xia, W., and Hung, M. C. (2000). Inhibition of angiogenesis and induction of apoptosis are involved in E1A-mediated bystander effect and tumor suppression. *Cancer Res.* **60,** 3123–3126.

Shay, J. W., and Bacchetti, S. (1997). A survey of telomerase activity in human cancer. *Eur. J. Cancer* **33,** 787–791.

Shimada, H., Matsubara, H., and Ochiai, T. (2002). p53 gene therapy for esophageal cancer. *J. Gastroenterol.* **37**(Suppl. 14), 87–91.

Song, J. S. (2004). Activity of the human telomerase catalytic subunit (hTERT) gene promoter could be increased by the SV40 enhancer. *Biosci. Biotechnol. Biochem.* **68,** 1634–1639.

Song, J. S., and Kim, H. P. (2004). Adenovirus-mediated HSV-TK gene therapy using the human telomerase promoter induced apoptosis of small cell lung cancer cell line. *Oncol. Rep.* **12,** 443–447.

Song, J. S., Kim, H. P., Yoon, W. S., Lee, K. W., Kim, M. H., Kim, K. T., Kim, H. S., and Kim, Y. T. (2003). Adenovirus-mediated suicide gene therapy using the human telomerase catalytic subunit (hTERT) gene promoter induced apoptosis of ovarian cancer cell line. *Biosci. Biotechnol. Biochem.* **67,** 2344–2350.

Spitz, F. R., Nguyen, D., Skibber, J. M., Meyn, R. E., Cristiano, R. J., and Roth, J. A. (1996). Adenoviral-mediated wild-type p53 gene expression sensitizes colorectal cancer cells to ionizing radiation. *Clin. Cancer Res.* **2,** 1665–1671.

Swisher, S. G., Roth, J. A., Komaki, R., Gu, J., Lee, J. J., Hicks, M., Ro, J. Y., Hong, W. K., Merritt, J. A., Ahrar, K., Atkinson, N. E., Correa, A. M., Dolormente, M., Dreiling, L., El-Naggar, A. K., Fossella, F., Francisco, R., Glisson, B., Grammer, S., Herbst, R., Huaringa, A., Kemp, B., Khuri, F. R., Kurie, J. M., Liao, Z., McDonnell, T. J., Morice, R., Morello, F., Munden, R., Papadimitrakopoulou, V., Pisters, K. M., Putnam, J. B., Jr., Sarabia, A. J., Shelton, T., Stevens, C., Shin, D. M., Smythe, W. R., Vaporciyan, A. A., Walsh, G. L., and Yin, M. (2003). Induction of p53-regulated genes and tumor regression in lung cancer patients after intratumoral delivery of adenoviral p53 (INGN 201) and radiation therapy. *Clin. Cancer Res.* **9,** 93–101.

Swisher, S. G., Roth, J. A., Nemunaitis, J., Lawrence, D. D., Kemp, B. L., Carrasco, C. H., Connors, D. G., El-Naggar, A. K., Fossella, F., Glisson, B. S., Hong, W. K., Khuri, F. R., Kurie, J. M., Lee, J. J., Lee, J. S., Mack, M., Merritt, J. A., Nguyen, D. M., Nesbitt, J. C., Perez-Soler, R., Pisters, K. M., Putnam, J. B., Jr., Richli, W. R., Savin, M., Schrump, D. S., Shin, D. M., Shulkin, A., Walsh, G. L., Wait, J., and Waugh, M. K. (1999). Adenovirus-mediated p53 gene transfer in advanced non-small-cell lung cancer. *J. Natl. Cancer Inst.* **91,** 763–771.

Tai, Y. T., Strobel, T., Kufe, D., and Cannistra, S. A. (1999). *In vivo* cytotoxicity of ovarian cancer cells through tumor-selective expression of the BAX gene. *Cancer Res.* **59,** 2121–2126.

Terazaki, Y., Yano, S., Yuge, K., Nagano, S., Fukunaga, M., Guo, Z. S., Komiya, S., Shirouzu, K., and Kosai, K. (2003). An optimal therapeutic expression level is crucial for suicide gene therapy for hepatic metastatic cancer in mice. *Hepatology* **37,** 155–163.

Thorburn, A. (2004). Death receptor-induced cell killing. *Cell Signal.* **16,** 139–144.

Ueno, N. T., Bartholomeusz, C., Xia, W., Anklesaria, P., Bruckheimer, E. M., Mebel, E., Paul, R., Li, S., Yo, G. H., Huang, L., and Hung, M. C. (2002). Systemic gene therapy in human xenograft tumor models by liposomal delivery of the E1A gene. *Cancer Res.* **62,** 6712–6716.

Ueno, N. T., Yu, D., and Hung, M. C. (1997). Chemosensitization of HER-2/neu-overexpressing human breast cancer cells to paclitaxel (Taxol) by adenovirus type 5 E1A. *Oncogene* **15,** 953–960.

Ueno, N. T., Yu, D., and Hung, M. C. (2001). E1A: Tumor suppressor or oncogene? Preclinical and clinical investigations of E1A gene therapy. *Breast Cancer* **8,** 285–293.

Van Tendeloo, V. F., Ponsaerts, P., Lardon, F., Nijs, G., Lenjou, M., Van Broeckhoven, C., Van Bockstaele, D. R., and Berneman, Z. N. (2001). Highly efficient gene delivery by mRNA electroporation in human hematopoietic cells: Superiority to lipofection and passive pulsing of mRNA and to electroporation of plasmid cDNA for tumor antigen loading of dendritic cells. *Blood* **98,** 49–56.

Villaret, D., Glisson, B., Kenady, D., Hanna, E., Carey, M., Gleich, L., Yoo, G. H., Futran, N., Hung, M. C., Anklesaria, P., and Heald, A. E. (2002). A multicenter phase II study of tgDCC-E1A for the intratumoral treatment of patients with recurrent head and neck squamous cell carcinoma. *Head Neck* **24,** 661–669.

Viniegra, J. G., Losa, J. H., Sanchez-Arevalo, V. J., Cobo, C. P., Soria, V. M., Ramon y Cajal, S., and Sanchez-Prieto, R. (2002). Modulation of PI3K/Akt pathway by E1a mediates sensitivity to cisplatin. *Oncogene* **21,** 7131–7136.

Vogelstein, B., Lane, D., and Levine, A. J. (2000). Surfing the p53 network. *Nature* **408,** 307–310.

Wang, S., and El-Deiry, W. S. (2004). The p53 pathway: Targets for the development of novel cancer therapeutics. *Cancer Treat. Res.* **119,** 175–187.

Wang, S. C., and Hung, M. C. (2001). HER2 overexpression and cancer targeting. *Semin. Oncol.* **28,** 115–124.

Wen, Y., Giri, D., Yan, D. H., Spohn, B., Zinner, R. G., Xia, W., Thompson, T. C., Matusik, R. J., and Hung, M. C. (2003). Prostate-specific antitumor activity by probasin promoter-directed p202 expression. *Mol. Carcinog.* **37,** 130–137.

Wen, Y., Yan, D. H., Spohn, B., Deng, J., Lin, S. Y., and Hung, M. C. (2000). Tumor suppression and sensitization to tumor necrosis factor alpha-induced apoptosis by an interferon-inducible protein, p202, in breast cancer cells. *Cancer Res.* **60,** 42–46.

Wen, Y., Yan, D. H., Wang, B., Spohn, B., Ding, Y., Shao, R., Zou, Y., Xie, K., and Hung, M. C. (2001). p202, an interferon-inducible protein, mediates multiple antitumor activities in human pancreatic cancer xenograft models. *Cancer Res.* **61,** 7142–7147.

Wills, K. N., Maneval, D. C., Menzel, P., Harris, M. P., Sutjipto, S., Vaillancourt, M. T., Huang, W. M., Johnson, D. E., Anderson, S. C., Wen, S. F., Bookstein, R., Shepard, H. M., and Gregory, R. J. (1994). Development and characterization of recombinant adenoviruses encoding human p53 for gene therapy of cancer. *Hum. Gene Ther.* **5,** 1079–1088.

Wolf, J. K., Bodurka, D. C., Gano, J. B., Deavers, M., Ramondetta, L., Ramirez, P. T., Levenback, C., and Gershenson, D. M. (2004). A phase I study of Adp53 (INGN 201; ADVEXIN) for patients with platinum- and paclitaxel-resistant epithelial ovarian cancer. *Gynecol. Oncol.* **94,** 442–448.

Xing, X., Yujiao Chang, J., and Hung, M. (1998). Preclinical and clinical study of HER-2/neu-targeting cancer gene therapy. *Adv. Drug Deliv. Rev.* **30,** 219–227.

Yan, D. H., Shao, R., and Hung, M. C. (2002). E1A Cancer Gene Therapy. *In* "Gene Therapy of Cancer" (E. C. Lattime and S. L. Gerson, eds.), 2nd Ed., pp. 465–477. Academic Press, San Diego.

Yan, D. H., Wen, Y., Spohn, B., Choubey, D., Gutterman, J. U., and Hung, M. C. (1999). Reduced growth rate and transformation phenotype of the prostate cancer cells by an interferon-inducible protein, p202. *Oncogene* **18,** 807–811.

Yang, L., Cao, Z., Li, F., Post, D. E., Van Meir, E. G., Zhong, H., and Wood, W. C. (2004). Tumor-specific gene expression using the survivin promoter is further increased by hypoxia. *Gene Ther.* **11,** 1215–1223.

Yoo, G. H., Hung, M. C., Lopez-Berestein, G., LaFollette, S., Ensley, J. F., Carey, M., Batson, E., Reynolds, T. C., and Murray, J. L. (2001). Phase I trial of intratumoral liposome E1A gene therapy in patients with recurrent breast and head and neck cancer. *Clin. Cancer Res* **7**, 1237–1245.

Yu, D., Hamada, J., Zhang, H., Nicolson, G. L., and Hung, M. C. (1992). Mechanisms of c-erbB2/neu oncogene-induced metastasis and repression of metastatic properties by adenovirus 5 E1A gene products. *Oncogene* **7**, 2263–2270.

Yu, D., and Hung, M. C. (2000). Overexpression of ErbB2 in cancer and ErbB2-targeting strategies. *Oncogene* **19**, 6115–6121.

Yu, D., Jia, W. W., Gleave, M. E., Nelson, C. C., and Rennie, P. S. (2004a). Prostate-tumor targeting of gene expression by lentiviral vectors containing elements of the probasin promoter. *Prostate* **59**, 370–382.

Yu, D., Matin, A., Xia, W., Sorgi, F., Huang, L., and Hung, M. C. (1995). Liposome-mediated in vivo E1A gene transfer suppressed dissemination of ovarian cancer cells that overexpress HER-2/neu. *Oncogene* **11**, 1383–1388.

Yu, D., Suen, T. C., Yan, D. H., Chang, L. S., and Hung, M. C. (1990). Transcriptional repression of the neu protooncogene by the adenovirus 5 E1A gene products. *Proc. Natl. Acad. Sci. USA* **87**, 4499–4503.

Yu, D., Wolf, J. K., Scanlon, M., Price, J. E., and Hung, M. C. (1993). Enhanced c-erbB-2/neu expression in human ovarian cancer cells correlates with more severe malignancy that can be suppressed by E1A. *Cancer Res.* **53**, 891–898.

Yu, D. H., Scorsone, K., and Hung, M. C. (1991). Adenovirus type 5 E1A gene products act as transformation suppressors of the neu oncogene. *Mol. Cell. Biol.* **11**, 1745–1750.

Yu, L., Yamamoto, N., Kadomatsu, K., Muramatsu, T., Matsubara, S., Sakiyama, S., and Tagawa, M. (2004b). Midkine promoter can mediate transcriptional activation of a fused suicide gene in a broader range of human breast cancer compared with c-erbB-2 promoter. *Oncology* **66**, 143–149.

Zeimet, A. G., and Marth, C. (2003). Why did p53 gene therapy fail in ovarian cancer? *Lancet Oncol.* **4**, 415–422.

Zhang, J., Thomas, T. Z., Kasper, S., and Matusik, R. J. (2000). A small composite probasin promoter confers high levels of prostate-specific gene expression through regulation by androgens and glucocorticoids in vitro and in vivo. *Endocrinology* **141**, 4698–4710.

Zhang, S. W., Xiao, S. W., Liu, C. Q., Sun, Y., Su, X., Li, D. M., Xu, G., Cai, Y., Zhu, G. Y., Xu, B., and Lu, Y. Y. (2003). Treatment of head and neck squamous cell carcinoma by recombinant adenovirus-p53 combined with radiotherapy: A phase II clinical trial of 42 cases. *Zhonghua Yi Xue Za Zhi* **83**, 2023–2028.

Zhang, W. W., Fang, X., Branch, C. D., Mazur, W., French, B. A., and Roth, J. A. (1993). Generation and identification of recombinant adenovirus by liposome-mediated transfection and PCR analysis. *Biotechniques* **15**, 868–872.

Zhong, H., De Marzo, A. M., Laughner, E., Lim, M., Hilton, D. A., Zagzag, D., Buechler, P., Isaacs, W. B., Semenza, G. L., and Simons, J. W. (1999). Overexpression of hypoxia-inducible factor 1alpha in common human cancers and their metastases. *Cancer Res.* **59**, 5830–5835.

Zhou, Z., Jia, S. F., Hung, M. C., and Kleinerman, E. S. (2001). E1A sensitizes HER2/neu-over-expressing Ewing's sarcoma cells to topoisomerase II-targeting anticancer drugs. *Cancer Res.* **61**, 3394–3398.

Zou, Y., Peng, H., Zhou, B., Wen, Y., Wang, S. C., Tsai, E. M., and Hung, M. C. (2002). Systemic tumor suppression by the proapoptotic gene Bik. *Cancer Res.* **62**, 8–12.

11

DNA Vaccine

Zhengrong Cui

Department of Pharmaceutical Sciences, College of Pharmacy
Oregon State University, Corvallis, Oregon 97331

0065-2660/05 $35.00
DOI: 10.1016/S0065-2660(05)54011-2

ABSTRACT

The DNA vaccine has proven to be one of the most promising applications in the field of gene therapy. Due to its unique ability to readily induce humoral as well as cellular immune responses, it attracted great interest when the concept was first confirmed in the early 1990s. After thousands of articles related to the DNA vaccine were published, scientists began to realize that although the DNA vaccine is very effective in small animal models, its effectiveness in recent clinical trails is rather disappointing. Therefore, current effort has been shifted to understanding the different performance of the DNA vaccine in mouse and large animal models and on how to transfer the success of the DNA vaccine in small animals to large animals and humans.   © 2005, Elsevier Inc.

I. INTRODUCTION

With the exception of clean water, vaccines have been the most effective modalities in reducing human mortality from infectious diseases. With an effective vaccine, the World Health Organization (WHO) was able to declare the complete eradication of smallpox around the world in the 1970s. Similarly, the world is now almost free of polio, with the exception of only six countries (CDC). Effective vaccines have been successfully developed for some pathogens. However, lack of effective vaccines for others [e.g., the Bacilli-Calette-Guerin (BCG)-based vaccine for tuberculosis] and the emergence of new pathogens (e.g., severe acute respiratory syndrome [SARS] virus) warrant continuous effort in vaccine development. In addition, the recent bioterror threat makes it more urgent to develop new or alternative vaccines for those pathogens that can be potentially used for bioterror purposes.

Vaccines are traditionally prepared with either live attenuated or killed bacteria or viruses. In many cases, these approaches have proven to be successful. Vaccines based on live attenuated or killed pathogens are usually very potent and can induce all aspects of immune responses. However, serious safety concerns may preclude the use of these approaches in the development of vaccines for some pathogens such as HIV (Baba *et al.*, 1999). New generation vaccines, such as recombinant protein-based vaccines, synthetic peptide-based vaccines, lipid-based vaccines/antigens, and vaccines based on polysaccharides, are thought to be potentially safer than traditional vaccines. Unfortunately, these new generation vaccines are often very poorly immunogenic, partially because the components having adjuvant activity were discarded in the purification or synthesis process. Some less or well defined structures of pathogens, such as bacterial cell wall components, unmethylated DNA, or double-stranded (ds) RNA, may activate the host defense system by acting as potent "danger" signals. Elimination of these components leads to poor immunogenicity. The reduced or lack of ability for some new generation vaccines to induce a cell-mediated immune (CMI) response, especially a cytotoxic T lymphocyte response (CTL), is another limitation. For example, a recombinant protein-based vaccine elicits mainly humoral immune response (i.e., IgG and IgE production) if administered without proper adjuvants. The polysaccharide vaccine, without conjugation to an appropriate carrier protein, generates only a T cell independent (TI) response by the production of IgM (Lesinski and Westerink, 2001). Importantly, it is currently believed that CMI, especially the CTL response, is as critical as neutralizing antibodies for the effective control of intracellular pathogens such as HIV, hepatitis, tuberculosis, and malaria. In case of tumor vaccines, the ability to induce a tumor-specific CTL response is thought to be indispensable for them to be effective. Therefore, the discovery in the early 1990s that a vaccine based on plasmid DNA can induce both humoral and cellular immune responses produced great excitement in the vaccine and immunology community.

Many different terms, such as genetic vaccine, polynucleotide vaccine, and nucleic acid vaccine, have been used to name the DNA vaccine. In 1994 the WHO chose the term nucleic acid vaccine, subtermed into the DNA vaccine and RNA vaccine. The DNA vaccine was based on the finding that administration of recombinant plasmid DNA into an animal resulted in the expression of a foreign protein encoded by the plasmid (Wolff *et al.*, 1990). Soon after this initial finding, Tang *et al.* (1992) for the first time demonstrated the elicitation of an immune response against a foreign protein by introducing a plasmid encoding the interesting antigen protein directly into mouse skin with a gene gun. Then, almost simultaneously, Ulmer *et al.* (1993) and Fynan *et al.* (1993) showed that immunization with plasmid DNA could protect mice against a lethal influenza challenge. Subsequently, thousands of papers have been published demonstrating that the DNA vaccine is potentially effective for a wide variety of diseases, including infectious diseases, cancers, autoimmune diseases, and allergic diseases.

II. COMPOSITION OF DNA VACCINE

To generate a DNA vaccine, the interesting antigen-encoding gene is inserted into a bacterial plasmid under the control of an appropriate eukaryotic promoter (e.g., the CMV promoter from cytomegalovirus in most cases). Due to the difference in codon usage preference between bacteria and eukaryotic cells, the antigen gene is often modified by point mutation to improve the efficiency of gene expression. Purified and detoxified plasmid DNA from bacteria is then administered into the host animal. From those plasmids that were picked up by appropriate cells and made their way into nuclei, the host cell will use its own gene transcription and protein expression machines to produce the interesting antigen. The host regards the expressed antigen as foreign and will then mount an immune response against it.

In addition, appropriate numbers of unmethylated CpG motifs with the right flanking sequences are usually engineered into the backbone of the plasmid. As explained later, the CpG motif is immunostimulatory and induces the production of Th1 type cytokines (INF-γ, TNF-α, IL-12, etc) and the upregulation of costimulatory molecules such as CD80 and CD86 on antigen-presenting cells (APCs) (Akira et al., 2001; Shimada et al., 1985; Tokunaga et al., 1984; Yamamoto et al., 1988).

III. ADVANTAGES AND DISADVANTAGES OF DNA VACCINE

In addition to its ability to elicit both humoral and cellular immune responses, the DNA vaccine is thought to be potentially safer than the traditional vaccine. It is relatively more stable and potentially more cost-effective for manufacture and storage. Also, multiple antigens may be combined into one plasmid to target multiple pathogens or multiple components of a single pathogen. In addition, the unmethylated CpG motifs on bacterial plasmid DNA in the context of flanking sequences have proven to be immunostimulatory by acting as a pathogen associated molecular pattern (PAMP) molecule and interacting with the Toll-like receptor 9 (TLR9) (Akira et al., 2001). However, the DNA vaccine does have limitations. Specifically, the DNA vaccine tends to be relatively poorly immunogenic, often requiring a large dose to be effective. Since its discovery in the early 1990s, the DNA vaccine has mainly been administered by either intramuscular (im) injection of naked plasmid DNA or by gene gun-mediated administration (i.e., ballistic penetration of pDNA adsorbed on gold beads into skin). Intramuscular injection has proven to be very effective in small animal models. However, the effectiveness of the DNA vaccine in non-human primates and humans in recent studies has not been encouraging, especially in eliciting an antibody response (Calarota et al., 2001; Conry et al., 2002; Klencke et al., 2002; Le et al., 2000; Mincheff et al., 2000; Rosenberg et al., 2003; Tacket et al., 1999; Tagawa et al., 2003; Timmerman et al., 2002; Wang et al., 1998; Weber et al., 2001). Gene gun-mediated administration resulted in a better immune response than an intramuscular injection in clinical trials (Roy et al., 2000). However, the administration of gold beads into humans might be problematic in long term.

IV. MECHANISMS OF IMMUNE INDUCTION FROM DNA VACCINE

As stated by Hilleman (1998) a vaccinologist's knowledge about immunology should be essential and simple. In general, immune responses include antibody production (humoral immune response), CTL response, and cytokine-mediated type 1 (Th1) and type 2 (Th2) T helper responses. In addition, APCs and B lymphocytes (known as detector cells), CD8$^+$ toxic T cells and B lymphocytes (known as effectors), and CD4$^+$ T cells (known as facilitators) determine the type of immune response.

It is now known that in order to successfully induce a primary immune response, professional APCs are required. This is because only the professional APCs can provide both the first signal and the secondary signal required for successful antigen presentation. Almost all cells (can) express MHC class I molecules, but only the professional APCs express or can be induced to express the secondary signal molecules such as CD80 and CD86. An interaction between a peptide epitope-loaded MHC molecule and the T cell receptor (TCR) without the appropriate secondary signal will lead to anergy. Dendritic cells (DCs) are the most potent antigen-presenting cells (Banchereau and Steinman, 1998). Immature DCs, such as Langerhan's cells (LCs), are extremely well equipped for antigen capture, and the capture of antigens induces maturation and mobilization of DCs (Banchereau and Steinman, 1998). In light of these findings, a successful antigen must make its way to APC, especially to DC to initiate a primary immune response.

The outcome of an immune response is determined partially by how the antigen is presented by the APCs to the T and B cells, exogenously or endogenously. For a live vaccine such as the live viral vaccine, the virus is still infectious to the host cells, including APCs. The relevant antigen is produced by the host cells. Therefore, the antigen is generated inside the cells (endogenously) and is processed by the proteosome apparatus into small peptides, which are then transferred into the endoplasmic reticulum where they can bind to the newly synthesized MHC class I molecules (Harding and Song, 1994). The MHC class I molecule with peptide epitope on its groove is mobilized onto the cell surface where it may be recognized by antigen-specific CD8$^+$ T cells with the appropriate TCR. This endogenous presentation results in a CTL immune response. The fact that a DNA vaccine may elicit a CTL immune response is also due to the endogenous presentation of the encoding antigen. However, for the nonlive vaccine, such as the protein (subunit)-based vaccine, the antigen is taken up by APCs in the intercellular spaces by endocytosis. In this pathway, the antigen is processed inside the endosome and lysosome. Some of the endosomal proteolytic degradation products will occupy the epitope groove of the class II MHC molecule and be transported to the extracellular surface of the APCs.

This class II MHC molecule, together with the epitope, will be available for recognition by CD4$^+$ T cells. The CD4$^+$ T cells may develop into either Th1 type or Th2 type cells. The Th1 cell and its related cytokines [interleukin-2 (IL-2), interferon-γ (IFN-γ), etc.] help initiate CMI, whereas the Th2 helper and its related cytokines (IL-4, IL-10, etc.) direct a humoral immune response (Castellino et al., 1997).

As mentioned earlier, for the DNA vaccine, plasmid DNA has primarily been administered by either im injection or via the gene gun into skin. It was found that gene expression from the administered plasmid was predominantly in the myoblasts after im injection and in the keratinocytes and fibroblasts after gene gun-mediated administration. These patterns of expression raised questions on how the expressed antigen is presented to T cells. Although myoblasts and keratinocytes express MHC class I molecules, they do not express other secondary signal molecules (e.g., CD 80 and CD 86). Therefore, theoretically, presentation of expressed antigen by these nonprofessional cells is more likely to tolerize than stimulate T cells. In addition, experiments carried out by Iwasaki et al. (1997) and Corr et al. (1996) clearly ruled out the possibility that the somatic cells (transfected with plasmid expressing the antigen only) directly presented antigens to the T cells. In the study by Corr et al. (1996), parent→F1 bone marrow chimeras, in which H-2^{bxd} recipient mice received bone marrow that expressed only H-2^b or H-2^d MHC molecules, were injected (im) with naked plasmid DNA encoding the nucleoprotein from the A/PR/8/34 influenza strain, which has epitopes for both H-2D^b and H-2K^d on a single antigen. The resulting CTL responses were restricted to the MHC haplotype of the bone marrow alone and not to the other haplotype expressed by the myocytes of the recipient (Corr et al., 1996). These results showed that dendritic cells and other APCs that are differentiated from bone marrow cells, instead of somatic cells, were responsible for antigen presentation. This may explain why intradermal gene gun injection often induced better immune responses than im injection. The gene gun may target DCs in the viable skin epidermis more effectively and directly, resulting in a very potent immune response. Moreover, significantly less DNA is required to elicit an immune response by the gene gun compared to im needle injection. Within the skin, there is a high population of immature DCs or LCs. In contrast, much fewer DCs exist in the muscle.

Currently, there are three proposed mechanisms of antigen presentation by the DNA vaccine: (1) direct transfection of professional APCs (i.e., DCs) and the presentation of expressed antigen by the professional APCs; (2) direct priming by modified somatic cells (myocytes or keratinocytes); and (3) cross-priming in which plasmid DNA transfects a somatic cell and/or professional APC and the proteins secreted from the transfected cells are taken up by other professional APCs and presented to T cells (Gurunathan *et al.*, 2000; Liu, 2003; Takashima and Morita, 1999).

The first mechanism of the direct transfection of professional APCs and the presentation of the self-expressed antigen by the APCs is easily understandable and has been well documented. It was reported that as few as 500 transfected mouse DCs are sufficient to successfully elicit an immune response (Takashima and Morita, 1999). Many experiments have also demonstrated the existence of cross-priming. In cross-priming, the antigen may be expressed by somatic cells transfected with the plasmid DNA. The antigen protein or peptide will then be picked up by professional APCs and presented to T cells. As aforementioned, the secreted or exogenous protein undergoes endocytosis or phagocytosis to enter the MHC class II antigen processing pathway to stimulate $CD4^+$ T cells. Endogenously produced proteins are processed by the proteosome apparatus and are presented through the MHC class I pathway to stimulate naive $CD8^+$ T cells. Although peptides derived from exogenous sources generally cannot be presented on MHC class I molecules, there are now many examples showing that this does occur if proper adjuvants or delivery systems are used (Falo *et al.*, 1995; Raychaudhuri and Rock, 1998). During cross-priming, the antigen or peptide (both MHC class I and II) generated by somatic cells (myocytes or keratinocytes) can be taken up by professional APCs to prime the T-cell response. A study carried out by Ulmer *et al.* (1996) clearly demonstrated that cross-priming from myocytes to APC happens. In their study, influenza NP-expressing myoblasts ($H-2^k$) were injected intraperitoneally into F1 hybrid mice ($H-2^{dxk}$) (Ulmer *et al.*, 1996). It was found that NP CTL restricted by the MHC haplotype of both parental strains was induced. $H-2^d$-restricted epitopes must have somehow found their way to move from the myoblasts to the $H-2^d$-expressing APCs. It is, of course, assumed that the very unlikely case of direct transfer of the NP expression plasmid from the myoblast to APC did not happen.

As to direct priming by transfected somatic cells, this mechanism is theoretically very unlikely. However, Agadjanyan *et al.* (1999) pointed out that when mice were vaccinated with DNA encoding both antigen and CD86, transfected muscle cells can prime an antigen-specific CTL response (Agadjanyan *et al.*, 1999). In this case, the myoblast expresses both HIV-1 envelope protein and CD86. It is interesting to find that CD86 alone can provide the necessary secondary signal. Another study showed that the transfected fibroblasts are able to induce an antigen-specific MHC class I-restricted response if they are physically relocated to secondary lymphoid tissue (Kundig *et al.*, 1995). In lymphoid tissues, a secondary signal may be provided by other cells to the fibroblasts. Thus, nonprofessional cells may be able to prime an immune response when the appropriate secondary signal is provided.

V. THE IMMUNOSTIMULATORY ACTIVITY OF CpG MOTIF

As mentioned earlier, one of the advantages of bacterial plasmid DNA vaccine is that it has a built-in powerful adjuvant, the unmethylated CpG motif. Bacterial DNA was unknowingly used experimentally as a component of an adjuvant over 60 years when Freund used the whole mycobacterial extract as a major constituent in his adjuvant formula based on empirical results (Freund, 1951). It has been shown that the mycobacterial DNA in the Freund's adjuvant contributed to the adjuvant effect (Shimada et al., 1985; Tokunaga et al., 1984; Yamamoto et al., 1988). Yamamoto et al. (1988) found that a purified nucleic acid fraction from the BCG vaccine has a limited antitumor activity that appears to be mediated through its ability to activate NK cells and to induce the production of interferon. Treatment of this fraction with DNase substantially reduced this activity (Tokunaga et al., 1999). This confirmed the immunostimulatory property of mycobacterial DNA. It is now known that the unmethylated CpG motif with its flanking sequences, even in the form of oligonucleotide, is strongly immunostimulatory (Klinman et al., 1999; Krieg et al., 1995).

The unmethylated CpG motif has been found to be a ligand for TLR9 (Hemmi et al., 2000). TLR9 belongs to a group of TLRs. TLRs were identified as major recognition receptors for a pathogen-associated molecular patterns (PAMP) such as lipopolysaccharides (LPS), peptidoglycan, lipoteichoic acid, and CpG-containing oligonucleotides (CpG ODN) (Hemmi et al., 2000; Poltorak et al., 1998). The term Toll was originally referred to as a cell surface receptor governing dorsal/ventral orientation in early Drosophila larvae (Stein et al., 1991). It was later found to also play a crucial role in antifungal defense, together with other antimicrobial peptides (Lemaitre et al., 1996). In the 1990s, the first mammalian protein structurally related to Drosophila Toll was identified and is now called the human Toll-like receptor (Medzhitov et al., 1997). To date, 10 human and 9 murine transmembrane proteins have been shown to belong to the mammalian TLR family (Akira et al., 2001; Zarember and Godowski, 2002). Very recently, a TLR11 responding specifically to uropathogenic bacteria was discovered in mice (Zhang et al., 2004).

Toll and TLR family proteins are characterized by the presence of an extracellular domain with leucine-rich repeats and an intracytoplasmic region containing a Toll/interleukin-1 receptor homology (TIR) domain. Individual mammalian TLR appears to recognize distinct microbial components. For example, LPS (Poltorak et al., 1998), bacterial lipoproteins (sBLP) (Aliprantis et al., 1999) and yeast (Underhill et al., 1999), flagellin (Hayashi et al., 2001), dsRNA (Alexopoulou et al., 2001), small antiviral compounds (Hemmi et al., 2000), and bacterial DNA (CpG-DNA) (Hemmi et al., 2000) engage TLR4, TLR2, TLR5, TLR3, TLR7, and TLR9, respectively.

Different TLRs can exert distinct, but overlapping, sets of biological effects, and increasing evidence indicates that this can be attributed to both common and unique aspects of the signaling mechanisms. All functional studies characterize the TLR signal via the TLR/IL-1 receptor pathway because the presence of the TIR domain can interact with the adaptor protein MyD88. The adaptor molecule MyD88 is recruited to the receptor complex, followed by engagement of the IL-1R-associated kinases (IRAK). IRAK1 and IRAK4 are serine-threonine kinases involved in the phosphorylation and activation of tumor necrosis factor (TNF) receptor-associated factor 6 (TRAF6) (Medzhitov *et al.*, 1998; Muzio *et al.*, 1998; Wesche *et al.*, 1997). In contrast, IRAK-M lacks kinase activity and regulates TLR signaling negatively by preventing the dissociation of phosphorylated IRAK1 and IRAK4 from MyD88. Phosphorylated TRAF6 leads to the activation of downstream kinases, such as the stress kinase JNK1/2 and the IκB kinase (IKK) complex (Baud *et al.*, 1999). This event frees nuclear factor-κB (NF-κB) from IκB and allows its nuclear translocation and the subsequent transcriptional activation of many proinflammatory genes, which encode cytokines, chemokines, adhesion molecules, and immune receptors. All of these molecules are involved in engaging and controlling the innate immune response and in orchestrating the transition to an adaptive immune response (Medzhitov, 2001).

TLR signaling stimulates the maturation of DCs, which migrate to the lymph nodes where they stimulate T cells by the presentation of MHC complexes. Antigen presentation alone can stimulate pathogen-specific T-cell clones, but is not sufficient to trigger efficient T-cell expansion. Clonal T-cell expansion requires an additional signal delivered by costimulatory molecules such as CD80/86. TLR signaling functions to trigger adaptive immunity by enhancing the expression of not only MHC molecules, but also of these costimulatory molecules in DCs (Kaisho and Akira, 2001).

Stimulation by bacterial CpG motifs generally leads to the production of Th1-type cytokines and to the upregulation of costimulatory molecules on lymphocytes and APCs. The interaction of CpG motifs with TLR9 was further confirmed by the fact that in TLR9 knockout mice, all CpG DNA-induced effects, including cytokine production, B-cell proliferation, and DC maturation, were completely abolished (Hemmi *et al.*, 2000). In fact, the immunostimulatory effect of CpG motif is so strong that CpG motifs containing oligonucleotides are now being used as a vaccine adjuvant. In tumor vaccine development, the CpG motif is an adjuvant of special interest, as generally, the CpG motif can skew the immune response to be Th1 biased, which favors the induction of CTL for tumor killing. However, caution needs to be applied in the repeated administration of CpG motifs. Heikenwalder *et al.* (2004) reported that repeated CpG oligodeoxynucleotide administration led to lymphoid follicle destruction and immunosuppression (Heikenwalder *et al.*, 2004). They reported that daily intraperitoneal injection of 60 μg CpG-ODN dramatically alters the morphology and functionality of mouse lymphoid organs. By day 7, lymphoid follicles were poorly defined; follicular dendritic cells and germinal center B lymphocytes were suppressed. Accordingly, CpG-ODN treatment for more than 7 days strongly reduced the primary humoral immune response and immunoglobulin class switching. By day 20, mice developed multifocal liver necrosis and hemorrhagic ascites. Of course, it is very unlikely for anybody to develop a vaccine that contains so many CpG motifs and to administer it daily for up to 7–20 days intraperitoneally. The finding is mentioned here simply to point out that high doses of CpG motifs could be potentially toxic.

VI. ROUTE OF ADMINISTRATION

The DNA vaccine was originally administered by an intramuscular injection or via a gene gun to the skin. These two have continued to be the main means/ routes of administration. Meanwhile, many other routes have been tried. These include subcutaneous injection, intradermal injection, noninvasive topical application onto the skin (Tang et al., 1997), intranasal and intravaginal applications (Livingston et al., 1998; Park et al., 2003), and vaccination via oral mucosa or gastrointestinal (GI) mucosa (oral). Depending on the route, different immune responses have been elicited. For example, due to the fact that most of the pathogens enter the host through either the mucosal surface or the skin, an immune response that can neutralize or kill the pathogens on the mucosal surface prior to their entrance is therefore desired. Also, due to the common mucosal system, dosing of vaccine from one mucosal site (e.g., nasal) may lead to a mucosal response on another mucosal site (e.g., vaginal) (McGhee et al., 1992). To achieve a mucosal immune response, the DNA vaccine has been applied on the nasal, oral, GI, or vaginal mucosal surface with specially designed devices or delivery systems. The following paragraphs provide a few examples for administration of the DNA vaccine by mucosal routes.

The oral buccal mucosa is covered by a network of DCs analogous to LCs in the skin. LCs represent the major APCs in human buccal mucosa epithelium. In addition, a high density of T cells and mucosal-associated lymphoid tissue (MALT) are present in the buccal mucosa. Therefore, immunization via the buccal mucosa may represent an attractive mucosal immunization approach. In fact, the efficacy of oral buccal mucosal DNA immunization has been confirmed by many using transepithelial needle injection, jet injection, or gene gun. For example, Lundholm et al. (1999) found that jet injection of plasmid DNA into the oral cheek of mice induced a very strong IgA mucosal response specific to the encoding HIV-1 proteins (gp160, p24, and TAT) (Lundholm et al., 1999). To avoid the inconvenience posed by those means of

DNA administration, Cui and Mumper (2002a) developed a buccal mucoadhesive film based on polymers Noveon and Eudragit S-100. When plasmid DNA (encoding β-galactosidase protein as a model antigen)-loaded film was applied to rabbit buccal mucosa, a serum-specific IgG level comparable to that induced by a subcutaneous injection of aluminum hydroxide (Alum) adjuvanted β-galactosidase protein was induced (Cui and Mumper, 2002a). In addition, strong serum IgA and splenocyte proliferative responses were induced. Although data are promising, application of vaccine on oral buccal mucosa has the limitation that tolerance instead of immunity might be very often induced. The buccal mucosa is in contact with many foreign antigens daily. However, rarely is an immune response induced.

Oral administration of vaccines is desired from both an immunological aspect and a patient compliance point of view. Oral administration of vaccines is convenient for patients and may make large population immunization more feasible. Moreover, oral administration of vaccines has been shown to induce both systemic and mucosal immune responses (Russell-Jones, 2000). The intestinal mucosa is rich in DCs and other MALT. However, the polio vaccine is still the only marketed vaccine administered orally, perhaps illustrating the difficulty in developing an effective oral vaccine. The extensive enzymatic systems and harsh physical and chemical environments in the GI tract make it difficult to develop an active oral vaccine. Therefore, a DNA vaccine for oral administration (at experimental level) is often encapsulated into particles. Encapsulation of DNA inside microparticles may provide protection to DNA.

Jones et al. (1997) reported the first oral DNA vaccine by encapsulating a plasmid-expressing insect luciferase protein into PLGA microparticles and administering it to mice by oral gavage (Jones et al., 1997). The authors observed good serum IgG, IgA, and IgM antibody responses and, most importantly, a significant level of mucosal IgA in saliva and stool samples. In contrast, unencapsulated DNA gave undetectable responses.

Soon after, Herrmann's group used this method to orally immunize mice with more relevant antigens, such as the capsid proteins of rotavirus, VP4, VP6, and VP7 (Chen et al., 1997, 1998; Herrmann et al., 1996, 1999). Using rotavirus VP6 DNA vaccine-encapsulated PLGA microparticles, Chen et al. (1998) reported that one dose of vaccine given to BALB/c mice elicited both rotavirus-specific serum antibodies and intestinal IgA. Moreover, after challenge with homologous rotaviruses, virus shedding was reduced significantly compared to control mice, which were immunized with PLGA microparticles encapsulated with VP6-free plasmid (Chen et al., 1998). Similar results were observed when the VP6 gene was replaced with VP4 and/or VP7 (Herrmann et al., 1999). These studies represented the first demonstration of protection against an infectious agent after oral administration of a DNA vaccine.

VII. IMMUNOLOGY OF THE IMMUNE RESPONSES FROM DNA VACCINE

Many studies have demonstrated the preclinical efficacy of the DNA vaccine in disease models of infectious disease, cancer, allergy, and autoimmune disease. Interested readers may refer to the supplemental tables tabulated by Gurunathan *et al.* (2000) for a comprehensive list of allergy, autoimmune disease, bacterial infection, and tumor models for which DNA vaccines have been attempted. Many more have been added since 2000. CTL, antibody, and different types of T-cell helper responses have been generated depending on the disease, antigen, animal, and route of administration.

A. Humoral response

Immunization with plasmid DNA can induce antibody responses in a variety of proteins in animal species, particularly in mouse. Moreover, the humoral response generated by DNA vaccination has been shown to be protective in several animal models *in vivo*. However, as detailed later, the antibody response in humans from the DNA vaccine has not been encouraging. The antibody response from the DNA vaccine is weak at the beginning then peaks and reaches a plateau between 1 and 3 months after a single DNA immunization in mice (Deck *et al.*, 1997). Furthermore, within a certain dose range, antibody production is generally increased in a dose-responsive manner with either a single injection or multiple injections of DNA by various routes of immunization (Deck *et al.*, 1997). The resulting antibody response can be long lived (e.g., significant serum levels were present up to 1.5 years after vaccination) (Deck *et al.*, 1997; Raz *et al.*, 1994). Unfortunately, when antibody responses from DNA, protein, and live virus vaccines were compared, the response from the DNA vaccine is generally weaker than that from protein-based vaccine and that from the liver virus vaccine. For example, in a study comparing the antibody responses to DNA encoding the hemagglutinin (HA) antigen and live influenza infection, the antibody titer in mice vaccinated with live influenza was significantly higher than that in DNA-vaccinated mice (Deck *et al.*, 1997). In comparing the antibody responses elicited by vaccination with DNA encoding a malarial surface protein and the protein itself, both antibody titers and avidity were significantly lower in mice vaccinated with DNA than in those vaccinated with the protein (Kang *et al.*, 1998). In contrast, in one study directly comparing the kinetics of an antibody response after vaccination with DNA-encoding ovalbumin (OVA) and OVA protein, there did not appear to be a difference in total OVA-specific antibody production at 2 or 4 weeks postvaccination when DNA was administered intradermally (Boyle *et al.*, 1997). In this study, antibody induced by DNA had a higher avidity than that induced by protein. It is worthwhile to point out that many parameters can affect the results from these comparisons. First of all, it is difficult to choose the doses of DNA and protein to compare. Therefore, one has to compare the highest antibody response from the DNA vaccine with the highest antibody response from other vaccines. Second, it is well known that when a protein is injected alone, usually no or very weak immune responses will be induced. Thus, an adjuvant such as Alum is often used. In the case of the DNA vaccine, intramuscular injection of "naked" DNA alone can lead to an antibody response. Third, the adjuvant or delivery system makes a difference in terms of the resulting responses. For example, previous studies found that the antibody response from the DNA vaccine was weaker than that from a protein-based vaccine adjuvanted with Alum when the DNA vaccine is administered as a "naked" unformulated plasmid. However, when the plasmid is formulated with nanoparticles as the carrier, the resulting antibody response was then stronger than that from the same protein-based vaccine (Cui and Mumper, 2002b). Finally, the route of administration has a significant effect. As mentioned earlier, when plasmid DNA was applied to rabbit buccal mucosa via the mucoadhensive films we developed, a strong antibody response was observed. However, when the protein antigen encoded by the plasmid was applied to rabbits in the same route with the same device, no detectable antibody response was induced (Cui and Mumper, 2002a). Therefore, caution should be applied when carrying out these comparisons. However, in general, especially in clinical trials, the antibody response from the DNA vaccine is not encouraging.

DNA vaccination induces the production of many subtypes of antibodies, including IgG, IgM, and IgA. Moreover, in most cases, antibodies generated by DNA vaccines are skewed toward IgG2a due to the fact that the CpG motifs on plasmid DNA stimulate production of the Th1 cytokine. However, exceptions do exist because DNA vaccination with a gene gun on skin was preferentially bias toward IgG1 production (Feltquate *et al.*, 1997). One reason to explain this is that when the DNA vaccine is administrated via the gene gun on skin, the DNA dose is much lower than in the case of intramuscular injection. Therefore, there may not be a sufficient amount of CpG motifs being administered. This has yet to be confirmed.

Finally, regarding the memory humoral response, it has been shown that mice vaccinated with DNA encoding an influenza viral HA antigen had levels of anti-HA antibodies comparable to or greater than those from convalescent sera of previously infected mice that persisted over 1 year (Martins *et al.*, 1995; Torres *et al.*, 1997). However, in other studies, plasmid DNA encoding a nucleoprotein of the LCMV virus administered intramuscularly failed to give appreciable antibody responses before viral challenge (Deck *et al.*, 1997). Thus, depending on the type of antigen used, DNA vaccination may be effective at inducing a long-term antibody response in some animal species.

B. Cellular immune response

As mentioned earlier, one of the advantages of the DNA vaccine is that it can induce cellular immune responses, including a CTL response, while it does not have the safety concern posed by other replicable vaccines, such as live virus. Both the $CD4^+$ T-cell response and the $CD8^+$ T-cell response (i.e., CTL) from DNA vaccination are discussed here.

Functionally, $CD4^+$ T cells may be divided into Th1 and Th2. Some researchers proposed a Th0 cell population, which has the function of both Th1 and Th2 cells. However, others believe that Th0 cells may just be the physical mixture of these two. Th1 cells produce IFN-γ exclusively, whereas Th2 cells produce IL-4, IL-5, and IL-13 exclusively. Also, the presence of Th1 cytokines facilitates the differentiation of $CD4^+$ T cells toward a Th1 phenotype and prevents the development of Th2, whereas the presence of Th2 cytokines allows for Th2 differentiation and prevents the differentiation of Th1 T cells. $CD4^+$ T cells can mediate at least three major functions. (1) Activated $CD4^+$ T cells promote B-cell survival and antibody production (Banchereau et al., 1994). (2) $CD4^+$ T cells, through production of IL-2 and/or through CD40L-CD40 costimulation, provide helper function to $CD8^+$ T cells (Bennett et al., 1998; Ridge et al., 1998; Schoenberger et al., 1998). (3) $CD4^+$ T cells secrete a variety of cytokines that regulate the resulting immune response as mentioned previously.

Because the CpG motif in bacterial DNA induces the production of a variety of proinflammatory cytokines, including IL-12, TNF-α, and INF-γ, it is understandable that the DNA vaccine generally skews the response toward Th1. Scientists have taken advantage of this property of the DNA vaccine to develop vaccines for tumor, which needs a strong Th1, especially CTL response to kill. Also, the DNA vaccine has been used to treat allergy, which is associated with a high-level production of the IgE antibody and Th2 cytokines. For example, Roy et al. (1999) developed a chitosan-encapsulated DNA vaccine for peanut allergy. Mice receiving chitosan-encapsulated DNA containing a dominant peanut allergen gene (pCMVArah2) produced secretory IgA and serum IgG2a (Roy et al., 1999). Compared to nonimmunized mice or mice treated with "naked" DNA, mice immunized with chitosan-encapsulated DNA showed a substantial reduction in allergen-induced anaphylaxis associated with reduced levels of IgE, plasma histamine, and vascular leakage (Roy et al., 1999).

The DNA vaccine is able to generate antigens endogenously, making them accessible to $CD8^+$ T cells via an MHC class I pathway. Numerous publications have demonstrated the induction of CTL response by DNA vaccines. Although the CTL response can also be induced by a live vaccine, it is difficult to induce with a protein-based vaccine. Of course, with a proper delivery system such as particles, a protein-based vaccine occasionally induces a CTL response (Falo et al., 1995). Depending on the disease models used, the magnitude of the CTL response from the DNA vaccine is, in most cases, comparable to that from a live viral vaccine. Also, the DNA vaccine can induce a CTL response against both dominant and subdominant epitopes. This may be useful for the development of a DNA vaccine for tumor immunotherapy. In their development process, tumor cells often become tolerated to the CTL response against the dominant epitopes of tumor-specific or tumor-associated antigens. Therefore, successful induction of CTL responses to the subdominant epitopes on these antigens should be helpful for tumor killing.

As to the memory cellular immune response, it has been shown that the frequency of antigen-specific CD4[+] T cells measured by proliferation remained elevated for 40 weeks postvaccination. Also, Raz et al. (1994) reported that a single intradermal needle injection of 0.3–15 μg of naked plasmid DNA induced an anti-influenza nucleoprotein-specific antibody and CTL that persisted for at least 68–70 weeks after vaccination (Raz et al., 1994). In a separate study, Davis et al. (1995) reported that a CTL response from the DNA vaccine to hepatitis B virus envelope proteins could be detected 4 months postvaccination. Also, Chen et al. (1999) showed that CTL activity specific for both dominant and subdominant epitopes on the Sendai virus nucleoprotein gene could be recalled readily 1 year after DNA vaccination and that the frequencies of CTL precursors specific for both of these epitopes were relatively high (Chen et al., 1999). Chattergoon et al. (2004) reported that coadministration of an IL-12 encoding plasmid pIL-12 with plasmid encoding HIV gp160 and influenza A hemagglutinin not only significantly enhanced the resulting CTL response, but also led to a more persistent CTL response.

In summary, depending on the route, antigen, species, and so on, a DNA vaccine may induce a very comprehensive and potent immune response.

VIII. CLINICAL TRIALS OF DNA VACCINES

The success of the DNA vaccine in a variety of small animal models propelled it into a number of human clinical trials for diseases or pathogens, including HIV (Boyer et al., 2000; MacGregor et al., 1998, 2000; Ugen et al., 1998; Weber et al., 2001), malaria (Epstein et al., 2004; Le et al., 2000; McConkey et al., 2003; Moorthy et al., 2003; Wang et al., 1998, 2004a), hepatitis B (Roy et al., 2000; Tacket et al., 1999), human papillomavirus (HPV) 16-associated anal dysplasia (Klencke et al., 2002), and many other cancers (Mincheff et al., 2000; Rosenberg et al., 2003; Tagawa et al., 2003; Timmerman et al., 2002). The selection of disease models for human clinical trials clearly showed that researchers were taking advantage of the ability of the DNA vaccine to induce cellular immune responses. As mentioned earlier, for a vaccine to be effective against those

Table 11.1. A Summary of Recent Human Trials with DNA Vaccines

Antigen	Pathogen or disease	Subjects	Route	Safety	Immune response	Clinical response	Reference
Env, Rev	HIV	15 asymptomatic HIV patients	im	Safe	Ab, CTL, T proliferative, IFN-γ, β-chemokine release	N/A	Boyle et al. (1997); MacGregor et al. (1998, 2000, 2002); Ugen et al. (1998)
gp120, gp160	HIV	4 asymptomatic HIV patients	im	Safe	No Ab response, Cellular response not reported	N/A	Weber et al. (2001)
Nef, Rev, Tat	HIV	9 asymptomatic patients	im	Safe	CTL	N/A	Calarota et al. (2001)
PfCSP	Malaria	20 healthy people	im	Few mild reaction	No Ab, CTL positive	N/A	Le et al. (2000); Wang et al. (1998)
PfCSP then RTS/AS02A	Malaria	24 healthy	im	Safe, well tolerated	Ab, CTL, Th	N/A	Epstein et al. (2004); Wang et al. (2004a)
TRAP and MVA	Malaria	63 people	im id	Safe	CTL, IFN-γ, Th1	Partial protection	McConkey et al. (2003); Moorthy et al. (2003)
HBsAg	HBV	7 healthy subjects	Powder-Ject	Well tolerated	No response	N/A	Tacket et al. (1999)

Antigen	Disease	Patients	Delivery	Safety	Immune response	Clinical response	Reference
HBsAg	HBV	12 healthy subjects	Powder-Ject	Safe, well tolerated	Ab in 12/12 subjects, CTL, Th	N/A	Roy et al. (2000)
HPV 16 E7	HPV	12 anal dysplasia patients	im	Well tolerated	10/12 CTL	3/12 partial histological response	Klencke et al. (2002)
PSMA with CD86	Prostate cancer	26 prostate cancer patients	im	Safe	Partially shown signs of immunization	Cannot conclude	Mincheff et al. (2000)
CEA, HBsAg	Colorectal carcinoma	17 patients	im	Grade I injection toxicity	0/17 anti-CEA, 4/17 proliferative response to CEA, 6/8 anti-HBsAg	No	Conry et al. (2002)
Id linked to MsIg	Lymphoma	12 patients	im id	Good	6/12 humoral and T-cell responses	No	Timmerman et al. (2002)
gp100	Melanoma	22 patients	im id	Good	No response	No	Rosenberg et al. (2003)
Tyrosinase	Melanoma	26 stage IV patients	Intra-nodal infusion	Well tolerated	11/26 tetramer response	No but longer survival time	Tagawa et al. (2003)

diseases or pathogens, it is believed that a strong specific CTL response, as well as antibody response, is required. In general, the DNA vaccine was found to be safe in humans. It was well tolerated. Only very weak and minor local side effects were observed in a small portion of the participants. Depending on different routes of administration, humoral and cellular immune responses can be generated. However, the immune responses, especially antibody response, have not been encouraging in most of the trials. The following findings are from some reported DNA vaccine clinical trials. Table 11.1 provides a brief summary of some clinical trial data.

A. HIV vaccine

The first human trial of the DNA vaccine for the treatment of HIV infection was reported by MacGregor *et al.* (1998). In their study, a DNA-based vaccine encoding HIV-1 env and rev genes was tested for safety and host immune response in 15 asymptomatic HIV-infected patients. Successive groups received three doses of vaccine intramuscularly (30, 100, or 300 μg) at 10-week intervals in a dose-escalation trial. The vaccine induced no local or systemic reactions, and no laboratory abnormalities were detected. Specifically, no patient developed anti-DNA antibody or showed muscle enzyme elevations. No consistent change occurred in CD4 or CD8 lymphocyte counts or in plasma HIV concentration. Antibody against gp120 increased in individual patients in the 100- and 300-μg groups. Some increases were noted in CTL activity against gp160-bearing targets and in lymphocyte proliferative activity. For the 100-μg treatment, ELISA showed that binding of antisera from the subjects to gp120 was increased significantly after vaccination. When the binding of pre- and postvaccination antisera against a V3 loop peptide derived from gp130MN was measured, an enhancement of binding (at 1:4500 dilution) of 47.6% 8-week postvaccination was observed. These data demonstrated that the DNA vaccine boosted the antibody response in humans. In case of cellular immune responses, some changes were noted in CTL activity against gp160-bearing targets. Enhanced specific lymphocyte proliferative activity against the HIV-1 envelope was observed in multiple patients. The majority of patients who exhibited an increase in any immune parameters were within the 300-μg dose group, the group with the highest dose. In at least one of multiple assays, the six subjects who received the 300-μg dose had DNA vaccine-induced antigen-specific lymphocyte proliferative responses and antigen-specific production of both IFN-γ and β-chemokine. Furthermore, four of five subjects in the 300-μg dose group responded to both rev and env components of the vaccine. The responses did not persist within inoculated individuals and scored in different individuals at different times in the trial (Boyer *et al.*, 2000; MacGregor *et al.*, 1998, 2000, 2002; Ugen *et al.*, 1998). Taken together, these studies show that the safety profile for the DNA vaccine is excellent. Both humoral and cellular immune responses, including CTL, are inducible in humans. Also, the vaccine can stimulate multiple immune responses in vaccine-naive subjects when multiple antigens were encoded in the plasmid DNA.

In addition to the work by Weiner and colleagues, Weber *et al.* (2001) reported a phase I study of a HIV-1 gp160 DNA vaccine. Asymptomatic HIV-1-infected subjects with CD4$^+$ lymphocyte counts >500/μl were injected intramuscularly four times with 400 μg of HIV-1-modified gp160 env and rev coding DNA vaccine at 0, 4, 10, and 28 weeks. The DNA vaccine was safe and did not induce anti-DNA autoimmune antibodies. Vaccination had no long-term effects on the CD4$^+$/CD8$^+$ lymphocyte counts, plasma HIV-1 RNA concentrations, or disease progression. However, anti-gp120 and anti-gp160 antibody titers did not change significantly over a time period of 28 weeks and did not increase in response to vaccination compared to the baseline value. In other words, this DNA vaccination did not cause any significant antibody response (Weber *et al.*, 2001). Calarota *et al.* (2001), however, reported that vaccination with a gene combination raised a broad HIV-specific CTL response (Calarota *et al.*, 2001). In their trial, the efficacy of a combination of DNA plasmids encoding the nef, rev, and tat HIV-1 regulatory genes in inducing cellular immune responses was analyzed in asymptomatic HIV-1-infected patients. Patients initially selected for having low or no detectable immune responses to Nef, Rev, or Tat antigens developed MHC class I-restricted cytolytic activities as well as enhanced bystander effects. The most remarkable change observed after immunization with the gene combination was an increase in CTL precursors to target T cells infected with the whole HIV-1 genome (Calarota *et al.*, 2001). An *in vitro* assessment of the expression of single and combined gene products showed that this was consistent with the induction of CTL responses *in vivo*.

B. Malaria vaccine

Malaria is an increasingly uncontrolled public health problem; 1–3 million people die annually from *Plasmodium falciparum* infection. A preventative vaccine is likely to be among the most effective means for its control. In 1998, Wang and colleagues reported the induction of antigen-specific CTL responses in humans by a malaria DNA vaccine. In their study, 20 healthy adult volunteers were enrolled in a phase I safety and tolerability clinical study of a DNA vaccine encoding a malaria antigen. Volunteers received three intramuscular injections of one of four different dosages (20, 100, 500, and 2500 μg) of the *P. falciparum* circumsporozoite protein (PfCSP) encoding plasmid DNA at monthly intervals and were followed for up to 12 months. Local reactogenicity and systemic symptoms were few and mild. There were no severe or serious adverse events, clinically significant biochemical or hematologic changes, or detectable anti-DNA antibodies. The volunteers developed antigen-specific, genetically restricted, CD8$^+$ T-cell-dependent CTLs. Responses were directed against all 10 peptides tested and were restricted by six human lymphocyte antigen (HLA) class I alleles. This was the first demonstration in healthy naïve humans of the induction of CD8$^+$ CTLs by a DNA vaccine, including CTLs, that were restricted by multiple HLA alleles in the same individual. However, very disappointingly, despite the induction of excellent CTL responses, the DNA vaccination failed to induce detectable antigen-specific antibodies in any of the volunteers (Le et al., 2000; Wang et al., 1998). More recently, this group reported that patients who received the PfCSP DNA vaccine followed by a recombinant protein vaccine have antibody and CD8$^+$ and CD4$^+$ T-cell responses, suggesting that this heterologous prime-boosting approach might be viable (Epstein et al., 2004; Wang et al., 2004a). Hill and colleagues at the University of Oxford approached this by alternative dosing with a DNA vaccine and recombinant modified vaccinia virus Ankara (MVA) (McConkey et al., 2003; Moorthy et al., 2003). They showed that a heterologous prime-boost vaccination regime of DNA either intramuscularly or epidermally, followed by intradermal recombinant MVA, induces high frequencies of IFN-γ-secreting, antigen-specific T-cell responses in humans to a preerythrocytic malaria antigen, thrombospondin-related adhesion protein (TRAP). These responses are 5- to 10-fold higher than the T-cell responses induced by the DNA vaccine or recombinant MVA vaccine alone and produce partial protection manifest as delayed parasitemia after sporozoite challenge with a different strain of P. falciparum. DNA ME-TRAP and MVA ME-TRAP are safe and immunogenic for effector and memory T-cell induction. MVA ME-TRAP, with or without prior DNA ME-TRAP immunization, was more immunogenic and more cross-reactive in malaria-exposed individuals than in malaria-naive individuals. Both CD4$^+$ and CD8$^+$ T cells were induced by these vaccines (McConkey et al., 2003; Moorthy et al., 2003).

C. Hepatitis B vaccine

Tacket et al. (1999) reported a HBV DNA vaccine human trial in 1998. The study was designed to determine the safety and immunogenicity of a DNA vaccine consisting of a plasmid-encoding hepatitis B surface antigen (HBsAg) delivered by the PowderJect XR1 gene delivery system into human skin. Seven healthy adult volunteers received two immunizations on days 0 and 56. The vaccine was well tolerated. However, only one out of six seronegative volunteers developed high titers of persistent anti-HBsAg Ab after a single immunization (Tacket et al., 1999). They reasoned that the lack of immune response might be due to the extremely low DNA dose (0.25 μg) used.

A similar trial carried out by the PowderJect Vaccines Inc. produced quite different and encouraging results. Roy et al. (2000) reported an induction of antigen-specific CD8$^+$ T cells, T helper cells, and protective levels of antibody in humans by particle-mediated administration of a hepatitis B virus DNA vaccine into the skin. The needle-free PowderJect system was used to deliver gold particles coated with DNA directly into cells of the skin of 12 healthy, hepatitis-naive human volunteers. Three groups of four volunteers received three administrations of DNA encoding the surface antigen of HBV at one of the three dose levels (1, 2, or 4 μg). The vaccine was safe and well tolerated, causing only transient and mild to moderate responses at the site of administration. All the volunteers developed protective antibody responses of at least 10 mIU/ml. In volunteers who were positive for the HLA class I A2 allele, the vaccine also induced antigen-specific CD8$^+$ cells that bound HLA-A2/HBsAg (335–343) tetramers, secreted IFN-γ, and lysed target T cells presenting a HBsAg CTL epitope (Roy et al., 2000). These results demonstrated that the DNA vaccine may induce protective antibody titers in humans, depending on how the DNA vaccine is administrated.

D. HPV 16-associated anal dysplasia

Klencke and colleagues (2002) at UCSF tested the PLGA microsphere-encapsulated DNA vaccine developed by Zycos, Inc. (Lexington, MA) in treating HPV 16-associated anal dysplasia. High-grade dysplasia induced by high-risk types of human papillomavirus (HPV) precedes invasive cancer in anal squamous epithelium just as it does in the cervix. In the trial, each subject was treated with four im injections of 50–400 μg of ZYC101 at 3-week intervals. The plasmid DNA in the ZYC101 encodes for multiple HLA-A2-restricted epitopes derived from the HPV-16 E7 protein, one of two HPV oncoproteins (E6 and E7) consistently expressed in neoplastic cells. Twelve eligible subjects with HPV-16 anal infection and a HLA-A2 haplotype were enrolled in the study. ZYC101 was well tolerated in all subjects at all dose levels tested. Three subjects experienced partial histological responses, including 1 of 3 subjects receiving the 200-μg dose and 2 subjects at the 400-μg dose level. Using a direct Elispot, 10 of 12 subjects demonstrated an increased immune response to the peptide epitopes encoded within ZYC101; each continued to show elevated immune responses 6 months after the initiation of therapy.

E. DNA vaccine for cancer immunotherapy

Because it requires a CTL response to kill tumor cells, the ability to induce a cellular immune response by the DNA vaccine makes it attractive in developing cancer vaccines. Advancement in biochemical and genetic techniques in the last decade has facilitated the discovery of a great number of tumor-specific antigens (TSAs) and tumor-associated antigens (TAAs). In most animal studies, the tumor DNA vaccine has proven to be very effective, although the tumors were grafted artificially in almost all the cases. These successful preclinical data propelled some tumor DNA vaccines into human trials.

In 2000, Mincheff and co-workers published their results from a trial of naked DNA and adenoviral immunizations for immunotherapy of prostate cancer. The prostate-specific membrane antigen (PSMA) was used as the tumor antigen. Immunizations included extracellular human PSMA DNA as well as human CD86 DNA into separate expression vectors (PSMA and CD86 plasmids) and into a combined PSMA/CD86 plasmid. In addition, the PSMA gene was inserted into a replication-deficient adenoviral expression vector. Twenty-six patients with prostate cancer were entered into a phase I/II toxicity-dose escalation study. Immunizations were performed intradermally at weekly intervals. Doses of DNA between 100 and 800 μg and of recombinant virus at 5×10^8 PFUs per application were used. They found no immediate or long-term side effects following immunizations. All patients who received the initial inoculation with the viral vector followed by PSMA plasmid boosts showed signs of immunization, as evidenced by the development of a delayed-type hypersensitivity reaction after the PSMA plasmid injection. In contrast, of the patients who received a PSMA plasmid and CD86 plasmid, only 50% showed signs of successful immunization. Of the patients who received PSMA plasmid and soluble granulocyte–macrophage colony-stimulating factor (GM-CSF), 67% were immunized. However, all patients who received the PSMA/CD86 plasmid and soluble GM-CSF became immunized. Patients who did not immunize during the first round were later successfully immunized after a boost with the viral vector. The heterogeneity of the medical status and the presence of concomitant hormone therapy in many patients did not permit unequivocal interpretation of data with respect to the effectiveness of the therapy. However, several responders, as evidenced by a change in the local disease, distant metastases, and PSA levels, were identified.

In 2002, Conry et al. reported safety and immunogenicity results from a dose-escalation clinical trial of a dual expression plasmid encoding a carcinoembryonic antigen (CEA) and HBsAg in 17 patients with metastatic colorectal carcinoma (Conry et al., 2002). CEA was selected as a prototypic tumor-associated self-antigen, and the HBsAg cDNA was included as a positive control for an immune response to the DNA vaccine without relying on breaking tolerance to a self-antigen. Groups of 3 patients received escalating single im doses of the DNA vaccine at 0.1, 0.3, and 1.0 mg. Subsequent groups of 3 patients received three repetitive 0.3- or 1.0-mg doses at 3-week intervals. A final group of 2 patients received three repetitive 2.0-mg doses at 3-week intervals. Toxicity was limited to transient grade 1 injection site tenderness, fatigue, and creatine kinase elevations, each affecting a minority of patients in a nondose-related manner. Repetitive dosing of the DNA vaccine induced HBsAg antibodies in 6 out of 8 patients, with protective antibody levels achieved in 4 of these patients. Although 4 of 17 patients developed lymphoproliferative responses to CEA after vaccination, CEA-specific antibody responses were not observed in any subject, echoing the results from other human clinical trials mentioned earlier. Also, this study shows that the antibody response is dependent on the antigen used. HBsAg is known to be very antigenic. This is probably why anti-HBsAg Ab was detected in all patients while anti-CEA Ab was not detected in any individual. No objective clinical responses to a DNA vaccine were observed among this population of patients with widely metastatic colorectal carcinoma.

Levy and colleagues reported results from a DNA vaccine encoding a chimeric idiotype in patients with B-cell lymphoma (Timmerman et al., 2002). B-cell lymphomas express tumor-specific immunoglobulin, the variable regions of which [idiotype (Id)] can serve as a target for active immunotherapy. The safety and immunogenicity of naked DNA Id vaccine in 12 patients with follicular B-cell lymphoma were investigated. The DNA encoded a chimeric immunoglobulin molecule containing variable heavy and light chain immunoglobulin sequences derived from each patient's tumor, linked to the IgG2a and κ mouse immunoglobulin (MsIg) heavy and light chain constant region chains, respectively. Patients in remission after chemotherapy received three monthly im injections of the DNA in three dose-escalation cohorts of 4 patients each (200, 600, and 1800 μg). After vaccination, 7 of 12 patients mounted either humoral ($n = 4$) or T-cell-proliferative ($n = 4$) responses to the MsIg component of the vaccine. In one patient, a T-cell response specific to autologous Id was also measurable. A second series of vaccinations were then administered using a needle-free injection device (Biojector) to deliver 1800 μg both im and intradermally (id); 9 of 12 patients had humoral ($n = 6$) and/or T-cell ($n = 4$) responses to MsIg. Six of 12 patients exhibited humoral and/or T-cell anti-Id responses; however, these were cross-reactive with Id proteins from other patient's tumors. Subsequently, a third series of vaccinations were carried out using 500 μg of human granulocyte–macrophage colony-stimulating factor DNA mixed with 1800 μg of Id DNA. The proportion of patients responding to MsIg remained essentially unchanged (8 of 12), although humoral or T-cell responses were boosted in some cases. Throughout the study, no significant side effects or toxicities were observed.

The result from Rosenberg et al. (2003) in NCI on a DNA vaccine encoding gp100 melanoma–melanocyte antigen in patients with metastatic melanoma reported in 2003 was very disappointing. In their study, 22 patients with metastatic melanoma were randomized to receive plasmid DNA either intradermally ($n = 10$) or intramuscularly ($n = 12$). One patient (4.5%)

exhibited a partial response of several subcentimeter cutaneous nodules. All other patients had progressive disease. Of 13 patients with cells available before and after immunization, no patient exhibited evidence of the development of an anti-gp100 cell response using *in vitro* boost assays. They were unable to demonstrate significant immunologic or clinical responses to plasmid DNA encoding the "self" nonmutated gp100 tumor antigen.

The study carried out by Tagawa *et al.* on melanoma DNA vaccine encoding tyrosinase epitopes was more encouraging (Tagawa *et al.*, 2003). Groups of eight stage IV melanoma patients each received 200, 400, or 800 μg of DNA intranodally by pump over 96 h every 14 days for four cycles. Blood was collected for immunologic assays and to measure plasmid in serum prior to treatment and 4 and 8 weeks later. Scans and X-rays were performed at baseline and after 8 weeks. It was found that the treatment was well tolerated, with only five patients demonstrating grade 1–2 toxicity. Vaccination by 96-h infusions of plasmid into a groin lymph node resulted in only one episode of catheter leakage in 107 cannulations. Detection of plasmid in serum was rare and transient in two patients. Immune responses by a peptide-tetramer assay to tyrosinase amino acids 207–216 were detected in 11 of 26 patients. Although clinical responses were not seen, survival of the heavily pretreated patients on this trial was unexpectedly long, with 16 of 26 patients alive at a median follow-up of 12 months.

The failure of DNA vaccine human trials for tumor agrees well with the reported tumor immunotherapy human trials with other vaccines. Possible reasons include the crippled immune system of cancer patients, especially after their chemotherapy and/or radiotherapy. In addition, tumors have developed a great number of escaping mechanisms, such as the loss of MHC I molecules, which makes the tumor cells unresponsive to CTL killing.

In summary, many human clinical trials have been reported in the past several years on DNA vaccine against quite a few disease models. Although they have all shown that the DNA vaccine is safe or causes very minor side effects, the resulted immunological responses have not been encouraging with no or low antibody response detected in most of the cases, often no clinical response, and somewhat weak cellular response in some trials. Therefore, research on how to transfer the success of DNA vaccine in small animals into large animals and humans is urgently needed.

IX. SAFETY ISSUES

Although the DNA vaccine was thought to be safer than the traditional live (viral) vaccine and phase I clinical trials reported no serious side effects from the DNA vaccine, there still are several concerns, including the integration of plasmid DNA into the host genome, induction of autoimmune response, induction of tolerance instead of immunity, and the effect of the CpG motif on the overall long-term immune response. Organizations such as FDA, WHO, and the European Union have already had their guidelines on the regulation of the DNA vaccine. The documents listed the manufacturing, preclinical, and human clinical issues relevant to the development of the DNA vaccine and described potential safety concerns that vaccine developers should address prior to the initiation of clinical trials. Several of these issues mentioned earlier are discussed briefly.

A. Plasmid integration

Similar to gene therapy, there is widespread concern that plasmid DNA might integrate into the host genome and increase the chance of malignant transformation, genomic instability, or cell growth dysregulation. However, for many years, the integration of plasmid DNA in host genome was not reported, making it difficult to reach a regulatory consensus concerning the magnitude to the problem (Gurunathan et al., 2000). It is well known that it is not technically easy to remove free plasmid DNA from the host genomic DNA. Wang et al. (2004b) clearly showed the integration of plasmid DNA into mouse genome, especially after intramuscular DNA electroporation (Wang et al., 2004b). For needle-injected mice, only about 17 copies of plasmid DNA were detected in 1 μg of genomic DNA. This agrees well with previous reports. It was indicated early that 3–30 copies of a DNA vaccine plasmid were associated with host genomic DNA 2 months after intramuscular needle injection (Ledwith et al., 2000; Martin et al., 1999). The calculated mutation rate from 3 to 30 copies per genome is 3000 times lower than the spontaneous mutation rate of 10^{-5} per cell. However, because the plasmid DNA copies in genome DNA were detected by polymerase chain reaction (PCR), PCR may not be able to detect all the integrations such as those short fragments. In addition, the effect from the integration of long stretches of plasmid DNA containing strong regulatory sequences might be more significant than that from the small spontaneous point mutations.

This integration rate, however, was sharply increased to about 980 copies/μg genomic DNA in the case of electroporated DNA (Wang et al., 2004b). Electroporation increased the plasmid tissue level by approximately 34-fold. Using a quantitative gel purification assay for integration, electroporation was found to markedly increase the level of plasmid associated with genomic DNA. To confirm the integration and to identify the insertion sites, Wang et al. (2004b) developed a new assay referred to as repeat-anchored integration capture (RAIC) PCR, which is capable of detecting rare integration events in a complex mixture in vivo. Using this assay, they identified four independent integration events. Sequencing of the insertion sites suggested a random integration process, but with short segments of homology between the vector breakpoint and the insertion site in three of the four cases. This highlights the dilemma scientists are facing. Ways to improve the efficiency of the DNA vaccine such as electroporation might at the same time increase the risk (integration) of the DNA vaccine. Therefore, it is necessary to continue monitoring the integration rate of the DNA administration approaches prior to entrance into a human clinical trial.

B. CpG effect and autoimmune response

As noted earlier, the CpG motif in the plasmid DNA biases the immune response toward Th1. This could be deleterious when a humoral response is required and therefore increases the susceptibility of the host to infections that need Th2 responses. In addition, the Th1-biased response in long term might lead to the development of Th1-mediated organ-specific autoimmune diseases. Segal et al. (1997) reported that in a murine model, the CpG motif, by enhancing the production of IL-12, promoted the development of experimental allergic encephalomyelitis, a Th1-dependent organ-specific autoimmune disease. Klinman et al. (1999) reported that when repeatedly administering (ip) CpG ODN two to four times a month, the animals remained healthy and developed neither macroscopic nor microscopic evidence of tissue damage or inflammation. In contrast, Heikenwalder et al. (2004) reported that repeated CpG ODN administration led to lymphoid follicle destruction and immunosuppression. These conflicting data clearly show that more thorough studies need to be carried out to investigate the long-term effect of the CpG motif.

C. Tolerance

In 1997, Mor and colleagues reported that the DNA vaccine encoding the circumsporozoite protein of the malaria induces tolerance rather than immunity when administered to 2- to 5-day-old mice, although it induced a strong protective immune response against live sporozoite challenge in adult BALB/c mice (Mor et al., 1996). Neonatally tolerized animals were unable to mount antibody, cytokine, or cytotoxic responses when rechallenged with the DNA vaccine in vitro or in vivo. Tolerance was specific for immunogenic epitopes expressed by the vaccine-encoded, endogenously produced antigen. This confirmed that in neonatals, an endogenously presented antigen from the DNA vaccine might be viewed by the neonatal as self. However, further studies pointed out that whether tolerance or immunity will develop is dependent on many parameters, such as dose, recipient's age, and the nature of antigen (Ichino et al., 1999). For example, for influenza, rabies, and HBV, no tolerance was observed, even in neonatals. Finally, a decreased response was also observed in aged mice (over 2 years old), suggesting that the DNA vaccine might be not effective in elderly people (Bender *et al.*, 1998). This is probably due to the generally weaker immune system in elderly people.

X. CONCLUSION

Since its first discovery in 1992, great advances have been made in the DNA vaccine field. The DNA vaccine has proven to be very successful for many diseases in small animal models. However, recent clinical trials have shown that it is far from being effective in humans. Research on how to transfer the success of DNA vaccines in small animal to human is thus needed.

References

Agadjanyan, M. G., Kim, J. J., Trivedi, N., Wilson, D. M., Monzavi-Karbassi, B., Morrison, L. D., Nottingham, L. K., Dentchev, T., Tsai, A., Dang, K., Chalian, A. A., Maldonado, M. A., Williams, W. V., and Weiner, D. B. (1999). CD86 (B7-2) can function to drive MHC-restricted antigen-specific CTL responses. *in vivo. J. Immunol.* **162**, 3417–3427.

Akira, S., Takeda, K., and Kaisho, T. (2001). Toll-like receptors: Critical proteins linking innate and acquired immunity. *Nature Immunol.* **2**, 675–680.

Alexopoulou, L., Holt, A. C., Medzhitov, R., and Flavell, R. A. (2001). Recognition of double-stranded RNA and activation of NF-kappaB by Toll-like receptor 3. *Nature* **413**, 732–738.

Aliprantis, A. O., Yang, R. B., Mark, M. R., Suggett, S., Devaux, B., Radolf, J. D., Klimpel, G. R., Godowski, P., and Zychlinsky, A. (1999). Cell activation and apoptosis by bacterial lipoproteins through toll-like receptor-2. *Science* **285**, 736–739.

Baba, T. W., Liska, V., Khimani, A. H., Ray, N. B., Dailey, P. J., Penninck, D., Bronson, R., Greene, M. F., McClure, H. M., Martin, L. N., and Ruprecht, R. M. (1999). Live attenuated, multiply deleted simian immunodeficiency virus causes AIDS in infant and adult macaques. *Nature Med.* **5**, 194–203.

Banchereau, J., Bazan, F., Blanchard, D., Briere, F., Galizzi, J. P., van Kooten, C., Liu, Y. J., Rousset, F., and Saeland, S. (1994). The CD40 antigen and its ligand. *Annu. Rev. Immunol.* **12**, 881–922.

Banchereau, J., and Steinman, R. M. (1998). Dendritic cells and the control of immunity. *Nature* **392**, 245–252.

Baud, V., Liu, Z. G., Bennett, B., Suzuki, N., Xia, Y., and Karin, M. (1999). Signaling by proin-flammatory cytokines: Oligomerization of TRAF2 and TRAF6 is sufficient for JNK and IKK activation and target gene induction via an amino-terminal effector domain. *Genes Dev.* **13**, 1297–1308.

Bender, B. S., Ulmer, J. B., DeWitt, C. M., Cottey, R., Taylor, S. F., Ward, A. M., Friedman, A., Liu, M. A., and Donnelly, J. J. (1998). Immunogenicity and efficacy of DNA vaccines encoding influenza A proteins in aged mice. *Vaccine* **16**, 1748–1755.

Bennett, S. R., Carbone, F. R., Karamalis, F., Flavell, R. A., Miller, J. F., and Heath, W. R. (1998). Help for cytotoxic-T-cell responses is mediated by CD40 signalling. *Nature* **393**, 478–480.

Boyer, J. D., Cohen, A. D., Vogt, S., Schumann, K., Nath, B., Ahn, L., Lacy, K., Bagarazzi, M. L., Higgins, T. J., Baine, Y., Ciccarelli, R. B., Ginsberg, R. S., MacGregor, R. R., and Weiner, D. B. (2000). Vaccination of seronegative volunteers with a human immunodeficiency virus type

1 env/rev DNA vaccine induces antigen-specific proliferation and lymphocyte production of beta-chemokines. *J. Infect. Dis.* **181,** 476–483.

Boyle, J. S., Silva, A., Brady, J. L., and Lew, A. M. (1997). DNA immunization: Iinduction of higher avidity antibody and effect of route on T cell cytotoxicity. *Proc. Natl. Acad. Sci. USA* **94,** 14626–14631.

Calarota, S. A., Kjerrstrom, A., Islam, K. B., and Wahren, B. (2001). Gene combination raises broad human immunodeficiency virus-specific cytotoxicity. *Hum. Gene Ther.* **12,** 1623–1637.

Castellino, F., Zhong, G., and Germain, R. N. (1997). Antigen presentation by MHC class II molecules: Invariant chain function, protein trafficking, and the molecular basis of diverse determinant capture. *Hum. Immunol.* **54,** 159–169.

Chattergoon, M. A., Saulino, V., Shames, J. P., Stein, J., Montaner, L. J., and Weiner, D. B. (2004). Co-immunization with plasmid IL-12 generates a strong T-cell memory response in mice. *Vaccine* **22,** 1744–1750.

Chen, S. C., Fynan, E. F., Robinson, H. L., Lu, S., Greenberg, H. B., Santoro, J. C., and Herrmann, J. E. (1997). Protective immunity induced by rotavirus DNA vaccines. *Vaccine* **15,** 899–902.

Chen, S. C., Jones, D. H., Fynan, E. F., Farrar, G. H., Clegg, J. C., Greenberg, H. B., and Herrmann, J. E. (1998). Protective immunity induced by oral immunization with a rotavirus DNA vaccine encapsulated in microparticles. *J. Virol.* **72,** 5757–5761.

Chen, Y., Usherwood, E. J., Surman, S. L., Hogg, T. L., and Woodland, D. L. (1999). Long-term CD8+ T cell memory to Sendai virus elicited by DNA vaccination. *J. Gen. Virol.* **80**(Pt. 6), 1393–1399.

Conry, R. M., Curiel, D. T., Strong, T. V., Moore, S. E., Allen, K. O., Barlow, D. L., Shaw, D. R., and LoBuglio, A. F. (2002). Safety and immunogenicity of a DNA vaccine encoding carcinoembryonic antigen and hepatitis B surface antigen in colorectal carcinoma patients. *Clin. Cancer Res.* **8,** 2782–2787.

Corr, M., Lee, D. J., Carson, D. A., and Tighe, H. (1996). Gene vaccination with naked plasmid DNA: Mechanism of CTL priming. *J. Exp. Med.* **184,** 1555–1560.

Cui, Z., and Mumper, R. J. (2002a). Bilayer films for mucosal (genetic) immunization via the buccal route in rabbits. *Pharm. Res.* **19,** 947–953.

Cui, Z., and Mumper, R. J. (2002b). Genetic immunization using nanoparticles engineered from microemulsion precursors. *Pharm. Res.* **19,** 939–946.

Davis, H. L., Schirmbeck, R., Reimann, J., and Whalen, R. G. (1995). DNA-mediated immunization in mice induces a potent MHC class I-restricted cytotoxic T lymphocyte response to the hepatitis B envelope protein. *Hum. Gene Ther.* **6,** 1447–1456.

Deck, R. R., DeWitt, C. M., Donnelly, J. J., Liu, M. A., and Ulmer, J. B. (1997). Characterization of humoral immune responses induced by an influenza hemagglutinin DNA vaccine. *Vaccine* **15,** 71–78.

Epstein, J. E., Charoenvit, Y., Kester, K. E., Wang, R., Newcomer, R., Fitzpatrick, S., Richie, T. L., Tornieporth, N., Heppner, D. G., Ockenhouse, C., Majam, V., Holland, C., Abot, E., Ganeshan, H., Berzins, M., Jones, T., Freydberg, C. N., Ng, J., Norman, J., Carucci, D. J., Cohen, J., and Hoffman, S. L. (2004). Safety, tolerability, and antibody responses in humans after sequential immunization with a PfCSP DNA vaccine followed by the recombinant protein vaccine RTS,S/AS02A. *Vaccine* **22,** 1592–1603.

Falo, L. D., Jr., Kovacsovics-Bankowski, M., Thompson, K., and Rock, K. L. (1995). Targeting antigen into the phagocytic pathway *in vivo* induces protective tumour immunity. *Nature Med.* **1,** 649–653.

Feltquate, D. M., Heaney, S., Webster, R. G., and Robinson, H. L. (1997). Different T helper cell types and antibody isotypes generated by saline and gene gun DNA immunization. *J. Immunol.* **158,** 2278–2284.

Freund, J. (1951). The effect of paraffin oil and mycobacteria on antibody formation and sensitization; a review. *Am. J. Clin. Pathol.* **21,** 645–656.

Fynan, E. F., Webster, R. G., Fuller, D. H., Haynes, J. R., Santoro, J. C., and Robinson, H. L. (1993). DNA vaccines: Protective immunizations by parenteral, mucosal, and gene-gun inoculations. *Proc. Natl. Acad. Sci. USA* **90,** 11478–11482.

Gurunathan, S., Klinman, D. M., and Seder, R. A. (2000). DNA vaccines: Immunology, application, and optimization. *Annu. Rev. Immunol.* **18,** 927–974.

Harding, C. V., and Song, R. (1994). Phagocytic processing of exogenous particulate antigens by macrophages for presentation by class I MHC molecules. *J. Immunol.* **153,** 4925–4933.

Hayashi, F., Smith, K. D., Ozinsky, A., Hawn, T. R., Yi, E. C., Goodlett, D. R., Eng, J. K., Akira, S., Underhill, D. M., and Aderem, A. (2001). The innate immune response to bacterial flagellin is mediated by Toll-like receptor 5. *Nature* **410,** 1099–1103.

Heikenwalder, M., Polymenidou, M., Junt, T., Sigurdson, C., Wagner, H., Akira, S., Zinkernagel, R., and Aguzzi, A. (2004). Lymphoid follicle destruction and immunosuppression after repeated CpG oligodeoxynucleotide administration. *Nature Med.* **10,** 187–192.

Hemmi, H., Takeuchi, O., Kawai, T., Kaisho, T., Sato, S., Sanjo, H., Matsumoto, M., Hoshino, K., Wagner, H., Takeda, K., and Akira, S. (2000). A Toll-like receptor recognizes bacterial DNA. *Nature* **408,** 740–745.

Herrmann, J. E., Chen, S. C., Fynan, E. F., Santoro, J. C., Greenberg, H. B., Wang, S., and Robinson, H. L. (1996). Protection against rotavirus infections by DNA vaccination. *J. Infect. Dis.* **174**(Suppl. 1), S93–S97.

Herrmann, J. E., Chen, S. C., Jones, D. H., Tinsley-Bown, A., Fynan, E. F., Greenberg, H. B., and Farrar, G. H. (1999). Immune responses and protection obtained by oral immunization with rotavirus VP4 and VP7 DNA vaccines encapsulated in microparticles. *Virology* **259,** 148–153.

Hilleman, M. R. (1998). A simplified vaccinologists' vaccinology and the pursuit of a vaccine against AIDS. *Vaccine* **16,** 778–793.

Ichino, M., Mor, G., Conover, J., Weiss, W. R., Takeno, M., Ishii, K. J., and Klinman, D. M. (1999). Factors associated with the development of neonatal tolerance after the administration of a plasmid DNA vaccine. *J. Immunol.* **162,** 3814–3818.

Iwasaki, A., Stiernholm, B. J., Chan, A. K., Berinstein, N. L., and Barber, B. H. (1997). Enhanced CTL responses mediated by plasmid DNA immunogens encoding costimulatory molecules and cytokines. *J. Immunol.* **158,** 4591–4601.

Jones, D. H., Corris, S., McDonald, S., Clegg, J. C., and Farrar, G. H. (1997). Poly(DL-lactide-co-glycolide)-encapsulated plasmid DNA elicits systemic and mucosal antibody responses to encoded protein after oral administration. *Vaccine* **15,** 814–817.

Kaisho, T., and Akira, S. (2001). Bug detectors. *Nature* **414,** 701–703.

Kang, Y., Calvo, P. A., Daly, T. M., and Long, C. A. (1998). Comparison of humoral immune responses elicited by DNA and protein vaccines based on merozoite surface protein-1 from *Plasmodium yoelii,* a rodent malaria parasite. *J. Immunol.* **161,** 4211–4219.

Klencke, B., Matijevic, M., Urban, R. G., Lathey, J. L., Hedley, M. L., Berry, M., Thatcher, J., Weinberg, V., Wilson, J., Darragh, T., Jay, N., Da Costa, M., and Palefsky, J. M. (2002). Encapsulated plasmid DNA treatment for human papillomavirus 16-associated anal dysplasia: A phase I study of ZYC101. *Clin. Cancer Res.* **8,** 1028–1037.

Klinman, D. M., Conover, J., and Coban, C. (1999). Repeated administration of synthetic oligodeoxynucleotides expressing CpG motifs provides long-term protection against bacterial infection. *Infect. Immun.* **67,** 5658–5663.

Krieg, A. M., Yi, A. K., Matson, S., Waldschmidt, T. J., Bishop, G. A., Teasdale, R., Koretzky, G. A., and Klinman, D. M. (1995). CpG motifs in bacterial DNA trigger direct B-cell activation. *Nature* **374,** 546–549.

Kundig, T. M., Bachmann, M. F., DiPaolo, C., Simard, J. J., Battegay, M., Lother, H., Gessner, A., Kuhlcke, K., Ohashi, P. S., Hengartner, H., et al. (1995). Fibroblasts as efficient antigen-presenting cells in lymphoid organs. Science **268,** 1343–1347.

Le, T. P., Coonan, K. M., Hedstrom, R. C., Charoenvit, Y., Sedegah, M., Epstein, J. E., Kumar, S., Wang, R., Doolan, D. L., Maguire, J. D., Parker, S. E., Hobart, P., Norman, J., and Hoffman, S. L. (2000). Safety, tolerability and humoral immune responses after intramuscular administration of a malaria DNA vaccine to healthy adult volunteers. Vaccine **18,** 1893–1901.

Ledwith, B. J., Manam, S., Troilo, P. J., Barnum, A. B., Pauley, C. J., Griffiths, T. G., 2nd, Harper, L. B., Schock, H. B., Zhang, H., Faris, J. E., Way, P. A., Beare, C. M., Bagdon, W. J., and Nichols, W. W. (2000). Plasmid DNA vaccines: Assay for integration into host genomic DNA. Dev. Biol. (Basel) **104,** 33–43.

Lemaitre, B., Nicolas, E., Michaut, L., Reichhart, J. M., and Hoffmann, J. A. (1996). The dorsoventral regulatory gene cassette spatzle/Toll/cactus controls the potent antifungal response in Drosophila adults. Cell **86,** 973–983.

Lesinski, G. B., and Westerink, M. A. (2001). Vaccines against polysaccharide antigens. Curr. Drug Targets Infect. Disord. **1,** 325–334.

Liu, M. A. (2003). DNA vaccines: A review. J. Intern. Med. **253,** 402–410.

Livingston, J. B., Lu, S., Robinson, H., and Anderson, D. J. (1998). Immunization of the female genital tract with a DNA-based vaccine. Infect. Immun. **66,** 322–329.

Lundholm, P., Asakura, Y., Hinkula, J., Lucht, E., and Wahren, B. (1999). Induction of mucosal IgA by a novel jet delivery technique for HIV-1 DNA. Vaccine **17,** 2036–2042.

MacGregor, R. R., Boyer, J. D., Ciccarelli, R. B., Ginsberg, R. S., and Weiner, D. B. (2000). Safety and immune responses to a DNA-based human immunodeficiency virus (HIV) type I env/rev vaccine in HIV-infected recipients: Follow-up data. J. Infect. Dis. **181,** 406.

MacGregor, R. R., Boyer, J. D., Ugen, K. E., Lacy, K. E., Gluckman, S. J., Bagarazzi, M. L., Chattergoon, M. A., Baine, Y., Higgins, T. J., Ciccarelli, R. B., Coney, L. R., Ginsberg, R. S., and Weiner, D. B. (1998). First human trial of a DNA-based vaccine for treatment of human immunodeficiency virus type 1 infection: Safety and host response. J. Infect. Dis. **178,** 92–100.

MacGregor, R. R., Ginsberg, R., Ugen, K. E., Baine, Y., Kang, C. U., Tu, X. M., Higgins, T., Weiner, D. B., and Boyer, J. D. (2002). T-cell responses induced in normal volunteers immunized with a DNA-based vaccine containing HIV-1 env and rev. Aids **16,** 2137–2143.

Martin, T., Parker, S. E., Hedstrom, R., Le, T., Hoffman, S. L., Norman, J., Hobart, P., and Lew, D. (1999). Plasmid DNA malaria vaccine: The potential for genomic integration after intramuscular injection. Hum. Gene Ther. **10,** 759–768.

Martins, L. P., Lau, L. L., Asano, M. S., and Ahmed, R. (1995). DNA vaccination against persistent viral infection. J. Virol. **69,** 2574–2582.

McConkey, S. J., Reece, W. H., Moorthy, V. S., Webster, D., Dunachie, S., Butcher, G., Vuola, J. M., Blanchard, T. J., Gothard, P., Watkins, K., Hannan, C. M., Everaere, S., Brown, K., Kester, K. E., Cummings, J., Williams, J., Heppner, D. G., Pathan, A., Flanagan, K., Arulanantham, N., Roberts, M. T., Roy, M., Smith, G. L., Schneider, J., Peto, T., Sinden, R. E., Gilbert, S. C., and Hill, A. V. (2003). Enhanced T-cell immunogenicity of plasmid DNA vaccines boosted by recombinant modified vaccinia virus Ankara in humans. Nature Med. **9,** 729–735.

McGhee, J. R., Mestecky, J., Dertzbaugh, M. T., Eldridge, J. H., Hirasawa, M., and Kiyono, H. (1992). The mucosal immune system: From fundamental concepts to vaccine development. Vaccine **10,** 75–88.

Medzhitov, R. (2001). Toll-like receptors and innate immunity. Nature Rev. Immunol. **1,** 135–145.

Medzhitov, R., Preston-Hurlburt, P., and Janeway, C. A., Jr. (1997). A human homologue of the Drosophila Toll protein signals activation of adaptive immunity. Nature **388,** 394–397.

Medzhitov, R., Preston-Hurlburt, P., Kopp, E., Stadlen, A., Chen, C., Ghosh, S., and Janeway, C. A., Jr. (1998). MyD88 is an adaptor protein in the hToll/IL-1 receptor family signaling pathways. *Mol. Cell* **2**, 253–258.

Mincheff, M., Tchakarov, S., Zoubak, S., Loukinov, D., Botev, C., Altankova, I., Georgiev, G., Petrov, S., and Meryman, H. T. (2000). Naked DNA and adenoviral immunizations for immunotherapy of prostate cancer: A phase I/II clinical trial. *Eur. Urol.* **38**, 208–217.

Moorthy, V. S., Pinder, M., Reece, W. H., Watkins, K., Atabani, S., Hannan, C., Bojang, K., McAdam, K. P., Schneider, J., Gilbert, S., and Hill, A. V. (2003). Safety and immunogenicity of DNA/modified vaccinia virus ankara malaria vaccination in African adults. *J. Infect. Dis.* **188**, 1239–1244.

Mor, G., Yamshchikov, G., Sedegah, M., Takeno, M., Wang, R., Houghten, R. A., Hoffman, S., and Klinman, D. M. (1996). Induction of neonatal tolerance by plasmid DNA vaccination of mice. *J. Clin. Invest.* **98**, 2700–2705.

Muzio, M., Natoli, G., Saccani, S., Levrero, M., and Mantovani, A. (1998). The human toll signaling pathway: Divergence of nuclear factor kappaB and JNK/SAPK activation upstream of tumor necrosis factor receptor-associated factor 6 (TRAF6). *J. Exp. Med.* **187**, 2097–2101.

Park, J. S., Oh, Y. K., Kang, M. J., and Kim, C. K. (2003). Enhanced mucosal and systemic immune responses following intravaginal immunization with human papillomavirus 16 L1 virus-like particle vaccine in thermosensitive mucoadhesive delivery systems. *J. Med. Virol.* **70**, 633–641.

Poltorak, A., He, X., Smirnova, I., Liu, M. Y., Van Huffel, C., Du, X., Birdwell, D., Alejos, E., Silva, M., Galanos, C., Freudenberg, M., Ricciardi-Castagnoli, P., Layton, B., and Beutler, B. (1998). Defective LPS signaling in C3H/HeJ and C57BL/10ScCr mice: Mutations in Tlr4 gene. *Science* **282**, 2085–2088.

Raychaudhuri, S., and Rock, K. L. (1998). Fully mobilizing host defense: Building better vaccines. *Nature Biotechnol.* **16**, 1025–1031.

Raz, E., Carson, D. A., Parker, S. E., Parr, T. B., Abai, A. M., Aichinger, G., Gromkowski, S. H., Singh, M., Lew, D., Yankauckas, M. A., *et al.* (1994). Intradermal gene immunization: The possible role of DNA uptake in the induction of cellular immunity to viruses. *Proc. Natl. Acad. Sci. USA* **91**, 9519–9523.

Ridge, J. P., Di Rosa, F., and Matzinger, P. (1998). A conditioned dendritic cell can be a temporal bridge between a CD4+ T-helper and a T-killer cell. *Nature* **393**, 474–478.

Rosenberg, S. A., Yang, J. C., Sherry, R. M., Hwu, P., Topalian, S. L., Schwartzentruber, D. J., Restifo, N. P., Haworth, L. R., Seipp, C. A., Freezer, L. J., Morton, K. E., Mavroukakis, S. A., and White, D. E. (2003). Inability to immunize patients with metastatic melanoma using plasmid DNA encoding the gp100 melanoma-melanocyte antigen. *Hum. Gene Ther.* **14**, 709–714.

Roy, K., Mao, H. Q., Huang, S. K., and Leong, K. W. (1999). Oral gene delivery with chitosan: DNA nanoparticles generates immunologic protection in a murine model of peanut allergy. *Nature Med.* **5**, 387–391.

Roy, M. J., Wu, M. S., Barr, L. J., Fuller, J. T., Tussey, L. G., Speller, S., Culp, J., Burkholder, J. K., Swain, W. F., Dixon, R. M., Widera, G., Vessey, R., King, A., Ogg, G., Gallimore, A., Haynes, J. R., and Heydenburg Fuller, D. (2000). Induction of antigen-specific CD8+ T-Cells, T helper cells, and protective levels of antibody in humans by particle-mediated administration of a hepatitis B virus DNA vaccine. *Vaccine* **19**, 764–778.

Russell-Jones, G. J. (2000). Oral vaccine delivery. *J. Control Release* **65**, 49–54.

Schoenberger, S. P., Toes, R. E., van der Voort, E. I., Offringa, R., and Melief, C. J. (1998). T-cell help for cytotoxic T lymphocytes is mediated by CD40-CD40L interactions. *Nature* **393**, 480–483.

Segal, B. M., Klinman, D. M., and Shevach, E. M. (1997). Microbial products induce autoimmune disease by an IL-12-dependent pathway. *J. Immunol.* **158**, 5087–5090.

Shimada, S., Yano, O., Inoue, H., Kuramoto, E., Fukuda, T., Yamamoto, H., Kataoka, T., and Tokunaga, T. (1985). Antitumor activity of the DNA fraction from Mycobacterium bovis BCG. II. Effects on various syngeneic mouse tumors. *J. Natl. Cancer Inst.* **74**, 681–688.

Stein, D., Roth, S., Vogelsang, E., and Nusslein-Volhard, C. (1991). The polarity of the dorsoventral axis in the *Drosophila* embryo is defined by an extracellular signal. *Cell* **65**, 725–735.

Tacket, C. O., Roy, M. J., Widera, G., Swain, W. F., Broome, S., and Edelman, R. (1999). Phase 1 safety and immune response studies of a DNA vaccine encoding hepatitis B surface antigen delivered by a gene delivery device. *Vaccine* **17**, 2826–2829.

Tagawa, S. T., Lee, P., Snively, J., Boswell, W., Ounpraseuth, S., Lee, S., Hickingbottom, B., Smith, J., Johnson, D., and Weber, J. S. (2003). Phase I study of intranodal delivery of a plasmid DNA vaccine for patients with stage IV melanoma. *Cancer* **98**, 144–154.

Takashima, A., and Morita, A. (1999). Dendritic cells in genetic immunization. *J. Leukocyte Biol.* **66**, 350–356.

Tang, D. C., DeVit, M., and Johnston, S. A. (1992). Genetic immunization is a simple method for eliciting an immune response. *Nature* **356**, 152–154.

Tang, D. C., Shi, Z., and Curiel, D. T. (1997). Vaccination onto bare skin. *Nature* **388**, 729–730.

Timmerman, J. M., Singh, G., Hermanson, G., Hobart, P., Czerwinski, D. K., Taidi, B., Rajapaksa, R., Caspar, C. B., Van Beckhoven, A., and Levy, R. (2002). Immunogenicity of a plasmid DNA vaccine encoding chimeric idiotype in patients with B-cell lymphoma. *Cancer Res.* **62**, 5845–5852.

Tokunaga, T., Yamamoto, H., Shimada, S., Abe, H., Fukuda, T., Fujisawa, Y., Furutani, Y., Yano, O., Kataoka, T., Sudo, T., *et al.* (1984). Antitumor activity of deoxyribonucleic acid fraction from Mycobacterium bovis BCG. I. Isolation, physicochemical characterization, and antitumor activity. *J. Natl. Cancer Inst.* **72**, 955–962.

Tokunaga, T., Yamamoto, T., and Yamamoto, S. (1999). How BCG led to the discovery of immunostimulatory DNA. *Jpn. J. Infect. Dis.* **52**, 1–11.

Torres, C. A., Iwasaki, A., Barber, B. H., and Robinson, H. L. (1997). Differential dependence on target site tissue for gene gun and intramuscular DNA immunizations. *J. Immunol.* **158**, 4529–4532.

Ugen, K. E., Nyland, S. B., Boyer, J. D., Vidal, C., Lera, L., Rasheid, S., Chattergoon, M., Bagarazzi, M. L., Ciccarelli, R., Higgins, T., Baine, Y., Ginsberg, R., Macgregor, R. R., and Weiner, D. B. (1998). DNA vaccination with HIV-1 expressing constructs elicits immune responses in humans. *Vaccine* **16**, 1818–1821.

Ulmer, J. B., Deck, R. R., Dewitt, C. M., Donnhly, J. I., and Liu, M. A. (1996). Generation of MHC class I-restricted cytotoxic T lymphocytes by expression of a viral protein in muscle cells: Antigen presentation by non-muscle cells. *Immunology* **89**, 59–67.

Ulmer, J. B., Donnelly, J. J., Parker, S. E., Rhodes, G. H., Felgner, P. L., Dwarki, V. J., Gromkowski, S. H., Deck, R. R., DeWitt, C. M., Friedman, A., *et al.* (1993). Heterologous protection against influenza by injection of DNA encoding a viral protein. *Science* **259**, 1745–1749.

Underhill, D. M., Ozinsky, A., Hajjar, A. M., Stevens, A., Wilson, C. B., Bassetti, M., and Aderem, A. (1999). The Toll-like receptor 2 is recruited to macrophage phagosomes and discriminates between pathogens. *Nature* **401**, 811–815.

Wang, R., Doolan, D. L., Le, T. P., Hedstrom, R. C., Coonan, K. M., Charoenvit, Y., Jones, T. R., Hobart, P., Margalith, M., Ng, J., Weiss, W. R., Sedegah, M., de Taisne, C., Norman, J. A., and Hoffman, S. L. (1998). Induction of antigen-specific cytotoxic T lymphocytes in humans by a malaria DNA vaccine. *Science* **282**, 476–480.

Wang, R., Epstein, J., Charoenvit, Y., Baraceros, F. M., Rahardjo, N., Gay, T., Banania, J. G., Chattopadhyay, R., de la Vega, P., Richie, T. L., Tornieporth, N., Doolan, D. L., Kester, K. E., Heppner, D. G., Norman, J., Carucci, D. J., Cohen, J. D., and Hoffman, S. L. (2004a). Induction in humans of CD8+ and CD4+ T cell and antibody responses by sequential immunization with malaria DNA and recombinant protein. *J. Immunol.* **172,** 5561–5569.

Wang, Z., Troilo, P. J., Wang, X., Griffiths, T. G., Pacchione, S. J., Barnum, A. B., Harper, L. B., Pauley, C. J., Niu, Z., Denisova, L., Follmer, T. T., Rizzuto, G., Ciliberto, G., Fattori, E., Monica, N. L., Manam, S., and Ledwith, B. J. (2004b). Detection of integration of plasmid DNA into host genomic DNA following intramuscular injection and electroporation. *Gene Ther.* **11,** 711–721.

Weber, R., Bossart, W., Cone, R., Luethy, R., and Moelling, K. (2001). Phase I clinical trial with HIV-1 gp160 plasmid vaccine in HIV-1-infected asymptomatic subjects. *Eur. J. Clin. Microbiol. Infect. Dis.* **20,** 800–803.

Wesche, H., Korherr, C., Kracht, M., Falk, W., Resch, K., and Martin, M. U. (1997). The interleukin-1 receptor accessory protein (IL-1RAcP) is essential for IL-1-induced activation of interleukin-1 receptor-associated kinase (IRAK) and stress-activated protein kinases (SAP kinases). *J. Biol. Chem.* **272,** 7727–7731.

Wolff, J. A., Malone, R. W., Williams, P., Chong, W., Acsadi, G., Jani, A., and Felgner, P. L. (1990). Direct gene transfer into mouse muscle *in vivo*. *Science* **247,** 1465–1468.

Yamamoto, S., Kuramoto, E., Shimada, S., and Tokunaga, T. (1988). *In vitro* augmentation of natural killer cell activity and production of interferon-alpha/beta and -gamma with deoxyribonucleic acid fraction from Mycobacterium bovis BCG. *Jpn. J. Cancer Res.* **79,** 866–873.

Zarember, K. A., and Godowski, P. J. (2002). Tissue expression of human Toll-like receptors and differential regulation of Toll-like receptor mRNAs in leukocytes in response to microbes, their products, and cytokines. *J. Immunol.* **168,** 554–561.

Zhang, D., Zhang, G., Hayden, M. S., Greenblatt, M. B., Bussey, C., Flavell, R. A., and Ghosh, S. (2004). A toll-like receptor that prevents infection by uropathogenic bacteria. *Science* **303,** 1522–1526.

12

Airway Gene Therapy

Jane C. Davies and Eric W. F. W Alton
Department of Gene Therapy
Imperial College London
National Heart and Lung Institute
London SW3 6LR, United Kingdom

Advances in Genetics, Vol. 54
Copyright 2005, Elsevier Inc. All rights reserved.

0065-2660/05 $35.00
DOI: 10.1016/S0065-2660(05)54012-4

ABSTRACT

Given both the accessibility and the genetic basis of several pulmonary diseases, the lungs and airways initially seemed ideal candidates for gene therapy. Several routes of access are available, many of which have been refined and optimized for nongene drug delivery. Two respiratory diseases, cystic fibrosis (CF) and α_1-antitrypsin (α_1-AT) deficiency, are relatively common; the single gene responsible has been identified and current treatment strategies are not curative. This type of inherited disease was the obvious initial target for gene therapy, but it has become clear that nongenetic and acquired diseases, including cancer, may also be amenable to this approach. The majority of preclinical and clinical studies in the airway have involved viral vectors, although for diseases such as CF, likely to require repeated application, non-viral delivery systems have clear advantages. However, with both approaches a range of barriers to gene expression have been identified that are limiting success in the airway and alveolar region. This chapter reviews these issues, strategies aimed at overcoming them, and progress into clinical trials with non-viral vectors in a variety of pulmonary diseases.   © 2005, Elsevier Inc.

I. ROUTES OF ADMINISTRATION FOR AIRWAY EXPRESSION

The respiratory tract, as one of the hollow body organs, is well suited to several routes of access, from either airway or adventitial surfaces. Conventional therapeutics are routinely administered topically, refinement of which has been achieved by an increased understanding of the strengths and limitations of nebulizers and inhalers. Substances can be targeted to various sites within the airway by modifications in size and, to a lesser extent, delivery rates. The systemic route is also used with success, for example, in the administration of antibiotics, some of which reach high levels in the airway surface liquid and overlying mucus. For gene therapy, the route of access chosen will depend on the nature of the disease being treated, the desired cellular site of gene expression, and, to a certain extent, the limitations of gene transfer agents (GTAs) and delivery systems.

A. Topical administration

Administration directly onto the airway surface may be most appropriate for epithelial membrane-bound proteins such as the cystic fibrosis transmembrane conductance regulator (CFTR) protein (mutated in cystic fibrosis) and would also be suitable for diseases where protein was to be secreted into the airway, e.g., α_1-antitrypsin deficiency. Widespread expression is likely to be best achieved by either inhalation or nebulization. To date, no studies of dry powder or aerosol delivery of gene transfer agents via inhalation have been reported, although ultimately this would be an ideal delivery system. Nebulization is an attractive method, which has a long history in the delivery of nongene-based drugs to the airways. It has been used with success in several clinical gene therapy studies, although it has disadvantages with respect to naked DNA; the majority of the currently available nebulizer systems destroy the DNA due to the physical forces exerted upon it during multiple passes through the equipment. Newer technologies, achieving nebulization by a single pass through a porous membrane, appear to offer more promise in this respect. Bronchoscopic techniques, such as instillation or injection, are probably most appropriate for single application localized delivery, such as in the context of endobronchial tumors. They would probably be impractical for diseases requiring repeated, widespread delivery such as CF.

B. Percutaneous injection

This route of administration, into a tumor or pleural space, may be appropriate for certain cancers. Radiological techniques such as ultrasound or computed tomography (CT) allow accurate placement of injections and already form part of mainstream clinical practice. Such an approach is generally well tolerated and, being somewhat less invasive than bronchoscopic injection, could be performed on a repeated basis. Major limitations would, however, arise if only a small proportion of the tumor was accessible.

C. Systemic injection

Intravascular injection could be appropriate if one wished to target the alveolar epithelium, but it is currently an inefficient method of targeting conducting airways. This likely relates both to the number of barriers the GTA needs to cross and to the distribution of the bronchial and pulmonary circulations in the human. Alternatively, if the disease required a systemically secreted protein (e.g., α_1-AT deficiency or certain inflammatory diseases), gene transfer to a distant organ, such as muscle or liver, which would serve as a "factory," could be an attractive option.

II. BARRIERS TO GENE TRANSFER AND EXPRESSION

It is now well recognized that a range of barriers hinder gene transfer to the lungs and airway (Zabner et al., 1995). These organs, open to the air and thus constantly at risk from foreign, potentially pathogenic material, have developed complex and highly successful exclusion mechanisms. Unfortunately, this extends to exogenously applied substances, such as gene transfer agents, leading investigators in the field to develop strategies aimed at overcoming these. In identifying each of these barriers, it may be helpful to consider the route taken by a gene from outside the host through to the nucleus of the desired cell. This route will vary depending on the administration technique being utilized, although there is likely a common pathway more distally.

A. Topical application

1. Reaching the cell surface

The conducting airways are lined with both ciliated and mucus-secreting epithelial cells, with the major concerted function of these being the rapid removal of inhaled material via a mechanism termed mucociliary clearance (MCC) (Houtmeyers et al., 1999). The mucus barrier prevents direct contact between inhaled particles and the cell surface, which in the case of airborne infection, for example, serves to limit the epithelial inflammatory response to pathogens. Even in normal health, MCC is likely to pose a significant barrier to topical gene delivery and, in certain disease states, such as CF, this problem may be even more pronounced. CF sputum, which is abnormal both in volume and in composition, has been shown to inhibit gene transfer by both viral and liposomal vectors (Perricone et al., 2000; Stern et al., 1998). Adjunctive therapies, such as the use of recombinant human DNase (Stern et al., 1998) or Nacystelyn (Ferrari et al., 2001) to reduce mucus viscosity, have shown promise in preclinical studies and could be clinically applicable. Several investigators have explored the use of surfactant (Raczka et al., 1998), other low surface tension liquids (Weiss et al., 1999), or thixotropic agents (Seiler, 2002) to facilitate the spreading of gene transfer vectors throughout the respiratory tract and increase contact time with the airway surface. However, because these compounds may increase the distal airway and alveolar expression of therapeutic genes at the expense of more proximal airway expression, the target area needs to be clearly identified before this approach could be considered applicable.

2. Entering the cell

Gene transfer agents that successfully evade MCC and come into direct contact with the cell surface need to enter the cell via a route that will not lead to their destruction. The cell surface glycolcalyx has been shown to be quite resistant to entry, although strategies have been explored to overcome this, such as enzymatic removal of its sialic acid moieties (Pickles et al., 2000). Agents that break down intercellular tight junctions, allowing access of the GTA to the basolateral surfaces of the cell, have been shown to increase the uptake of both viral and synthetic vectors (Croyle et al., 2001; Das and Niven, 2001; Meng et al., 2004), although the safety of these strategies and thus their clinical applicability remain to be determined. Some viral vectors utilize specific cell surface receptors in order to enter the cell. In an attempt to mimic the potential efficiency of this process, ligands for such receptors can be added to synthetic vectors. One success in this area involved targeting the serpin-enzyme complex receptor (sec-R), which is normally involved in the uptake and processing of serine proteases bound to their inhibitors. Sec-R targeted CFTR led to a degree of correction of chloride ion transport in the CF mouse nose, whereas nontargeted complexes were unsuccessful (Ziady et al., 2002). Another group targeted integrins, located on the apical membrane of respiratory epithelia, and demonstrated an increased uptake of plasmid coupled to an integrin-binding motif via receptor-mediated endocytosis (Scott et al., 2001). The repeatability of other approaches, incorporating viral peptides, remains to be determined.

B. Systemic administration: reaching and entering the cell

Attempts have been made to circumvent barriers at the mucosal surface by delivering transgenes systemically. The intravenous route might make it possible to access the basolateral membrane of airway epithelial cells, characterized by a higher rate of endocytosis and an increased density of receptors. However, most studies have demonstrated that cells transfected via the systemic circulation are pulmonary endothelial cells or alveolar epithelial cells (Griesenbach et al., 1998), with only a minority reporting airway epithelial cell transfection (Koehler et al., 2001; Zhu et al., 1993). This is likely to be related to a number of barriers, including reticuloendothelial system clearance, inhibition by serum proteins, escape from the vasculature, the interstitium, and the epithelial basement membrane. Various attempts to overcome these barriers have been reviewed (Fenske, 2001; Niidome and Huang, 2002).

C. Common pathway: crossing the cytoplasm and the nuclear membrane

Once inside the cell, GTAs must cross the cytoplasm, while at risk of destruction from cytoplasmic DNases (Lechardeur et al., 1999), and penetrate the nuclear membrane. Potential methods to improve endosomal escape include coupling the polycation with peptides similar to those used by viruses for this purpose, the use of proton sponges with intrinsic endosomolytic activity, and pH-responsive histidine-based systems (Cho et al., 2003). Alternative approaches include conjugation of non-viral vectors with specific sugar moieties (Grosse et al., 2002), the addition of peptides to facilitate cytoskeletal transport, or the breakdown of cytoskeletal elements, such as microtubules or microfila-ments (Kitson et al., 1999). Once at the nucleus, the nuclear envelope prevents the entry of most foreign material. In the nondividing cell, such as most of those in the airway, this barrier is likely to be a major limiting step to successful gene transfer. Molecules needed within the nucleus, e.g., transcription factors, enter through the nuclear pore complex, facilitated by the possession of specific nuclear localizing sequences (NLS). Ways around this barrier could be the incorporation of specific NLS peptide sequences, e.g., from the HIV TAT protein (Snyder and Dowdy, 2001) (although issues of repeatability need to be considered once again), and the identification of novel NLS (Munkonge et al., 1998). Alternatively, this barrier can be overcome completely by the use of cytoplasmic expression systems such as those incorporating the T7 promoter and RNA polymerase (Brisson et al., 1999).

III. NOVEL STRATEGIES TO INCREASE THE UPTAKE OF GENE TRANSFER AGENTS

In vitro, animal and early human studies have confirmed the difficulties inherent in airway gene transfer. Some of the strategies used to overcome the identi-fied barriers were outlined earlier. In addition, a variety of methods aimed at adding "energy" to the system have been considered, including electroporation, ultrasound, and the use of magnetism.

A. Electroporation

This is the application of electric pulses across the tissue in an attempt to increase uptake of the GTA, probably via both cell membrane permeabilization and DNA electrophoresis. The differential effects on these two mechanisms appear to depend on the type of electric pulse administered (Satkauskas et al., 2002). In animal models, this results in surprisingly little tissue damage, and the technique has been used with some success on skeletal muscle, a variety of tumors, blood vessels, skin, and neural tissue (Bigey et al., 2002). In the majority of these organs, the DNA has been injected directly into the tissue. Dean et al. (2003) have shown dose-dependent transfer of naked DNA in response to electroporation through the chest wall in mice. Levels of expression were significantly lower than in comparative ex vivo experiments, due most probably to the multiple layers of tissue through which the electrical field had to travel. When sensitive detection techniques were employed, gene transfer was seen in both alveolar and airway cells. Importantly, there was no macroscopic or microscopic evidence of damage at the doses used. Reporter gene expression has been demonstrated in the lungs of mice following thoracotomy and direct placement of electrodes on the lung surface (Pringle *et al.*, 2004), and another group has reported that the electroporation-mediated transfer of naked DNA encoding the Na^+-K^+/ATPase led to enhanced fluid clearance in the rat lung (Machado-Aranda *et al.*, 2005). The translation of this technique into the larger chest of humans poses obvious challenges, although it may be particularly applicable to isolated areas such as tumors. Development of bronchoscopic techniques may further increase the applicability of this approach (Pringle *et al.*, 2004).

B. Ultrasound

Ultrasound has been used in both preclinical and clinical settings to increase the percutaneous absorption of drugs, such as insulin, local anesthetics, and non-steroidal anti-inflammatory agents (Machet and Boucaud, 2002). Success has been mixed, with some studies demonstrating convincing evidence of enhanced uptake. Mechanisms of action would appear to include heating and cavitation, both of which lead to increased tissue permeability (Miller *et al.*, 2002). The latter mechanism seems to account for the observed enhanced uptake and expression of DNA, both *in vitro* and *in vivo* in animal models. To date, these experiments have been performed on joints (Kim *et al.*, 1996), tumors (Manome *et al.*, 2000), and the vascular tree (Amabile *et al.*, 2001), but not on the lung or airways. Intuitively, one might consider this form of energy to be poorly suited to such an air-filled organ. However, low-frequency ($<100\,MHz$) ultrasound, which has been used in clinical trials on skin, is able to traverse air-filled spaces and may therefore be applicable in the respiratory tract.

C. Magnetism

Magnetofection is the term used to describe the application of a magnetic field to the organ or animal during the administration of GTA-linked magnetic nanoparticles. The purpose of this approach is to increase contact time between the host cells and the DNA, increasing the likelihood of uptake. *In vitro* experiments have shown up to 100% of cells to have vector particles bound to their surface within a few minutes using this technique. While those working in the field have not claimed that magnetofection increases overall gene transfer efficiency, they do argue for its beneficial effects on time, dose required, and the potential for targeting a specific organ. In addition, cells resistant to adenoviral gene transfer, because they lack the CAR receptor, were transfectable using this technique (Scherer *et al.*, 2002), apparently due to increased contact time and enhanced uptake. Using an *ex vivo* porcine airway model, Gersting *et al.* (2004) demonstrated that magnetofection led to significant increases in reporter gene expression in short time periods, in part as it reduced clearance by the mucociliary escalator.

IV. EXTENDING THE DURATION OF EXPRESSION

The duration of transgene expression is limited by a variety of mechanisms. Most GTAs utilize viral promoters, which may give high, early peak levels of expression, but which decline rapidly thereafter due to transcriptional silencing, probably mediated via inflammatory cytokines. Progress has been reported with human promoters such as polyubiquitin (Ub)C (Gill et al., 2001) or UbB (Yew et al., 2001). The former was shown to extend the expression of naked DNA from 2 weeks to 6 months in the murine lung (Gill et al., 2001). These types of promoters could clearly have major advantages in the clinical setting where prolonged expression is desirable. Another reason for suboptimal duration of expression relates to death of the transfected cell, either due to natural turnover or in response to host recognition. The cells of the conducting airways are terminally differentiated (Warburton et al., 1998) and are thought to have a life span of several months. Once such a cell dies, the transgene will be lost, and thus repeated applications will be required. One way around this could be the selective transfection of progenitor or stem cells. This approach is complicated by our limited understanding of exactly which subpopulation(s) of cells possesses these properties and how they could be targeted. Suggestions have included Clara and basal cells for the airways and type II cells and neuroepithelial bodies (NEBs) in the alveolar region. However, here most non-viral delivery systems possess significant disadvantages over the integrating viral vectors. The latter could give rise to generations of corrected progeny, whereas this is conceptually more difficult with episomal DNA, such as that delivered with the majority of non-viral vectors. Novel approaches such as the transposon system, whereby DNA is inserted directly into the host genome (Yant et al., 2000), may overcome this barrier, although, as with integrating viral systems, concerns over insertional mutagenesis (Hacein-Bey-Abina et al., 2003) must be overcome.

V. LIMITING THE HOST IMMUNE RESPONSE

Both innate and adaptive immune responses can limit the efficiency and duration of gene transfer and may pose problems with all routes of administration, including production of a secreted protein at a distant site. Within the respiratory tract, alveolar macrophages can act either directly as phagocytes or indirectly as antigen-presenting cells, resulting in fairly rapid clearance of a topically applied gene transfer vector (Worgall et al., 1997). Similarly, macrophages in the reticulendothelial system can mediate clearance of a systemically applied vector (Plank *et al.*, 1996). If vectors are not cleared by such innate immune mechanisms, the adaptive immune system has the potential to significantly impair gene transfer and expression. Both cell-mediated and humoral responses have been demonstrated with viral and synthetic vectors, although the major toxicity problems in clinical trials have been encountered with the former. Initiation of a cytotoxic T lymphocyte response may result in killing of the transduced cells, thus effectively limiting the duration of transgene expression. The presence and number of unmethylated CpG motifs on plasmid DNA have been suggested as causes for an observed inflammatory response with non-viral vectors (Schwarzt *et al.*, 1997; Yew *et al.*, 1999). Efforts are being aimed at selectively methylating such motifs, (although this can reduce expression levels) (McLachlan *et al.*, 2000), reducing their numbers, or inhibiting the pathways through which such they signal with agents such as chloroquine (Yew *et al.*, 2000). Although apparently not a significant problem in clinical trials to date, the possibility of antibody production to previously unencountered transgene-derived protein also exists.

VI. SPECIFIC LUNG DISEASES: PROGRESS TOWARD CLINICAL APPLICATION

A. Cystic fibrosis

Cystic fibrosis results from a variety of mutations in the gene encoding the CFTR protein, a cAMP-regulated chloride channel in the apical surface of epithelial cells (Welsh and Smith, 1993). Pulmonary disease results from impaired mucociliary clearance from the abnormal epithelial ion transport, chronic bacterial infection, and severe inflammation (Armstrong *et al.*, 1995). It is the cause of death in over 90% of patients. Treatment for CF patients has advanced greatly over the last few decades, but at best, it slows the inevitable progression of lung damage. Median survival is still only approximately 30 years. Clearly new approaches to treatment are required.

As the major site of pathology and leading cause of mortality, the airways have been the target in the majority of gene therapy trials. Topical gene delivery to the airway epithelium aims to normalize the functions of CFTR, including ion transport. There are several unresolved questions with this approach, including the types of cell to be transfected, the degree of correction required, and how success should be measured. With respect to the cell type targeted, maximal CFTR expression in non-CF airways appears to be in the submucosal glands (Engelhardt *et al.*, 1992, 1994) and in the surface epithelium of the distal small airways. It is in this latter site that clinically detectable disease begins. Topical application, for example, via inhalation, is likely to target the surface epithelium, but is less likely to reach the deeper submucosal gland cells. Whether gene transfer to these cells will be necessary for clinical effect remains to be determined.

With regard to what degree of transfection is required, different levels of expression may be required to restore the various functions of CFTR. For example, lower numbers of cells likely need to be corrected to restore chloride transport compared with those required for normalization of sodium absorption (Johnson et al., 1995), and differences have also been observed with regard to the correction of glycoconjugate sulfation and ion transport (Zhang et al., 1998). Which function(s) of CFTR is most important and whether all identified functions (and perhaps as yet unrecognised ones) need to be corrected to prevent disease initiation/progression are important questions that remain to be resolved. However, based on genetic studies, it would seem that as little as 5–10% of wild-type levels of CFTR in each cell may be sufficient for a normal disease-free phenotype (Gan et al., 1995). However, the sensitivity of epithelia to the effects of low levels of functional CFTR differs among organs; the male reproductive tract is highly sensitive, the pancreas much less so, and the airways are of intermediate sensitivity (Cutting, 2004). Thus, although most gene therapy efforts are focused on the airway, should other organs become therapeutic targets, the levels of transgene expression required for disease amelioration may differ. The final issue is one of the assessment of success. Detection of either mRNA or CFTR protein provides evidence of gene expression but no confirmation of *functional* correction. Ion transport was the first function of CFTR identified and is thought by most to be key in the pathophysiology of the disease. As such, it is the function assessed most commonly in both gene-based and pharmacological clinical studies. Ion transport is assessed most readily *in vivo* by measurement of the transepithelial potential difference (PD), both at baseline and in response to a variety of drugs, including those that block the sodium channel, such as amiloride, and stimulate chloride secretion (Knowles et al., 1981; Middleton et al., 1994). These measurements can now be obtained in the airway via the bronchoscope (Alton et al., 1999), as well as from the nasal epithelium. Additional *ex vivo* techniques, such as epifluorescence microscopy, have also proved useful in some studies (Stern et al., 1995), and alternative functions of the protein, such as those involved in bacterial adherence (Alton et al., 1999), have also been utilized as end points.

The majority of CF gene therapy trials have been conducted with viral vectors, which are outside the scope of this chapter. The interested reader is directed toward a recent review of the subject (Griesenbach et al., 2003). Several placebo-controlled clinical trials of liposome-mediated *CFTR* gene transfer to the nasal epithelium have confirmed safety and demonstrated variable degrees of functional correction (Caplen *et al.*, 1995; Gill *et al.*, 1997; Knowles *et al.*, 1998; Noone *et al.*, 2000; Porteous *et al.*, 1997; Sorscher *et al.*, 1994). Subsequently, our group conducted the first trial of liposome-mediated *CFTR* to the lower airways of CF patients (Alton *et al.*, 1999). Administration was well tolerated, although mild respiratory symptoms including chest tightness and cough were seen in both groups. This had not been observed with liposome alone in non-CF subjects (Chadwick *et al.*, 1997) and may relate to the pulmonary inflammation present in the lungs of CF patients. In addition, all patients in the treatment group reported mild influenza-like symptoms within the first 24 h. Importantly, these symptoms were not reported after nasal administration, which was included for comparison purposes, suggesting that for safety at least, the nasal epithelium may not be a good surrogate site for such trials. The reason for this was unclear, but it may relate to the presence of unmethylated CpG groups on the bacterially derived DNA. A similar clinical syndrome was seen in another lung trial using the same lipid vector in escalating doses (Ruiz *et al.*, 2001). There was an associated rise in serum interleukin (IL)-6, but not of other inflammatory markers. *In vitro* studies by this group demonstrated that the proinflammatory activity of the DNA alone was minimal, but a brisk response was induced once it was complexed with the cationic lipid, suggesting that CpG motifs were not the major cause.

The first of these two lung trials assessed efficacy in the lower airway by measuring lower airway PD (Alton *et al.*, 1999). In neither the treated nor the placebo (lipid alone) group was there any change in the parameters of sodium absorption (baseline or amiloride response). The treated group, however, demonstrated a significant response to perfusion with low chloride and isoprenaline of approximately 25% of non-CF values. Unlike the reported problems with readministration of adenoviral-mediated gene transfer, Hyde *et al.* (2000), using DC-Chol/DOPE, reported that repeated nasal administration was both well tolerated and could be effective. Most recently, a single dose escalation study has been conducted in the nasal epithelium of CF subjects using nanoparticles composed of a single plasmid DNA molecule and polyethylene glycol-substituted polylysine (Konstan *et al.*, 2003). The study confirmed safety and demonstrated partial correction of chloride transport in some patients. Longer term clinical trials are planned by this group.

In summary, clinical trials in CF have proved safe, but not without side effects. Future plasmid and vector design should be directed at improving efficiency and duration and limiting inflammation. With regard to efficacy, there has been a degree of correction of chloride secretion, but not of sodium hyperabsorption. How this relates to a clinical benefit is as yet unknown. Future trials may need to be designed to incorporate more clinically relevant assays, which are likely to require repeated application of therapy.

B. α_1-Antitrypsin deficiency

Deficiency of α_1-AT, the principal endogenous antiprotease, leads to pulmonary emphysema in adult life and liver disease, which can occur as early as infancy. The disease is inherited in an autosomal recessive fashion, with several mutations having been identified, resulting in either absent or severely low levels of circulating protein (Coakley et al., 2001). The principal action of α_1-AT in the lung is to counter the adverse effects of proteases such as neutrophil elastase in the distal conducting airways and alveoli. Current therapy involves avoidance of damaging environmental triggers such as cigarette smoke and symptomatic treatments. Plasma-derived α_1-AT can be administered intravenously, but is costly and has a short half-life, necessitating frequent administration (Pierce, 1997). Its purification from human serum also raises the possibility of viral transmission. Gene therapy has therefore been considered as an alternative approach.

In contrast to the situation with CF, the secreted nature of the deficient protein makes expression at a distant site a possibility for exogenous gene transfer and simplifies end point assays. The liver, as the natural site of synthesis, would seem the logical choice, although it is possible that were the protein produced at its site of desired action, in the lung, that lower levels of transfection might be sufficient. Several preclinical animal studies have been conducted using viral vectors (Rosenfeld et al., 1991; Stecenko and Brigham, 2003). In a study of aerosolized cationic liposome-mediated α_1-AT gene transfer to rabbit lung, protein was demonstrated both in the airway and in alveolar cells (Canonico et al., 1994). On the basis of these data, the first clinical trial of α_1-AT gene transfer was conducted (Brigham et al., 2000). Patients with α_1-AT deficiency received a single dose of cationic liposome–α_1-AT complex into one nostril, with the other nostril acting as a control. Protein was detected in nasal lavage fluid, with levels peaking at day 5 at approximately one-third of normal values. This rise was not seen in fluid from the control nostril. In addition, levels of the proinflammatory cytokine IL-8 were decreased in the treated nostril. Most interestingly, this anti-inflammatory effect was not observed when intravenously administered purified α_1-AT protein achieved levels within the normal range in nasal lavage, leading the authors to speculate that different routes of administration may lead to a variable response in different sites of expression. Future studies will address this issue and assess administration to the lower airway. One potential difficulty in later-phase studies will be the design of end points to assess clinical benefit in this disease, which progresses extremely slowly in nonsmoking patients.

C. Cancer

In comparison to the two previous single gene disorders, cancer may seem a less obvious choice for gene therapy. Although cancer is fundamentally a disorder of genes, mutation of more than one gene is usually necessary to produce disease.

However, this is a rapidly growing field, with several approaches demonstrating some success. The majority of the strategies reported to date have utilized viral vectors; these are discussed only briefly here, but have been well reviewed elsewhere (Gottesman, 2003; Mosca et al., 2000; Wu et al., 2001).

Mutation of the tumor supressor gene p53 is one of the most common findings in nonsmall cell lung cancers (Rom et al., 2000). Patients bearing such mutations on their tumor cells are less likely to respond to conventional treatment, either chemotherapy or radiotherapy. In vitro studies showed that p53-transfected lung cancer cells were more susceptible to apoptosis (Fujiwara et al., 1994), an effect that was enhanced in the presence of chemotherapeutic agents. Importantly, a bystander effect was observed in animal models in which a subset of cells expressing the gene inhibits the growth of neighboring nontransfected cells. Several groups have administered viral-mediated p53 to lung cancer patients, although the trials have generally been uncontrolled and variable in their success. Suicide gene therapy involves the transfer of a gene encoding an enzyme that converts a nontoxic agent to a toxic chemotherapeutic (Smythe, 2000). This limits the activity of the active drug to the site of gene expression and thus minimizes side effects. The most commonly used system has been the herpes simplex thymidine kinase (HSVtk) gene, which converts ganciclovir into a triphosphorylated derivative that inhibits cell replication. Two major advantages of this approach are a significant bystander effect due to local spread of toxic metabolites and the stimulation of a local immune response. The approach is being utilized in trials of malignancies in many sites, including patients with pleural mesothelioma (Sterman et al., 1998), although all studies are utlizing viral vectors. Immunogenetic therapy is based on the transfer of genes encoding molecules involved in the host immune response in an attempt to enhance immune recognition and destruction of tumor cells. Such genes include various cytokines, interferon-γ, granulocyte macrophage-colony stimulating factor, and heat shock proteins (Leroy et al., 1998). An alternative strategy has been to introduce a foreign gene, such as the bacterially derived β-galactosidase, in the hope that cells expressing this gene will be targeted for destruction.

Although conventional cancer therapies can be highly successful, a major drawback is the high incidence of severe systemic adverse effects. Gene therapy strategies to overcome this include designing of the construct such that transgene expression is dependent on the presence or absence of cancer cell-specific factors. Examples are adenoviral vectors able to replicate only in the absence of normal p53 function (Nemunaitis et al., 2001) or plasmids driven by the survivin promoter. The latter was used complexed to a liposomal vector in a preclinical model, where it was expressed at higher levels in the tumor than elsewhere throughout the body (Chen et al., 2004). Several alternative approaches are being attempted, which have been reviewed elsewhere (Moon et al., 2003). As with other diseases discussed in this chapter, if repeat application is required in order to kill all the malignant cells, viral vectors may be inappropriate.

D. Inflammatory and miscellaneous conditions

1. Acute lung injury (ALI)

Acute lung injury (adult/acute respiratory distress syndrome) is the end result of a variety of insults, including severe sepsis, aspiration, trauma, near-drowning, and pancreatitis (Dennehy et al., 1999). The clinical hallmarks of ALI include impaired oxygenation, which is often poorly responsive to invasive ventilation and patchy infiltrates on chest X-ray in the absence of a raised pulmonary arterial wedge pressure or left atrial hypertension. Despite the improvements in intensive care management, which have led to a recent reduction in mortality, ALI carries a poor prognosis, particularly in children. No specific therapy exists, and current management is largely supportive (Weinacker and Vaszar, 2001). The pathophysiology of ALI is a sequential process of (a) immediate injury, (b) exudative alveolar inflammation with edema and, finally, (c) fibro-proliferative repair (Dennehy et al., 1999). Recognition that the early stages of this process are characterized by generalized intravascular activation, endothelial damage, and high levels of proinflammatory cytokines has led to the development of novel gene and small molecule-based approaches to treatment. These have some theoretical advantages over the use of proteins, including the potential for cell specificity, the possibility of delivering intracellular proteins, duration of action, and possibly cost. The majority of approaches target either inflammation or oxidative stress, although other investigators have explored the potential for increasing fluid reabsorption and reducing pulmonary edema in this disease process. Inflammation can be targeted either by antagonizing proinflammatory cytokines or by increasing the levels or effects of anti-inflammatory cytokines. In the context of ALI, studies have been undertaken both with conventional gene transfer techniques and with antisense oligonucleotides, which reduce translation to protein by specific binding to mRNA. The vast majority of these studies have utilized viral vectors (reviewed by Brigham and Stecenko, 2000) and are not discussed further here. Conary et al. (1994) administered a gene encoding the prostaglandin synthase gene complexed to cationic lipid in a rabbit model of endotoxin-induced acute lung damage and reported decreased pulmonary edema and thromboxane B_2 release. An alternative target in this disease is pulmonary edema. In the healthy state, alveolar liquid is cleared by the basolaterally situated Na^+/K^+ ATPase, which is upregulated during the resolution phase of pulmonary edema (Factor, 2001), a recognized feature of acute lung injury. Our group has used cationic liposomes to overexpress this ATPase in a mouse model of acute lung injury and demonstrated a significant reduction in pulmonary edema (Stern *et al.*, 2000). Another approach has been to increase activity of the epithelial sodium ion channel, ENaC, levels and function of which are both upregulated by β_2-adrenoreceptor expression (Dumasius *et al.*, 2001), although to date this has not been performed with a non-viral vector.

2. Asthma

Asthma is a disease of high prevalence characterized by type 2 T helper (Th2) lymphocyte-mediated inflammation (Lee *et al.*, 2001a) and airway hyperreactivity. Current treatment with bronchodilators and anti-inflammatory agents is successful in treating wheeze, cough, and breathlessness in the vast majority of patients (Suissa and Ernst, 2001). However, there is a subgroup who fail to respond for whom novel therapeutic approaches may be relevant. Following success in animal models with Th1-type cytokine protein therapy, including IFN-γ. (Lack *et al.*, 1994) and IL-12 (Schwarze *et al.*, 1998), beneficial effects have been demonstrated with a variety of cytokine genes. Of these, non-viral vectors have been used to transfer an IL-12 variant to the airways of mice with dust mite-induced inflammation (Lee *et al.*, 2001b). This group demonstrated significant reductions in airway hyperresponsiveness, IL-5, and local eosinophils after treatment. Galectin-3 selectively downregulates IL-5 in several cell types. Plasmid DNA encoding this gene was administered intratracheally in a rat model of asthma and was shown to normalize eosinophil and T-cell counts in bronchoalveolar lavage fluid (del Pozo *et al.*, 2002). Success has also been reported by two groups in mouse models with lipid-complexed interferon-γ (Dow *et al.*, 1999; Li *et al.*, 1996). Thus, in line with the multifactorial nature of the disease, many angles are being explored for new gene therapies for asthma. Given the success of conventional treatment, however, it is likely that such therapies may only be useful for a small minority of patients who are currently failing standard regimes.

3. Fibrotic lung disease

Lung fibrosis can be idiopathic, part of a multisystem (e.g. autoimmune) disorder, or iatrogenic [following radiotherapy or drugs such as bleomycin (Fonseca *et al.*, 1999)]. The prognosis is often poor and available therapies are limited. A variety of growth factors, in particular TGF-β, are considered key in the progressive nature of the disease (Sime and O'Reilly, 2001), leading to these molecules as targets for novel therapies. Epperly *et al.* (1998) have reported prevention of radiation-induced lung fibrosis and improved survival with liposome-mediated-manganese superoxide dismutase. The design of radiation-induced promoters (Scott *et al.*, 2000) may help limit sites of expression of therapeutic transgenes in both radiation-related fibrosis and lung cancers.

Finally, a clinical study has shown success with interferon-γ 1b protein therapy, which, in combination with prednisolone, led to improvements in pulmonary function and oxygen saturation (Ziesche *et al.*, 1999); in contrast, the group treated with the corticosteroid alone deteriorated. This and other reports suggesting a role for Fas-mediated alveolar cell apoptosis (Kuwano *et al.*, 1999) may lead to further new gene-based strategies for this disease.

4. Lung transplantation

The major obstacle to organ transplantation programs worldwide remains the lack of sufficient donor organs. Medium- to long-term success is, however, further limited both by acute ischemia–reperfusion injury, which is particularly problematic for lung transplantation (Mal *et al.*, 1998), and by the host response leading to organ rejection (Ward and Muller, 2000). Given the shortage of available organs, strategies to attenuate these processes would be of major benefit. Organ transplantation theoretically creates a unique window of opportunity for gene therapy: in addition to administration to the host prior to removal of the organ, or the recipient after surgery, therapeutic genes could be administered to the organ *ex vivo* during the procedure. Cassivi *et al.* (1999) demonstrated the feasibility of Ad-β—gal transfection either before procurement (via tracheostomy) or *ex vivo* after surgical removal. At the critical time of reperfusion, they demonstrated significantly greater transgene levels in lungs transfected via the tracheostomy than in those transfected *ex vivo*. Levels in other organs were virtually absent, confirming limitation to the lungs. However, another study using liposomes demonstrated that *ex vivo* transfection was superior to intravenous injection of donors prior to organ harvesting (Boasquevisque *et al.*, 1999). Another concern may be the cold preservation of organs prior to transplantation, which has been shown to affect efficacy of gene transfer adversely (Boasquevisque *et al.*, 1998). The optimal route and timing of gene administration may therefore depend on the vector and the desired function and site of transgene expression. Various mediators have been implicated in ischemia–reperfusion injury, including stimulated leucocytes and platelets, complement, proinflammatory cytokines, and oxidants (Mal *et al.*, 1998). Attempts to combat this process with exogenous recombinant proteins have been limited, thought largely to be related to the inability to achieve and maintain high local levels. Gene transfer may therefore be a more useful approach. Schmid *et al.* (2000) administered the lipid-mediated Fas ligand retrogradely through the pulmonary venous system, prior to lung removal along with a single dose of cyclosporine. Compared with controls, rats receiving Fas-transfected lungs had better day 5 gas exchange and demonstrated significantly less histological evidence of acute rejection. Other groups have reported success with TGF-β. Daddi *et al.* (2003) administered naked DNA to lung transplanted rats and reported that the active form of TGF-β (but not the latent form) led to better postoperative oxygenation and significantly lower levels of inflammatory mediators. Although this approach looks promising, data have yet to be translated into clinical settings, which will be complicated by the clinical heterogeneity of the patient groups studied.

VII. CONCLUSIONS

The principle of non-viral gene therapy for lung disease has now been proved, with clinical trials showing success in a variety of genetic and acquired disorders. One of the major benefits of the non-viral approach is the ability to repeat dose, which for most diseases is likely to be essential. The major problem currently is with low levels of efficiency and short duration. Much research is focusing on both overcoming the identified barriers to gene transfer and improving the design of GTAs to both enhance efficiency and reduce toxicity.

References

Alton, E. W. F. W., Stern, M., Farley, R., Jaffe, A., Chadwick, S. L., Phillips, J., Davies, J., Smith, S. N., Browning, J., Davies, M. G., Hodson, M. E., Durham, S. R., Li, D., Jeffery, P. K., Scallan, M., Balfour, R., Eastman, S. J., Cheng, S. H., Smith, A. E., Meeker, D., and Geddes, D. M. (1999). Cationic lipid-mediated CFTR gene transfer to the lungs and nose of patients with cystic fibrosis: A double-blind placebo-controlled trial. *Lancet* **353,** 947–954.

Amabile, P. G., Waugh, J. M., Lewis, T. N., Elkins, C. J., Janas, W., and Dake, M. D. (2001). High-efficiency endovascular gene delivery via therapeutic ultrasound. *J. Am. Coll. Cardiol.* **37,** 1975–1980.

Armstrong, D. S., Grimwood, K., Carzino, R., Carlin, J. B., Olinsky, A., and Phelan, P. D. (1995). Lower respiratory infection and inflammation in infants with newly diagnosed cystic fibrosis. *BMJ* **310,** 1571–1572.

Bigey, P., Bureau, M. F., and Scherman, D. (2002). *In vivo* plasmid DNA electrotransfer. *Curr. Opin. Biotechnol.* **13,** 443–447.

Boasquevisque, C. H., Mora, B. N., Bernstein, M., Osburn, W. O., Nietupski, J., Scheule, R. K., Cooper, J. D., Botney, M., and Patterson, G. A. (1998). *Ex vivo* liposome-mediated gene transfer to lung isografts. *J. Thorac. Cardiovasc. Surg.* **115,** 38–44.

Boasquevisque, C. H., Mora, B. N., Boglione, M., Ritter, J. K., Scheule, R. K., Yew, N., Debruyne, L., Qin, L., Bromberg, J. S., and Patterson, G. A. (1999). Liposome-mediated gene transfer in rat lung transplantation: A comparison between the *in vivo* and *ex vivo* approaches. *J. Thorac. Cardiovasc. Surg.* **117,** 8–14.

Brigham, K. L., Lane, K. B., Meyrick, B., Stecenko, A. A., Strack, S., Cannon, D. R., Caudill, M., and Canonico, A. E. (2000). Transfection of nasal mucosa with a normal α_1-antitrypsin gene in α_1-antitrypsin-deficient subjects: Comparison with protein therapy. *Hum. Gene Ther.* **11,** 1023–1032.

Brigham, K. L., and Stecenko, A. A. (2000). Gene therapy for acute lung injury. *Intensive Care Med.* **26**(Suppl. 1), S119–S123.

Brisson, M., He, Y., Li, S., Yang, J. P., and Huang, L. (1999). A novel T7 RNA polymerase autogene for efficient cytoplasmic expression of target genes. *Gene Ther.* **6**, 263–270.

Canonico, A. E., Conary, J. T., Meyrick, B. O., and Brigham, K. L. (1994). Aerosol and intravenous transfection of human alpha 1-antitrypsin gene to lungs of rabbits. *Am. J. Respir. Cell Mol. Biol.* **10**, 24–29.

Caplen, N. J., Alton, E. W., Middleton, P. G., Dorin, J. R., Stevenson, B. J., Gao, X., Durham, S. R., Jeffery, P. K., Hodson, M. E., Coutelle, C., et al. (1995). Liposome-mediated CFTR gene transfer to the nasal epithelium of patients with cystic fibrosis. *Nature Med.* **1**, 39–46.

Cassivi, S. D., Cardella, J. A., Fischer, S., Liu, M., Slutsky, A. S., and Keshavjee, S. (1999). Transtracheal gene transfection of donor lungs prior to organ procurement increases transgene levels at reperfusion and following transplantation. *J. Heart Lung Transplant.* **18**, 1181–1188.

Chadwick, S. L., Kingston, H. D., Stern, M., Cook, R. M., O'Connor, B. J., Lukasson, M., Balfour, R. P., Rosenberg, M., Cheng, S. H., Smith, A. E., Meeker, D. P., Geddes, D. M., and Alton, E. W. (1997). Safety of a single aerosol administration of escalating doses of the cationic lipid GL-67/DOPE/DMPE-PEG$_{5000}$ formulation to the lungs of normal volunteers. *Gene Ther.* **4**, 937–942.

Chen, J. S., Liu, J. C., Shen, L., Rau, K. M., Kuo, H. P., Li, Y.M, Shi, D., Lee, Y. C., Chang, K. J, Hung, M. C. (2004). Cancer-specific activation of the survivin promoter and its potential in gene therapy. *Cancer Gene Therapy.* **11**, 740–747.

Cho, Y. W., Kim, J. D., and Park, K. (2003). Polycation gene delivery systems: Escape from endosomes to cytosol. *J. Pharm. Pharmacol.* **55**, 721–734.

Coakley, R. J., Taggart, C., O'Neill, S., and McElvaney, N. G. (2001). Alpha1-antitrypsin deficiency: Biological answers to clinical questions. *Am. J. Med. Sci.* **321**, 33–41.

Conary, J. T., Parker, R. E., Christman, B. W., Faulks, R. D., King, G. A., Meyrick, B. O., and Brigham, K. L. (1994). Protection of rabbit lungs from endotoxin injury by *in vivo* hyperexpression of the prostaglandin G/H synthase gene. *J. Clin. Invest.* **93**, 1834–1840.

Croyle, M. A., Cheng, X., Sandhu, A., and Wilson, J. M. (2001). Development of novel formulations that enhance adenoviral-mediated gene expression in the lung *in vitro* and *in vivo*. *Mol. Ther.* **4**, 22–28.

Cutting, G. R. (2004). What we have learned from correlating genotype to phenotype. *Pediatr. Pulmonol.* **S27**, 94.

Daddi, N., Kanaan, S. A., Suda, T., Tagawa, T., D'Ovidio, F., Grapperhaus, K., Kozower, B. D., Ritter, J. H., Mohanakumar, T., and Patterson, G. A. (2003). Recipient intramuscular administration of naked plasmid TGF-beta1 attenuates lung graft reperfusion injury. *J. Heart Lung Transplant.* **22**, 1323–1334.

Das, A., and Niven, R. (2001). Use of perfluorocarbon (fluorinert) to enhance reporter gene expression following intratracheal instillation into the lungs of Balb/c mice: Implications for nebulized delivery of plasmids. *J. Pharm. Sci.* **90**, 1336–1344.

Dean, D. A., Machado-Aranda, D., Blair-Parks, K., Yeldandi, A. V., and Young, J. L. (2003). Electroporation as a method for high-level non-viral gene transfer to the lung. *Gene Ther.* **10**, 1608–1615.

del Pozo, V., Rojo, M., Rubio, M. L., Cortegano, I., Cardaba, B., Gallardo, S., Ortega, M., Civantos, E., Lopez, E., Martin-Mosquero, C., Peces-Barba, G., Palomino, P., Gonzalez-Mangado, N., Lahoz, C. (2002) Gene therapy with galectin-3 inhibits bronchial obstruction and inflammation in antigen-challenged rats through interleukin-5 gene down regulation. *Am. J. Respir. Crit. Care Med.* **166**(5) 732–737.

Dennehy, K. C., and Bigatello, L. M. (1999). Pathophysiology of the acute respiratory distress syndrome. *Int. Anesthesiol. Clin.* **37**, 1–13.

Dow, S. W., Schwarze, J., Heath, T. D., Potter, T. A., and Gelfand, E. W. (1999). Systemic and local interferon gamma gene delivery to the lungs for treatment of allergen-induced airway hyperresponsiveness in mice. *Hum. Gene Ther.* **10**, 1905–1914.

Dumasius, V., Sznajder, J. I., Azzam, Z. S., Boja, J., Mutlu, G. M., Maron, M. B., and Factor, P. (2001). beta(2)-adrenergic receptor overexpression increases alveolar fluid clearance and responsiveness to endogenous catecholamines in rats. *Circ. Res.* **89,** 907–914.

Engelhardt, J. F., Yankaskas, J. R., Ernst, S. A., *et al.* (1992). Submucosal glands are the predominant site of CFTR expression in the human bronchus. *Nature Genet.* **2,** 240–248.

Engelhardt, J. F., Zepeda, M., Cohn, J. A., Yankaskas, J. R., and Wilson, J. M. (1994). Expression of the cystic fibrosis gene in adult human lung. *J. Clin. Invest.* **93,** 737–749.

Epperly, M., Bray, J., Kraeger, S., Zwacka, R., Engelhardt, J., Travis, E., and Greenberger, J. (1998). Prevention of late effects of irradiation lung damage by manganese superoxide dismutase gene therapy. *Gene Ther.* **5,** 196–208.

Factor, P. (2001). Role and regulation of lung Na,K-ATPase. *Cell. Mol. Biol.* **47,** 347–361.

Fenske, D. B., MacLachlan, I., and Cullis, P. R. (2001). Long-circulating vectors for the systemic delivery of genes. *Curr. Opin. Mol. Ther.* **3,** 153–158.

Ferrari, S., Kitson, C., Farley, R., Steel, R., Marriott, C., Parkins, D. A., Scarpa, M., Wainwright, B., Evans, M. J., Colledge, W. H., Geddes, D. M., and Alton, E. W. (2001). Mucus altering agents as adjuncts for non-viral gene transfer to airway epithelium. *Gene Ther.* **8,** 1380–1386.

Fonseca, C., Abraham, D., and Black, C. M. (1999). Lung fibrosis. *Spring. Semin. Immunopathol.* **21,** 453–474.

Fujiwara, T., Grimm, E. A., Mukhopadhyay, T., Zhang, W. W., Owen-Schaub, L. B., and Roth, J. A. (1994). Induction of chemosensitivity in human lung cancer cells *in vivo* by adenovirus-mediated transfer of the wild-type p53 gene. *Cancer Res.* **54,** 2287–2291.

Gan, K. H., Veeze, H. J., van den Ouweland, A. M., Halley, D. J., Scheffer, H., van der Hout, A., Overbeek, S. E., de Jongste, J. C., Bakker, W., and Heijerman, H. G. (1995). A cystic fibrosis mutation associated with mild lung disease. *N. Engl. J. Med.* **333,** 95–99.

Gersting, S. W., Schillinger, U., Lausier, J., Nicklaus, P., Rudolph, C., Plank, C., Reinhardt, D., and Rosenecker, J. (2004). Gene delivery to respiratory epithelial cells by magnetofection. *J. Gene Med.* **6,** 913–922.

Gill, D. R., Smyth, S. E., Goddard, C. A., Pringle, I. A., Higgins, C. F., Colledge, W. H., and Hyde, S. C. (2001). Increased persistence of lung gene expression using plasmids containing the ubiquitin C or elongation factor 1alpha promoter. *Gene Ther.* **8,** 1539–1546.

Gill, D. R., Southern, K. W., Mofford, K. A., Seddon, T., Huang, L., Sorgi, F., Thomson, A., MacVinish, L. J., Ratcliff, R., Bilton, D., Lane, D. J., Littlewood, J. M., Webb, A. K., Middleton, P. G., Colledge, W. H., Cuthbert, A. W., Evans, M. J., Higgins, C. F., and Hyde, S. C. (1997). A placebo-controlled study of liposome-mediated gene transfer to the nasal epithelium of patients with cystic fibrosis. *Gene Ther* **4,** 199–209.

Gottesman, M. M. (2003). Cancer gene therapy: An awkward adolescence. *Cancer Gene Ther.* **10,** 501–508.

Griesenbach, U., Chonn, A., Cassady, R., Hannam, V., Ackerley, C., Post, M., Tanswell, A. K., Olek, K., O'Brodovich, H., and Tsui, L. C. (1998). Comparison between intratracheal and intravenous administration of liposome-DNA complexes for cystic fibrosis lung gene therapy. *Gene Ther.* **5,** 181–188.

Griesenbach, U., Geddes, D. M., and Alton, E. W. (2003). Update on gene therapy for cystic fibrosis *Curr. Opin. Molec. Ther.* **5,** 489–494.

Grosse, S., Tremeau-Bravard, A., Aron, Y., Briand, P., and Fajac, I. (2002). Intracellular rate-limiting steps of gene transfer using glycosylated polylysines in cystic fibrosis airway epithelial cells. *Gene Ther.* **9,** 1000–1007.

Hacein-Bey-Abina, S., von Kalle, C., Schmidt, M., Le Deist, F., Wulffraat, N., McIntyre, E., Radford, I., Villeval, J. L., Fraser, C. C., Cavazzana-Calvo, M., and Fischer, A. (2003). A serious adverse event after successful gene therapy for X-linked severe combined immunodeficiency. *N. Engl. J. Med.* **348,** 255–256.

Houtmeyers, E., Gosselink, R., Gayan-Ramirez, G., and Decramer, M. (1999). Regulation of mucociliary clearance in health and disease. *Eur. Respir. J.* **13**, 1177–1188.

Hyde, S. C., Southern, K. W., Gileadi, U., *et al.* (2000). Repeat administration of DNA/liposomes to the nasal epithelium of patients with cystic fibrosis. *Gene Ther.* **7**, 1156–1165.

Johnson, L. G., Boyles, S. E., Wilson, J., and Boucher, R. C. (1995). Normalization of raised sodium absorption and raised calcium-mediated chloride secretion by adenovirus-mediated expression of cystic fibrosis transmembrane conductance regulator in primary human cystic fibrosis airway epithelial cells. *J. Clin. Invest.* **95**, 1377–1382.

Kim, H. J., Greenleaf, J. F., Kinnick, R. R., Bronk, J. T., and Bolander, M. E. (1996). Ultrasound-mediated transfection of mammalian cells. *Hum. Gene Ther.* **7**, 1339–1346.

Kitson, C., Angel, B., Judd, D., Rothery, S., Severs, N. J., Dewar, A., Huang, L., Wadsworth, S. C., Cheng, S. H., Geddes, D. M., and Alton, E. W. (1999). The extra- and intracellular barriers to lipid and adenovirus-mediated pulmonary gene transfer in native sheep airway epithelium. *Gene Ther.* **6**, 534–546.

Knowles, M., Gatzy, J., and Boucher, R. (1981). Increased bioelectric potential difference across respiratory epithelia in cystic fibrosis. *N. Engl. J. Med.* **305**, 1489–1495.

Knowles, M. R., Noone, P. G., Hohneker, K., Johnson, L. G., Boucher, R. C., Efthimiou, J., Crawford, C., Brown, R., Schwartzbach, C., and Pearlman, R. (1998). A double-blind, placebo controlled, dose ranging study to evaluate the safety and biological efficacy of the lipid-DNA complex GR213487B in the nasal epithelium of adult patients with cystic fibrosis. *Hum. Gene Ther.* **9**, 249–269.

Koehler, D. R., Hannam, V., Belcastro, R., Steer, B., Wen, Y., Post, M., Downey, G., Tanswell, A. K., and Hu, J. (2001). Targeting transgene expression for cystic fibrosis gene therapy. *Mol. Ther.* **4**, 58–65.

Konstan, M. W., Wagener, J. S., Hilliard, K. A., Kowalczyk, T. H., Hyatt, S. L., Davis, P. B., Moen, R. C., and Cooper, M. J. (2003). Single dose escalation study to evaluate safety of nasal administration of CFTR001 gene transfer vector to subjects with cystic fibrosis. *Mol. Ther.* **7**, S386.

Kuwano, K., Hagimoto, N., Kawasaki, M., Yatomi, T., Nakamura, N., Nagata, S., Suda, T., Kunitake, R., Maeyama, T., Miyazaki, H., and Hara, N. (1999). Essential roles of the Fas-Fas ligand pathway in the development of pulmonary fibrosis. *J. Clin. Invest.* **104**, 13–19.

Lack, G., Renz, H., Saloga, J., Bradley, K. L., Loader, J., Leung, D. Y., Larsen, G., and Gelfand, E. W. (1994). Nebulized but not parenteral IFN-gamma decreases IgE production and normalizes airways function in a murine model of allergen sensitization. *J. Immunol.* **152**, 2546–2554.

Lechardeur, D., Sohn, K. J., Haardt, M., Joshi, P. B., Monck, M., Graham, R. W., Beatty, B., Squire, J., O'Brodovich, H., and Lukacs, G. L. (1999). Metabolic instability of plasmid DNA in the cytosol: A potential barrier to gene transfer. *Gene Ther.* **6**, 482–497.

Lee, N. A., Gelfand, E. W., and Lee, J. J. (2001a). Pulmonary T cells and eosinophils: Co-conspirators or independent triggers of allergic respiratory pathology? *J. Allergy Clin. Immunol.* **107**, 945–957.

Lee, Y. L., Ye, Y. L., Yu, C. I., Wu, Y. L., Lai, Y. L., Ku, P. H., Hong, R. L., and Chiang, B. L. (2001b). Construction of single-chain interleukin-12 DNA plasmid to treat airway hyperresponsiveness in an animal model of asthma. *Hum. Gene. Ther* **12**, 2065–2079.

Leroy, P., Slos, P., Homann, H., Erbs, P., Poitevin, Y., Regulier, E., Colonna, F. Q., Devauchelle, P., Roth, C., Pavirani, A., and Mehtali, M. (1998). Cancer immunotherapy by direct *in vivo* transfer of immunomodulatory genes. *Res. Immunol.* **149**, 681–684.

Li, X. M., Chopra, R. K., Chou, T. Y., Schofield, B. H., Wills-Karp, M., and Huang, S. K. (1996). Mucosal IFN-gamma gene transfer inhibits pulmonary allergic responses in mice. *J. Immunol.* **157**, 3216–3219.

Machado-Aranda, D., Adir, Y., Young, J. L., Briva, A., Budinger, G. R., Yeldandi, A. V., Sznajder, J. I., and Dean, D. A. (2005). Gene transfer of the Na+, K+-ATPase β 1 subunit using electroporation increases lung liquid clearance. *Am. J. Respir. Crit. Care Med.* **171,** 204–211.

Machet, L., and Boucaud, A. (2002). Phonophoresis: Efficiency, mechanisms and skin tolerance. *Int. J. Pharm.* **243,** 1–15.

Mal, H., Dehoux, M., Sleiman, C., Boczkowski, J., Leseche, G., Pariente, R., and Fournier, M. (1998). Early release of proinflammatory cytokines after lung transplantation. *Chest* **113,** 645–651.

Manome, Y., Nakamura, M., Ohno, T., and Furuhata, H. (2000). Ultrasound facilitates transduction of naked plasmid DNA into colon carcinoma cells *in vitro* and *in vivo. Hum. Gene Ther.* **11,** 1521–1528.

McLachlan, G., Stevenson, B. J., Davidson, D. J., and Porteous, D. J. (2000). Bacterial DNA is implicated in the inflammatory response to delivery of DNA/DOTAP to mouse lungs. *Gene Ther.* **7,** 384–392.

Meng, Q. H., Robinson, D., Jenkins, R. G., McAnulty, R. J., and Hart, S. L. (2004). Efficient transfection of non-proliferating human airway epithelial cells with a synthetic vector system. *J. Gene Med.* **6,** 210–221.

Middleton, P. G., Geddes, D. M., and Alton, E. W. F. W. (1994). Protocols for *in vivo* measurement of the ion transport defects in cystic fibrosis nasal epithelium. *Eur. Respir. J.* **7,** 2050–2056.

Miller, D. L., Pislaru, S. V., and Greenleaf, J. E. (2002). Sonoporation: Mechanical DNA delivery by ultrasonic cavitation. *Somat. Cell Mol. Genet.* **27,** 115–134.

Moon, C., Oh, Y., and Roth, J. A. (2003). Current status of gene therapy for lung cancer and head and neck cancer. *Clin. Cancer Res.* **9,** 5055–5067.

Mosca, P. J., Morse, M. A., D'Amico, T. A., Crawford, J., and Lyerly, H. K. (2000). Gene therapy for lung cancer. *Clin. Lung Cancer* **1,** 218–226.

Munkonge, F. M., Hillery, E., Griesenbach, U., Geddes, D. M., and Alton, E. W. F. W. (1998). Isolation of a putative nuclear import DNA shuttle protein. *Mol. Biol. Cell* **9,** 187a.

Nemunaitis, J., Cunningham, C., Buchanan, A., Blackburn, A., Edelman, G., Maples, P., Netto, G., Tong, A., Randlev, B., Olson, S., and Kirn, D. (2001). Intravenous infusion of a replication-selective adenovirus (ONYX-015) in cancer patients: safety, feasibility and biological activity. *Gene Ther.* **8,** 746–759.

Niidome, T., and Huang, L. (2002). Gene therapy progress and prospects: Non-viral vectors. *Gene Ther.* **9,** 1647–1652.

Noone, P. G., Hohneker, K. W., Zhou, Z., Johnson, L. G., Foy, C., Gipson, C., Jones, K., Noah, T. L., Leigh, M. W., Schwartzbach, C., Efthimiou, J., Pearlman, R., Boucher, R. C., and Knowles, M. R. (2000). Safety and biological efficacy of a lipid-CFTR complex for gene transfer in the nasal epithelium of adult patients with cystic fibrosis. *Mol. Ther.* **1,** 105–114.

Perricone, M. A., Rees, D. D., Sacks, C. R., Smith, K. A., Kaplan, J. M., and St George, J. A. (2000). Inhibitory effect of cystic fibrosis sputum on adenovirus-mediated gene transfer in cultured epithelial cells. *Hum. Gene Ther.* **11,** 1997–2008.

Pickles, R. J., Fahrner, J. A., Petrella, J. M., Boucher, R. C., and Bergelson, J. M. (2000). Retargeting the coxsackievirus and adenovirus receptor to the apical surface of polarized epithelial cells reveals the glycocalyx as a barrier to adenovirus-mediated gene transfer. *J. Virol.* **74,** 6050–6057.

Pierce, J. A. (1997). Alpha1-antitrypsin augmentation therapy. *Chest* **112,** 872–874.

Plank, C., Mechtler, K., Szoka, F. C., Jr., and Wagner, E. (1996). Activation of the complement system by synthetic DNA complexes: A potential barrier for intravenous gene delivery. *Hum. Gene Ther.* **7,** 1437–1446.

Porteous, D. J., Dorin, J. R., McLachlan, G., Davidson-Smith, H., Davidson, H., Stevenson, B. J., Carothers, A. D., Wallace, W. A., Moralee, S., Hoenes, C., Kallmeyer, G., Michaelis, U., Naujoks, K., Ho, L. P., Samways, J. M., Imrie, M., Greening, A. P., and Innes, J. A. (1997). Evidence for safety and efficacy of DOTAP cationic liposome mediated CFTR gene transfer to the nasal epithelium of patients with cystic fibrosis. *Gene Ther.* 4, 210–218.

Pringle, I. A., Davies, L. A., McLachlan, G., Collie, D. D., Gill, D. R., and Hyde, S. C. (2004). Duration of reporter gene expression from naked DNA in the mouse lung following direct electroporation and development of wire electrodes for sheep lung electroporation studies. *Mol. Ther.* 9, S1–S56.

Raczka, E., Kukowska-Latallo, J. F., Rymaszewski, M., Chen, C., and Baker, J. R., Jr. (1998). The effect of synthetic surfactant Exosurf on gene transfer in mouse lung *in vivo. Gene Ther.* 5, 1333–1339.

Rom, W. N., Hay, J. G., Lee, T. C., Jiang, Y., and Tchou-Wong, K.-M. (2000). Molecular and genetic aspects of lung cancer. *Am. J. Respir. Crit. Care Med.* 161, 1355–1367.

Rosenfeld, M. A., Siegfried, W., Yoshimura, K., Yoneyama, K., Fukayama, M., Stier, L. E., Paakko, P. K., Gilardi, P., Stratford-Perricaudet, L. D., Perricaudet, M., *et al.* (1991). Adenovirus-mediated transfer of a recombinant alpha 1-antitrypsin gene to the lung epithelium *in vivo. Science* 252, 431–434.

Ruiz, F. E., Clancy, J. P., Perricone, M. A., Bebok, Z., Hong, J. S., Cheng, S. H., Meeker, D. P., Young, K. R., Schoumacher, R. A., Weatherly, M. R., Wing, L., Morris, J. E., Sindel, L., Rosenberg, M., van Ginkel, F. W., McGhee, J. R., Kelly, D., Lyrene, R. K., and Sorscher, E. J. (2001). A clinical inflammatory syndrome attributable to aerosolized lipid-DNA administration in cystic fibrosis. *Hum. Gene Ther.* 12, 751–761.

Satkauskas, S., Bureau, M. F., Puc, M., Mahfoudi, A., Scherman, D., Miklavcic, D., and Mir, L. M. (2002). Mechanisms of *in vivo* DNA electrotransfer: Respective contributions of cell electropermeabilization and DNA electrophoresis. *Mol. Ther.* 5, 133–140.

Scherer, F., Anton, M., Schillinger, U., Henke, J., Bergemann, C., Kruger, A., Gansbacher, B., and Plank, C. (2002). Magnetofection: Enhancing and targeting gene delivery by magnetic force *in vitro* and *in vivo. Gene Ther.* 9, 102–109.

Schmid, R. A., Stammberger, U., Hillinger, S., Gaspert, A., Boasquevisque, C. H., Malipiero, U., Fontana, A., and Weder, W. (2000). Fas ligand gene transfer combined with low dose cyclosporine A reduces acute lung allograft rejection. *Transpl. Int.* 13(Suppl. 1), S324–S328.

Schwarze, J., Hamelmann, E., Cieslewicz, G., Tomkinson, A., Joetham, A., Bradley, K., and Gelfand, E. W. (1998). Local treatment with IL-12 is an effective inhibitor of airway hyperresponsiveness and lung eosinophilia after airway challenge in sensitized mice. *J. Allergy Clin. Immunol.* 102, 86–93.

Schwarzt, D. A., Quinn, T. J., Thorne, P. S., *et al.* (1997). CpG motifs in bacterial DNA cause inflammation in the lower respiratory tract. *J. Clin. Invest.* 100, 68–73.

Scott, E. S., Wiseman, J. W., Evans, M. J., and Colledge, W. H. (2001). Enhanced gene delivery to human airway epithelial cells using an integrin-targeting lipoplex. *J. Gene Med.* 3, 125–134.

Scott, S. D., Marples, B., Hendry, J. H., Lashford, L. S., Embleton, M. J., Hunter, R. D., Howell, A., and Margison, G. P. (2000). A radiation-controlled molecular switch for use in gene therapy of cancer. *Gene Ther.* 7, 1121–1125.

Seiler, M. P., Luner, P., Moninger, T. O., Karp, P. H., Keshavjee, S., and Zabner, J. (2002). Thixotropic solutions enhance viral-mediated gene transfer to airway epithelia. *Am. J. Respir. Cell Mol. Biol.* 27, 133–140.

Sime, P. J., and O'Reilly, K. M. (2001). Fibrosis of the lung and other tissues: New concepts in pathogenesis and treatment. *Clin. Immunol.* 99, 308–319.

Smythe, W. R. (2000). Prodrug/drug sensitivity gene therapy: Current status. *Curr. Oncol. Rep.* 2, 17–22.

Snyder, E. L., and Dowdy, S. F. (2001). Protein/peptide transduction domains: Potential to deliver large DNA molecules into cells. *Curr. Opin. Mol. Ther.* **3,** 147–152.

Sorscher, E. J., Logan, J. J., Frizzell, R. A., Lyrene, R. K., Bebok, Z., Dong, J. Y., Duvall, M. D., Felgner, P. L., Matalon, S., Walker, L., *et al.* (1994). Gene therapy for cystic fibrosis using cationic liposome mediated gene transfer: A phase I trial of safety and efficacy in the nasal airway. *Hum. Gene Ther.* **5,** 1259–1277.

Stecenko, A. A., and Brigham, K. L. (2003). Gene therapy progress and prospects: Alpha-1 antitrypsin. *Gene Ther.* **10,** 95–99.

Sterman, D. H., Treat, J., Litzky, L. A., Amin, K. M., Coonrod, L., Molnar-Kimber, K., Recio, A., Knox, L., Wilson, J. M., Albelda, S. M., and Kaiser, L. R. (1998). Adenovirus-mediated herpes simplex virus thymidine kinase/ ganciclovir gene therapy in patients with localised malignancy: Results of a phase I clinical trial in malignant mesothelioma. *Hum. Gene Ther.* **9,** 1083–1092.

Stern, M., Caplen, N. J., Browning, J. E., Griesenbach, U., Sorgi, F., Huang, L., Gruenert, D. C., Marriot, C., Crystal, R. G, Geddes, D. M., and Alton, E. W. (1998). The effects of mucolytic agents on gene transfer across a CF sputum barrier *in vitro*. *Gene Ther.* **5,** 91–98.

Stern, M., Munkonge, F. M., Caplen, N. J., *et al.* (1995). Quantitative fluorescence measurements of chloride secretion in native airway epithelium from CF and non-CF subjects. *Gene Ther.* **2,** 766–774.

Stern, M., Ulrich, K., Robinson, C., Copeland, J., Griesenbach, U., Masse, C., Cheng, S., Munkonge, F., Geddes, D., Berthiaume, Y., and Alton, E. (2000). Pretreatment with cationic lipid-mediated transfer of the Na+K+-ATPase pump in a mouse model *in vivo* augments resolution of high permeability pulmonary oedema. *Gene Ther.* **7,** 960–966.

Suissa, S., and Ernst, P. (2001). Inhaled corticosteroids: Impact on asthma morbidity and mortality. *J. Allergy Clin. Immunol.* **107,** 937–944.

Warburton, D., Wuenschell, C., Flores-Delgado, G., and Anderson, K. (1998). Commitment and differentiation of lung cell lineages. *Biochem. Cell Biol.* **76,** 971–995.

Ward, S., and Muller, N. L. (2000). Pulmonary complications following lung transplantation. *Clin. Radiol.* **55,** 332–339.

Weinacker, A. B., and Vaszar, L. T. (2001). Acute respiratory distress syndrome: Physiology and new management strategies. *Annu. Rev. Med.* **52,** 221–237.

Weiss, D. J., Strandjord, T. P., Liggitt, D., and Clark, J. G. (1999). Perflubron enhances adenovirus-mediated gene expression in lungs of transgenic mice with chronic alveolar filling. *Hum Gene Ther.* **10,** 2287–2293.

Welsh, M. J., and Smith, A. E. (1993). Molecular mechanisms of CFTR chloride channel dysfunction in cystic fibrosis. *Cell* **73,** 1251–1254.

Worgall, S., Leopold, P. L., Wolff, G., Ferris, B., Van Roijen, N., and Crystal, R. G. (1997). Role of alveolar macrophages in rapid elimination of adenovirus vectors administered to the epithelial surface of the respiratory tract. *Hum. Gene Ther.* **8,** 1675–1684.

Wu, Q., Moyana, T., and Xiang, J. (2001). Cancer gene therapy by adenovirus-mediated gene transfer. *Curr. Gene Ther.* **1,** 101–122.

Yant, S. R., Meuse, L., Chiu, W., Ivics, Z., Izsvak, Z., and Kay, M. A. (2000). Somatic integration and long-term transgene expression in normal and haemophilic mice using a DNA transposon system. *Nature Genet.* **25,** 35–41.

Yew, N. S., Przybylska, M., Ziegler, R. J., Liu, D., and Cheng, S. H. (2001). High and sustained transgene expression *in vivo* from plasmid vectors containing a hybrid ubiquitin promoter. *Mol. Ther.* **4,** 75–82.

Yew, N. S., Wang, K. X., Przybylska, M., Bagley, R. G., Stedman, M., Marshall, J., Scheule, R. K., and Cheng, S. H. (1999). Contribution of plasmid DNA to inflammation in the lung after administration of cationic lipid:pDNA complexes. *Hum. Gene Ther.* **10,** 223–234.

Yew, N. S., Zhao, H., Wu, I. H., Song, A., Tousignant, J. D., Przybylska, M., and Cheng, S. H. (2000). Reduced inflammatory response to plasmid DNA vectors by elimination and inhibition of immunostimulatory CpG motifs. *Mol. Ther.* **1,** 255–262.

Zabner, J., Fasbender, A. J., Moninger, T., Poellinger, K. A., and Welsh, M. J. (1995). Cellular and molecular barriers to gene transfer by a cationic lipid. *J. Biol. Chem.* **270,** 18997–19007.

Zhang, Y., Jiang, Q., Dudus, L., Yankaskas, J. R., and Engelhardt, J. F. (1998). Vector-specific profiles of two independent primary defects in cystic fibrosis airways. *Hum. Gene Ther.* **20,** 635–648.

Zhu, N., Liggitt, D., Liu, Y., and Debs, R. (1993). Systemic gene expression after intravenous DNA delivery into adult mice. *Science* **261,** 209–211.

Ziady, A. G., Kelley, T. J., Milliken, E., Ferkol, T., and Davis, P. B. (2002). Functional evidence of CFTR gene transfer in nasal epithelium of cystic fibrosis mice *in vivo* following luminal application of DNA complexes targeted to the serpin-enzyme complex receptor. *Mol. Ther.* **5,** 413–419.

Ziesche, R., Hofbauer, E., Wittmann, K., Petkov, V., and Block, L. H. (1999). A preliminary study of long-term treatment with interferon gamma-1b and low-dose prednisolone in patients with idiopathic pulmonary fibrosis. *N. Engl. J. Med.* **341,** 1264–1269.

13

Non-Viral Vector as Vaccine Carrier

Weihsu Claire Chen and Leaf Huang
Center for Pharmacogenetics
School of Pharmacy
University of Pittsburgh
Pittsburgh, Pennsylvania 15261

ABSTRACT

Over the last several years, advances in gene-based delivery technology arising from the field of gene therapy have helped revitalize the field of vaccine development. Genetic vaccination encoding antigen from bacteria, virus, and cancer has shown promise in protective humoral and cellular immunity; however, the potential disadvantages of naked DNA vaccine have reduced the value of the approach. To optimize antigen delivery efficiency as well as vaccine efficacy, the non-viral vector as vaccine carrier, for example, the cationic liposome, has shown particular benefits to circumvent the obstacles that both peptide/protein- and gene-based vaccines have encountered. Liposome-mediated vaccine delivery provides greater efficacy and safer vaccine formulation for the development of vaccine for human use. The success of the liposome-based vaccine has been demonstrated in clinical trials and further human trials are also in progress.   © 2005, Elsevier Inc.

I. INTRODUCTION

Vaccines historically represent one of the most established and cost-effective procedures in medicine, having perhaps the greatest impact on human health of any medical intervention. Jenner's smallpox vaccine, for example, eliminated human infection by the variola virus and led to eradication of the naturally transmitted virus with which mortality has exceeded 300 million lives in the last century alone. The impact of other vaccines, such as polio or influenza vaccines, has reinforced the value of the approach that prevents millions of deaths each year.

However, vaccines made of attenuated live organisms, although efficacious in producing diverse and persistent immune response by mimicking the natural infection without the symptom, can have certain disadvantages. Thus, there is a risk of reversion during the replication of live viruses or even mutation to a more pathogenic state, and with immune-compromised individuals, some of the attenuated viruses may still provoke disease. However, although the extracellular localization of killed virus vaccines and their subsequent phagocytosis by professional antigen-presenting cells (APCs) or antigen-specific B cells lead to MHC class II-restricted presentation and to T helper cell and humoral immunity, they do not elicit a significant cytotoxic T lymphocyte (CTL) response. Subunit and peptide vaccines produced recombinantly or synthetically are considered to be safer than the attenuated or killed virus vaccine. Unfortunately, they are weak immunogens and are often unable to induce adequate immune responses.

Recent developments in molecular and cellular immunology, genetics, and gene delivery have stimulated a renaissance in the field of vaccine research. In addition to a variety of modified viral vectors, researchers have developed novel adenovirus serotypes, alphaviruses, adeno-associated viruses, and non-viral vectors such as naked DNA that can be used as vaccine vectors for a variety of infectious diseases (Nabel, 2004).

Naked DNA has shown particular promise in a variety of different animal models of infectious disease, including HIV (Donnelly *et al.*, 1997), influenza virus (Ulmer *et al.*, 1993), malaria (Becker *et al.*, 1998; Doolan *et al.*, 1996; Gardner *et al.*, 1996; Sedegah *et al.*, 1994), tuberculosis (Tascon *et al.*, 1996), Ebola virus (Xu *et al.*, 1998), rabies (Lodmell *et al.*, 1998), lymphocytic choriomeningitis virus (Martins *et al.*, 1995; Yokoyama *et al.*, 1995), and herpes simplex virus (Manickan *et al.*, 1995). This technology has been evaluated in phase I human clinical trials for HIV, malaria, and influenza virus.

Naked DNA vaccines are easy and inexpensive to produce, as plasmid vectors can be constructed and tested rapidly. Plasmids are also more temperature stable than live vaccines and their storage in a lyophilized form is straightforward. In addition to eliciting a potent T-cell immunity, including the CTL response, a robust antibody response has been documented in a variety of animal models (e.g., rabies, West Nile virus, anthrax, hanta virus), including mammals much larger than humans (Fischer *et al.*, 2003a). These studies demonstrated that immunization with naked DNA can elicit both antibody and cell-mediated immunities. Furthermore, the protective immunity has also been documented in more than just mouse models of infectious disease; efficacy has been demonstrated in fish (Corbeil *et al.*, 1999), dogs (Fischer *et al.*, 2003b), birds (Turell *et al.*, 2003), horses (Fischer *et al.*, 2003a), and nonhuman primates (Lodmell *et al.*, 1998).

It also appeared that DNA immunization could be applied in cancer treatment: injection of a plasmid encoding tumor antigen resulted in the induction of an immune response (Bright *et al.*, 1996; Conry *et al.*, 1995), which was protective in an animal model (Manickan *et al.*, 1997). Following these pioneering studies, the concept of DNA immunization has now been adopted by vaccinologists worldwide using an ever-increasing number of plasmids encoding immunogens from bacterial, viral, and parasitic pathogens and a variety of tumors (Gregoriadis, 1998; Lewis and Babiuk, 1999).

II. POTENTIAL DISADVANTAGES OF NAKED DNA IMMUNIZATION

Disadvantages with naked DNA vaccination include uptake of DNA by only a minor fraction of muscle cells (Davis et al., 1993a) and the exposure of DNA to deoxyribonuclease in the interstitial fluid. Thus, the use of relatively large quantities of DNA is required (e.g., 50–200 µg per mouse) and, in some cases, the need to inject into regenerating muscle to enhance the immunity. Administration of antigen-encoding plasmid DNA via liposome can, however, circumvent the need for muscle involvement. Instead, the uptake is mediated by the APC, e.g., those infiltrating the site of injection or in the lymphatics. Liposome also protects DNA from the nuclease attack.

    Most of the immunization procedures carried out so far have opted for intramuscular and, to a lesser extent, intraepidermal routes (Gregoriadis et al., 2002). Other routes, such as oral, nasal, vaginal, intravenous, intraperitoneal, and subcutaneous, have also been used (Gregoriadis, 1998; Lewis and Babiuk, 1999). For intramuscular injections, it is not uncommon to pretreat the tissue with cardiotoxin and other drugs that cause muscle damage followed by regeneration (Davis et al., 1993b) or the anesthetic bupivacaine, which dilates local vessels, thus enhancing DNA uptake by myocytes (Wang et al., 1993). There are a number of other variables that need full evaluation for an optimal immune response (Spier, 1996).

III. VACCINE ADJUVANTS AND FORMULATION

A. Type of vaccine adjuvant

Most biologics formulated for use in humans contain additives that are included to improve solubility, viscosity, bioabsorption, and stability or, in the case of vaccines, are added to stimulate the immune system. This is especially true for most of the synthetic peptide and subunit vaccine antigens, which, in addition to being costly or only available in small quantities (e.g., recombinant DNA products), can be weakly or nonimmunogenic. A great variety of experimental immunological adjuvants (Gregoriadis et al., 2002) now available in rendering such vaccines stronger and more efficient, e.g., Freund's complete and incomplete adjuvants, bacterial endotoxins, polyanions, and mineral adsorbents, induce local or systemic toxicity. However, only two other adjuvants, liposomes (Gregoriadis, 1990) and MF59, have been approved for use in humans (Gluck et al., 1992) after the introduction of aluminum salts as an adjuvant.

    Attempts to organize adjuvants in grouped categories in order to facilitate adjuvant selection have met with difficulty because of the multiple and overlapping biological effects of many adjuvants. A practical categorization of three different types of immunostimulators has been proposed by Edelman and Tacket (1990): adjuvants per se, carriers, and vehicles. Included in the adjuvant category are aluminum salts, saponin, muramyl di- and tripeptides, monophosphoryl lipid A, Bordetella pertussis, cytokines, and many others. The carriers, which mainly provide T-cell help, include bacterial toxoids, fatty acids, and living vectors. The vehicle category includes mineral oil emulsions (e.g., incomplete Freunds adjuvant), biodegradable oil emulsions (e.g., emulsions containing peanut oil, squalene, or squalane), nonionic block copolymer surfactants, liposomes, and biodegradable polymer microspheres (Eldridge et al., 1991).

B. Vaccine entrapment in liposomes

The immunological adjuvant property of liposomes was first established (Allison and Gregoriadis, 1974) when strong humoral immune responses to diphtheria toxoid (entrapped in liposomes) were obtained in mice. Unlike other adjuvants, there were no granulomas at the site of injection (Allison and Gregoriadis, 1974; Manesis et al., 1978) and no hypersensitivity reactions were observed in pre-immunized animals when the antigen was given in the entrapped form (Gregoriadis and Allison, 1974). Furthermore, liposomes composed of the appropriate phospholipid do not develop antibodies against their phospholipid component (Alving, 1991) nor have they produced any side effects in repeatedly injected patients (Gregoriadis et al., 1982). Extensive work in this and other laboratories since the early 1980s has shown that liposome adjuvanticity applies to a large variety of bacterial, viral, protozoan, tumor, and other antigens (Alving, 1991; Gregoriadis, 1990). In many experiments, protection of the host was achieved by immunization with a relevant liposome-entrapped antigen.

IV. LIPOSOME-MEDIATED GENETIC VACCINATION

A. Naked DNA carried by liposome increases its uptake by antigen-presenting cells (APCs)

Vaccination with naked DNA by the intramuscular route relies on the ability of the myocytes to engulf the plasmid. Some of the DNA may also be endocytosed by APCs infiltrating the site of injection or in the lymph nodes following its migration to the lymphatics. The extent of DNA degradation by extracellular deoxyribonucleases is unknown, but depending on the time of its residence interstitially, degradation could be considerable. Approaches to protect DNA from the extracellular biological milieu, introducing it into cells more efficiently, or targeting it to immunologically relevant cells should improve the design of the DNA vaccine.

It has been proposed (Gregoriadis *et al.*, 1997) that as APCs are a preferred alternative to muscle cells as targets for DNA vaccine uptake and expression; liposomes would be a suitable means of delivery for the entrapped DNA to such cells. Locally injected liposomes are known (Gregoriadis, 1990) to be taken up avidly by APCs infiltrating the site of injection or in the lymphatics, an event that has been implicated (Gregoriadis, 1990) in their immunoadjuvant activity. Liposomes would also protect their DNA content from deoxyribonuclease attack. Because of the structural versatility of the system (Gregoriadis *et al.*, 1996), its transfection efficiency could be further improved by the judicial choice of vesicle surface charge, size, and lipid composition or by the coentrapment of cytokine genes and other adjuvants (e.g., immunostimulatory CpG sequences), together with the plasmid vaccine. Moreover, as a number of injectable liposome-based drug formulations, including vaccines against hepatitis A and influenza, have been already licensed in the United States and Europe for clinical use (Gregoriadis, 1995), clinical acceptance of the system would be less problematic.

B. Naked DNA carried by liposome enhances both humoral and cellular immunity

Plasmid-containing liposomes were tested in immunization experiments (Gregoriadis *et al.*, 1997; Perrie *et al.*, 2001) using plasmid (pRc/CMV HBS) as naked DNA, encoding the S region of the hepatitis B surface antigen (HBsAg). Figure 13.1 shows that mice (BALB/c) injected repeatedly by the intramuscular route with 5 or 10 μg plasmid entrapped in cationic liposomes (DOTAP, DC-Chol or SA) elicited at all times much greater up to a 100-fold antibody (IgG1) response than animals immunized with naked DNA alone. Responses for other subclasses (IgG2a and IgG2b) for the liposomal DNA were also greater, albeit to a lesser extent (up to 10-fold). Significantly, the IgG1 response for the liposome-entrapped DNA via the dehydration-rehydration method was also higher (up to 10-fold) than those obtained for DNA complexed with similar cationic liposomes (Gregoriadis *et al.*, 1997). This was also true for IFN-γ and IL-4 levels in the spleen of immunized mice (Fig. 13.2).

C. Naked DNA carried by liposome induces cytotoxic T lymphocyte response

In a subsequent study (Bacon *et al.*, 2002), the CTL component of the immune response was measured by the specific killing of syngeneic target cells pulsed with a recognized CTL epitope peptide derived from the antigen. To that end, the type and degree of immune response induced were analyzed following subcutaneous injection of DNA in liposomes and compared to that achieved by DNA alone delivered by the same route. For this purpose, 6- to 8-week-old

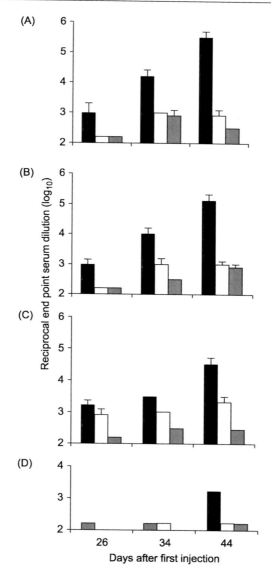

Figure 13.1. Comparison of immune responses in mice injected with naked or liposome-entrapped pRc/CMV HBS. BALB/c mice were injected intramuscularly on days 0, 10, 20, 27, and 37 with 5 μg of DNA entrapped in cationic liposomes composed of PC, DOPE, and DOTAP (A); DC-Chol (B); SA (C) (molar ratios 1:0.5:0.25); or in the naked form (D). Animals were bled 7, 15, 26, 34, and 44 days after the first injection and sera tested by ELISA for IgG1 (black bars), IgG2a (white bars), or IgG2b (gray bars) responses against the encoded hepatitis B surface antigen (HBsAg; S region, ayw subtype) (Gregoriadis et al., 1997).

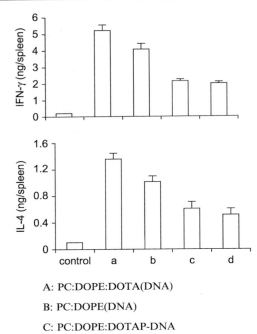

A: PC:DOPE:DOTA(DNA)

B: PC:DOPE(DNA)

C: PC:DOPE:DOTAP-DNA

D: Naked DNA

Figure 13.2. Cytokine levels in spleens of mice injected with naked, entrapped, or complexed pRc/
CMV HBS. BALB/c mice were immunized as in Fig. 13.1 with DNA entrapped into
either cationic (DOTAP) (a) or uncharged liposomes (b), mixed with cationic
(DOTAP) liposomes (c), or in the naked form (d). "Control" denotes cytokine levels
in normal unimmunized mice. Three weeks after the final injection, mice were killed
and their spleens subjected to IL-4 and IFN-γ analysis. Each bar represents the mean
$\pm$ SE of a group of four mice. Cytokine values in mice immunized with cationic
liposomes were significantly higher than those in the other groups ($P < 0.001 - 0.05$).
Modified from Gregoriadis et al. (1997).

female C57/BL6 (H-2d) mice received either one or two doses of 2.5 or 10 μg of
ovalbumin (OVA)-encoding plasmid DNA (pCI-OVA) either alone or in
liposomes. As a positive control, animals were immunized subcutaneously with
100 μg of OVA protein complexed with 1 μg of cholera toxin (CT). One week
after the last immunization, blood samples and spleens were collected from all
animals and tested for anti-OVA serum total IgG antibody levels, CTL activity,
and cytokine release. After a single dose of antigen, only animals immunized
with either protein or 10 μg of liposomal DNA showed significant anti-OVA
antibody titers by ELISA. After two doses of antigen, only animals immunized
with either protein or liposomal DNA (both 2.5 and 10 μg) showed significant

levels of seroconversion and antibody titers against OVA by ELISA (Bacon et al., 2002). Similarly, no anti-OVA CTL activity was detected in animals immunized with DNA alone. However, animals immunized with two doses of 10 μg of liposome-entrapped DNA (Lipodine) displayed a CTL response higher (60% cell killing versus 50%) than that obtained in the positive control group, immunized with OVA protein + adjuvant (CT) (Fig. 13.3). Thus, delivery of a small dose of liposomal plasmid DNA subcutaneously, a route of immunization not normally inducing significant plasmid DNA-mediated immune activation (Lewis and Babiuk, 1999), results in a strong antigen-specific cellular response that is achieved by higher doses of a conventional protein antigen together with a powerful adjuvant.

V. LIPOSOME-MEDIATED PEPTIDE VACCINATION

A. Peptide-based vaccine for cancer therapy

Vaccination with antigenic peptides derived from these antigens has the potential to be a powerful weapon for the treatment of cancer (van Driel et al., 1999). Unfortunately, most tumor antigens that have been identified are self-antigens, and to be effective, immunization strategies utilizing these antigens must overcome the formidable obstacle of self-tolerance (Houghton et al., 2001). Because of this, most vaccination approaches have met with only modest success (Dallal and Lotze, 2000). Methods to enhance peptide vaccine potency, including the coadministration of immunostimulatory cytokines such as interleukin 12 (IL-12) or granulocyte macrophage colony stimulating factor (GM-CSF) as an adjuvant or the delivery of peptide-pulsed dendritic cells, are currently under investigation (Babai et al., 2001; Schuler and Steinman, 1997).

For a peptide antigen delivery strategy to effectively induce cell-mediated antitumor immunity, at least two obstacles must be overcome. First, the antigen must make its way into organized lymph tissue such as draining lymph nodes or the spleen (Karrer et al., 1997). This could be accomplished either through effective delivery of peptide directly to the lymphoid organs or by effective targeting of peptide to APC. It has been shown that large amounts of immunogenic antigens are ignored by the immune system, as long as they remain outside of the lymphatic system (Ochsenbein et al., 1999; Zinkernagel et al., 1997). Studies involving genetic vaccinations have shown that despite the fact that the majority of antigen expression is restricted to peripheral tissues, immune responses are initiated by APCs in draining lymph nodes (Akbari et al., 1999). Direct administration of plasmid DNA encoding a tumor antigen into lymph nodes or the spleen results in a 100- to 1000-fold increase in vaccine potency (Maloy et al., 2001).

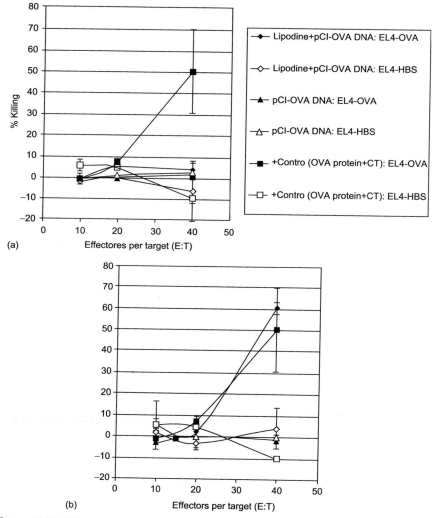

(a) Effectores per target (E:T)

(b) Effectors per target (E:T)

Legend:
- Lipodine+pCI-OVA DNA: EL4-OVA
- Lipodine+pCI-OVA DNA: EL4-HBS
- pCI-OVA DNA: EL4-OVA
- pCI-OVA DNA: EL4-HBS
- +Contro (OVA protein+CT): EL4-OVA
- +Contro (OVA protein+CT): EL4-HBS

Figure 13.3. CTL response to EL4 pulsed with an OVA CTL epitope peptide in animals immunized with 2.5 μg (a) and 10 μg (b) of pCI-OVA either alone or entrapped in cationic DOTAP liposomes. (a) CTL activity (two 2.5-μg pCI-OVA doses); splenocytes + OVA (peptide) + EL4 + IL-2 stimulation; (b) CTL activity (two 10 μg pCI-OVA doses); splenocytes + OVA (peptide) + EL4 + IL-2 stimulation.

Second, to induce effective cellular immunity, antigenic peptides in the lymphoid tissues must be presented to T cells in association with MHC molecules on the surface of APCs. In addition, the strength of priming depends on the duration of the antigenic stimulus (Mackay *et al.*, 1990; Zinkernagel *et al.*, 1997). While the extracellular peptide antigen can form stable complexes with MHC class I molecules on the surface of APCs *in vivo*, degradation of minimal epitopes by extracellular proteases generally makes this mechanism of association exceedingly inefficient. Furthermore, internalization of extracellular peptides by APCs generally results in delivery to lysosomal compartments, in which the minimal epitopes are degraded rapidly (Toes *et al.*, 1996). To circumvent these obstacles, antigen delivery strategies are being designed to deliver extracellular peptides to the cytoplasm of APCs from which they could access the MHC class I-restricted processing pathway. Wang and Wang (2002) demonstrated that *ex vivo* loading of antigen into the cytoplasm of dendritic cells extends the time that it is presented on MHC class I molecules and enhances CTL response.

B. Lipid–polycation–DNA (LPD) complex

A great number of antigen delivery systems have been designed to deliver antigenic peptides. The use of peptides associated with adjuvants or heat shock proteins (Liu *et al.* (2000), encapsulated in neutral liposomes or biodegradable polymers (Eldridge *et al.*, 1991), or coupled to synthetic beads (Cruz *et al.*, 2004) have all been investigated. While these methods have had some success, none has been specifically designed to deliver peptide antigens directly to the cytoplasm of APCs inside organized lymph tissue. Our laboratory has developed a novel liposome-based system, called LPD (lipid–polycation–DNA), for the delivery of plasmid DNA (Li and Huang, 1997). Polylysine was initially used as the polycation to condense DNA, but protamine was later shown to elicit an improved activity (Gao and Huang, 1996). LPD particles are formed by combining cationic liposomes and polycation-condensed DNA. Upon mixing, the components rearrange to form a virus-like structure with condensed DNA inside of lipid membranes (Li *et al.*, 1998). When injected systemically, LPD particles distribute to all major organs (including the spleen), where they are endocytosed and release their contents into the cytoplasm. Administration of these particles initiates the rapid production of several Th1 cytokines, most notably TNFα, IL-12, and IFN-γ, due to the presence of unmethylated CpG motifs in the plasmid DNA. This nonspecific immunostimulation is associated with tumoristatic effects (Whitmore *et al.*, 1999, 2001).

Because these particles can successfully entrap proteins and deliver their contents to the cytoplasm, it was hypothesized that they can be used to entrap and deliver antigenic peptides to the cytoplasm of APCs in the lymph node or the spleen (S. J. Han *et al.*, unpublished result). This delivery strategy could result in the extended presentation of antigen and enhanced immune response to peptide vaccines.

C. LPD/peptide complex accumulates in the spleen and is taken up by APCs

To investigate the antitumor immunity of LPD particles as a carrier of tumor-specific antigen, mouse TC-1 cells that were transformed by human papilloma virus (HPV) DNA were used as a model for cervical cancer cells. A major CTL epitope of the HPV antigen, E7 peptide (amino acids 49 to 57), was used as the immunogen.

As delivery to organized lymph tissue is important for successful vaccination, the distribution of LPD particles within the spleen was investigated (Dileo *et al.*, 2003). To this end, C57BL/6 mice were injected iv with LPD/E7 particles containing trace amounts of Cy3-labeled DNA. At 12, 24, and 48 h after injection, spleens were collected and sectioned and the distribution of fluorescence was observed. Mice injected with nonfluorescent LPD particles served as a negative control. After iv administration, LPD particles accumulate rapidly in the marginal zones of the spleen (Fig. 13.4). At 24 h, the majority of the fluorescence was still located in the marginal zones but began to distribute in the white pulp. By 48 h, fluorescence was less intense in a diffuse pattern. These results suggest that LPD particles are vividly taken up by the phagocytic cells in the spleen.

To determine if LPD/E7 particles are taken up by antigen-presenting cells, C57BL/6 mice were sc-injected with LPD/E7 particles containing trace amounts of FITC-labeled DNA. At 16 h after injection, cells in the draining lymph node were collected, stained for CD11c (to identify dendritic cells) and CD11b (to identify macrophages and myeloid lineage DCs) using PE-labeled antibodies, and subjected to flow cytometry analysis. Mice injected with nonfluorescent LPD particles served as control. At 12 h, 2.2% of total splenocytes were CD11b/DNA double positive and 1.9% were CD11c/DNA positive. However, these numbers represent approximately 17 and 26% of all CD11b and CD11c positive cells, respectively (S. J. Han *et al.*, unpublished result). Results indicated that LPD/E7 particles were taken up by a significant portion of the APCs in the lymph node.

D. LPD/E7 complexes induce an E7-specific immune response

To determine if LPD/E7 particles induce enhanced immunization versus traditional liposome/peptide vaccines, mice were vaccinated sc or iv with LPD particles containing 0, 1, 10, or 20 μg of E7 peptide on days 0 and 5 (Dileo

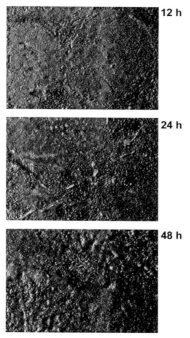

Figure 13.4. Distribution of LPD/E7 particles in the spleen following iv administration. LPD/E7 particles containing Cy3-labeled ODN were visualized at the indicated times after administration. (See Color Insert.)

et al., 2003). For comparison, mice were injected with 20 μg E7 peptide in phosphate-buffered saline (PBS) or encapsulated in SL liposomes (Ignatius et al., 2000). Five days after the final vaccination, splenocytes were collected and used as effector cells in a chromium release assay. Mouse EL4 cells pulsed with the antigenic peptide were used as the target cells. Consistent with previous reports, SL liposomes containing 20 μg E7 peptide induced a significant level of CTL activity (61% specific lysis) following sc injection (Fig. 13.5B). However, vaccination with LPD containing the same amount of E7 peptide produced a higher level of CTL activity by the same injection routes. Injection of a lower amount of LPD/E7 particles induced lower levels of CTL activity in both iv and sc routes (Fig. 13.5). These data indicated that LPD is an excellent carrier for the peptide antigen for inducing a CTL response and its activity is greater than that of the SL liposomes.

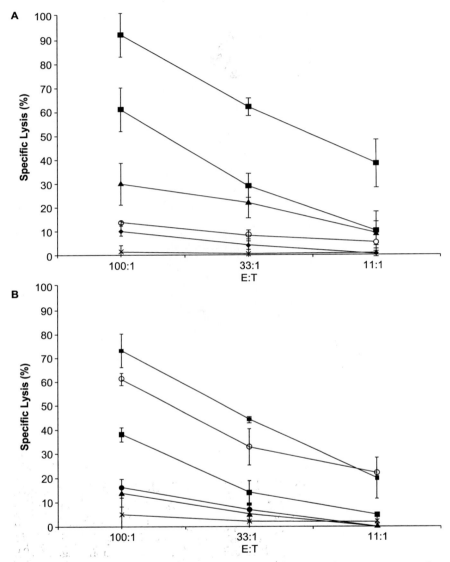

Figure 13.5. Tumor-specific CTL activity induced by LPD/E7 vaccination. Mice were injected iv (A) or sc (B) with LPD particles containing 0 (◆), 1 (▲), 10 (■), or 20 (●) μg of E7 peptide or SL liposome containing 20 μg E7 peptide (○) or 20 μg E7 peptide in PBS μg on days 0 and 5. Five days after the last injection, splenocytes were isolated, restimulated for 4 days, and used as effectors in a chromium release assay. Nonspecific lysis was <8% in all groups. Means ± SD, two mice/group. One representative of two experiments is shown.

E. LPD/E7 vaccination protects mice from HPV + tumor formation as well as eradicates established HPV + tumors

To determine if the induced immune response is adequate to provide protective immunity, mice were vaccinated iv or sc with 20 μg E7 peptide in PBS, empty LPD, or LPD containing 20 μg of E7 peptide on days 0 and 5 (Dileo et al., 2003). Untreated mice served as a control. Five days after the last vaccination, mice were sc challenged with 0.5×10^6 E7 expressing TC-1 cells. Mice that received LPD/E7 particles by either route failed to develop tumors, whereas control mice and mice that received LPD alone or free E7 peptide developed tumors within 12 days (Fig.13.6). Results clearly showed that LPD/E7 was a potent tumor vaccine that provided complete protection against the tumor challenge.

To determine the potential of the LPD/E7 complex for use as a therapeutic strategy, subcutaneous tumors were established in C57BL/6 mice by inoculation of 0.5×10^6 TC-1 cells (Dileo et al., 2003). On days 3 and 6 following inoculation, mice were injected iv or sc with LPD containing 10 μg of E7 peptide. To determine the importance of antigen delivery to the spleen in the generation of the observed immune response, a group of mice that had their spleens removed surgically prior to the experiment were included as a control. Untreated mice and mice receiving empty LPD served as additional controls. All mice that received iv LPD/E7 peptide showed steady tumor shrinkage and complete regression within 2 weeks. Subcutaneous treatment also resulted in complete regression but with a slower kinetics. As expected, empty LPD administration slowed tumor growth progression but failed to eradicate the tumors. The antitumor effect in asplenic mice depended on the route of administration. Intravenous treatment showed a tumor growth rate similar to those observed in mice treated with LPD alone, whereas sc delivery resulted in complete tumor regression. Untreated mice showed unimpeded tumor progression (Fig. 13.7). These results clearly demonstrated that LPD/E7 could be used as a therapeutic strategy against established tumors and that the activity depended on the intactness of the organized lymphoid tissue, i.e., the spleen for the iv injection route.

VI. LIPOSOME-MEDIATED RECOMBINANT PROTEIN VACCINATION

Liposomes have been shown to deliver encapsulated antigen directly to the cytoplasm and MHC class I pathway of antigen-presenting cells (Rao and Alving, 2000). Thus, liposomes were considered to be an excellent candidate as the basis for an adjuvanted vaccine against cancer because induction of CTL activity is necessary for maximum efficacy of a cancer vaccine in which cytoplasmic antigens are involved. Numerous studies have demonstrated that CTLs

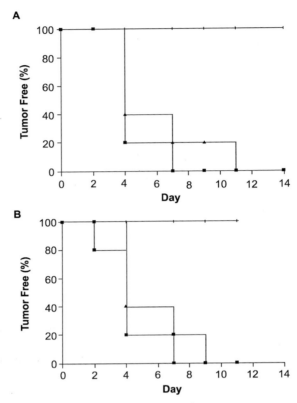

Figure 13.6. LPD/E7 vaccination prevents tumor establishment. Mice were injected iv (A) or sc
(B) with LPD particles with (●) or without (■) 20 μg E7 peptide or 20 μg E7 peptide
in PBS (◆) on days 0 and 5 or were left untreated (▲). On day 10, the mice were sc
challenged with 0.5×10^6 TC-1 cells. Tumor formation was monitored twice per
week by palpation. Five mice per group. One representative of three experiments
showing similar tumor formation kinetics is shown.

are induced readily by liposome/lipid A-encapsulated antigens in animals
(Alving and Wassef, 1994).

A. Liposome-encapsulated malaria vaccine

The malaria vaccine consists of liposome-encapsulated recombinant protein
(R32NS1), containing epitopes from the repeat region of the circumsporozoite
protein of *Plasmodium falciparum*. The whole formulation was adsorbed on
aluminum hydroxide with monophosphoryl lipid A in the lipid bilayer as an

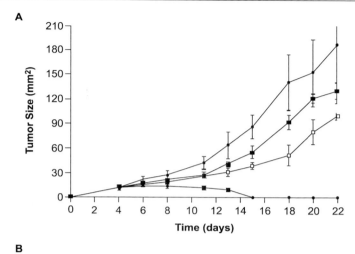

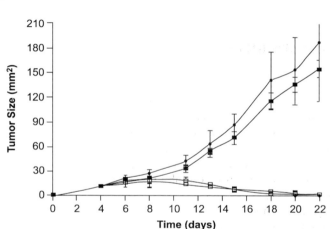

Figure 13.7. LPD/E7 treatment eradicates established tumors. Intact (closed symbols) or asplenic mice (open symbols) were inoculated sc with 0.5×10^6 TC-1 cells on day 0. On days 3 and 6, mice were injected iv (A) or sc (B) with LPD particles with (●) or without (■) 10 μg E7 peptide or left untreated (◆). Tumor growth was measured three times per week. Means ± SD, five mice per group. One representative of three experiments showing similar tumor growth kinetics is shown.

adjuvant (Alving *et al.*, 1999; Owens *et al.*, 2000). The formulation showed considerable promise for an induction of high titers of relevant antimalarial antibodies in human volunteers in a phase I dose-escalating study (Fries *et al.*, 1992). R32NS1, or a similar construct, was also tested as an antigen in a series of phase I trials in which other adjuvant formulations were employed. In all cases, the adjuvants being compared had previously been found safe in human trials with other antigens. When the results from the various studies were compared, the liposome formulation emerged as the most potent and least reactogenic of the adjuvants studied (Gordon, 1993).

As noted earlier, the liposomes, referred to as Walter Reed liposomes, exhibited no detectable toxicity despite the presence of huge concentrations of endotoxic monophosphoryl lipid A (determined in a dose-ranging study that started with 0.02 mg and went up to more than 2 mg of endotoxin) (Fries et al., 1992). The degree of reduced toxicity that was actually observed in humans was much greater than that predicted by any in vitro or animal models (Alving, 1999). The case of liposomal lipid A, therefore, clearly illustrates the principle that the expression of adjuvanticity can occur in humans completely independently from the expression of pyrogenicity. The Walter Reed liposomal formulation also exhibited low toxicity and high potency that was comparable or superior to any of the other formulations (McElrath, 1995).

B. Liposome-stabilized prostate cancer vaccine trials

A series of phase I human trials were undertaken to develop an immunotherapeutic vaccine against prostate cancer under a cooperative research and development agreement with Jenner Biotherapies Inc. using the liposome/lipid A-encapsulated recombinant prostate-specific antigen (PSA) (Harris et al., 1999). The strategy was to use the Walter Reed liposomes as the basis of an initial adjuvant and carrier formulation in a phase I trial in patients with advanced prostate cancer and then to add additional adjuvants to this basic formulation for comparison of the ability to break tolerance against PSA in subsequent trials. A total of six phase I trials were undertaken with small groups of patients (five or six per group) having advanced hormone therapy-unresponsive prostate cancer. The trials started in the first two groups of patients with the formulation consisting of Walter Reed liposomes containing lipid A and PSA as described previously. This Walter Reed liposome formulation by itself, although safe, failed to break immunological tolerance when given either intramuscularly or intravenously. Because of this, four additional cohorts were tested with the original liposome/lipid A/PSA formulation in combination with additional adjuvants. Each of the four adjuvant formulations in which liposome/lipid A/PSA was strengthened by the inclusion of additional adjuvants was successful in breaking tolerance. The advance of this strategy has resulted successfully in a new liposome/lipid A/PSA emulsion adjuvant formulation that shows greater safety and immunological potency than other formulations and has been transitioned to phase II trials.

VII. CONCLUSION

During the past decade, several immunization technologies once thought promising, but technically challenging, have proven tractable and are making significant inroads into the practice of medicine. DNA immunization is a promising approach for the design of vaccines in which the antigen is either ineffective or unavailable. However, plasmid DNA vaccines used as naked DNA are vulnerable to attack by deoxyribonuclease following their administration and do not normally target antigen-presenting cells. Such problems can be circumvented by entrapping the DNA within various liposome formulations, including cationic liposomes. Immunization studies have also shown that cationic liposomes promote a much greater humoral and cytotoxic T lymphocyte immune response against the antigen encoded by the entrapped DNA vaccine. The procedure of entrapment is simple (one step), and the formulation can be a freeze-dried preparation that can be stored for a prolonged period of time. There is a growing need for multivalent combination vaccines that are less complex to develop and manufacture, but are safer, more stable, and require fewer injections. In addition to the DNA vaccine, liposomes as a vaccine carrier have also shown promising results in other types of vaccine application, such as the peptide-based vaccine as well as the recombinant antigen vaccine against a variety of cancers and infection diseases. Some of the successes of the liposome-based vaccine formulations have been demonstrated in the either phase I or phase II clinical trials. Thus, liposomes and other non-viral vectors are excellent antigen carriers and/or adjuvants for immunization and play an important role in vaccine development.

Acknowledgments

The authors thank Stella Weidner and Nicole Sebula for their excellent assistance. The original work from this laboratory was supported by NIH Grants CA74918 and AI48851.

References

Akbari, O., Panjwani, N., Garcia, S., Tascon, R., Lowrie, D., and Stockinger, B. (1999). DNA vaccination: Transfection and activation of dendritic cells as key events for immunity. *J. Exp. Med.* **189**(1), 169–178.

Allison, A. G., and Gregoriadis, G. (1974). Liposomes as immunological adjuvants. *Nature* **252** (5480), 252.

Alving, C. R. (1991). Liposomes as carriers of antigens and adjuvants. *J. Immunol. Methods* **140**(1), 1–13.

Alving, C. R. (1999). Encapsulated high-concentration lipid A composition as immunogenic agents to produce human antibodies to prevent or treat gram-negative bacteria infection. U.S. Patent No. 5,888,519.

Alving, C. R., Owens, R. R., and Wassef, N. M. (1999). Process for making liposome preparation. U.S. Patent No. 6,007,838.

Alving, C. R., and Wassef, N. M. (1994). Cytotoxic T lymphocytes induced by liposomal antigens: Mechanisms of immunological presentation. *AIDS Res. Hum. Retrovir.* 10(Suppl. 2), S91–S94.

Babai, I., Barenholz, Y., Zakay-Rones, Z., Greenbaum, E., Samira, S., Hayon, I., Rochman, M., and Kedar, E. (2001). A novel liposomal influenza vaccine (INFLUSOME-VAC) containing hemagglutinin-neuraminidase and IL-2 or GM-CSF induces protective anti-neuraminidase antibodies cross-reacting with a wide spectrum of influenza A viral strains. *Vaccine* 20(3–4), 505–515.

Bacon, A., Caparros-Wanderley, W., Zadi, B., and Gregoriadis, G. (2002). Induction of a cytotoxic T lymphocyte (CTL) response to plasmid DNA delivered via Lipodine liposomes. *J. Liposome Res.* 12(1–2), 173–183.

Becker, S. I., Wang, R., Hedstrom, R. C., Aguiar, J. C., Jones, T. R., Hoffman, S. L., and Gardner, M. J. (1998). Protection of mice against *Plasmodium yoelii* sporozoite challenge with P. yoelii merozoite surface protein 1 DNA vaccines. *Infect. Immun.* 66(7), 3457–3461.

Bright, R. K., Beames, B., Shearer, M. H., and Kennedy, R. C. (1996). Protection against a lethal challenge with SV40-transformed cells by the direct injection of DNA-encoding SV40 large tumor antigen. *Cancer Res.* 56(5), 1126–1130.

Conry, R. M., LoBuglio, A. F., Loechel, F., Moore, S. E., Sumerel, L. A., Barlow, D. L., Pike, J., and Curiel, D. T. (1995). A carcinoembryonic antigen polynucleotide vaccine for human clinical use. *Cancer Gene Ther.* 2(1), 33–38.

Corbeil, S., Lapatra, S. E., Anderson, E. D., Jones, J., Vincent, B., Hsu, Y. L., and Kurath, G. (1999). Evaluation of the protective immunogenicity of the N, P, M, NV and G proteins of infectious hematopoietic necrosis virus in rainbow trout oncorhynchus mykiss using DNA vaccines. *Dis. Aquat. Organ* 39(1), 29–36.

Cruz, L. J., Iglesias, E., Aguilar, J. C., Gonzalez, L. J., Reyes, O., Albericio, F., and Andreu, D. (2004). A comparative study of different presentation strategies for an HIV peptide immunogen. *Bioconjug. Chem.* 15(1), 112–120.

Dallal, R. M., and Lotze, M. T. (2000). The dendritic cell and human cancer vaccines. *Curr. Opin. Immunol.* 12(5), 583–588.

Davis, H. L., Demeneix, B. A., Quantin, B., Coulombe, J., and Whalen, R. G. (1993a). Plasmid DNA is superior to viral vectors for direct gene transfer into adult mouse skeletal muscle. *Hum. Gene Ther.* 4(6), 733–740.

Davis, H. L., Whalen, R. G., and Demeneix, B. A. (1993b). Direct gene transfer into skeletal muscle *in vivo*: Factors affecting efficiency of transfer and stability of expression. *Hum. Gene Ther.* 4(2), 151–159.

Dileo, J., Banerjee, R., Whitmore, M., Nayak, J. V., Falo, L. D., Jr., and Huang, L. (2003). Lipid-protamine-DNA-mediated antigen delivery to antigen-presenting cells results in enhanced anti-tumor immune responses. *Mol. Ther.* 7(5 Pt 1), 640–648.

Donnelly, J. J., Ulmer, J. B., Shiver, J. W., and Liu, M. A. (1997). DNA vaccines. *Annu. Rev. Immunol.* 15, 617–648.

Doolan, D. L., Sedegah, M., Hedstrom, R. C., Hobart, P., Charoenvit, Y., and Hoffman, S. L. (1996). Circumventing genetic restriction of protection against malaria with multigene DNA immunization: CD8+ cell-, interferon gamma-, and nitric oxide-dependent immunity. *J. Exp. Med.* 183(4), 1739–1746.

Edelman, R., and Tacket, C. O. (1990). Adjuvants. *Int. Rev. Immunol.* 7(1), 51–66.

Eldridge, J. H., Staas, J. K., Meulbroek, J. A., McGhee, J. R., Tice, T. R., and Gilley, R. M. (1991). Biodegradable microspheres as a vaccine delivery system. *Mol. Immunol.* 28(3), 287–294.

Fischer, L., Minke, J., Dufay, N., Baudu, P., and Audonnet, J. C. (2003a). Rabies DNA vaccine in the horse: Strategies to improve serological responses. *Vaccine* 21(31), 4593–4596.

Fischer, L., Tronel, J. P., Minke, J., Barzu, S., Baudu, P., and Audonnet, J. C. (2003b). Vaccination of puppies with a lipid-formulated plasmid vaccine protects against a severe canine distemper virus challenge. *Vaccine* 21(11–12), 1099–1102.

Fries, L. F., Gordon, D. M., Richards, R. L., Egan, J. E., Hollingdale, M. R., Gross, M., Silverman, C., and Alving, C. R. (1992). Liposomal malaria vaccine in humans: A safe and potent adjuvant strategy. *Proc. Natl. Acad. Sci. USA* 89(1), 358–362.

Gao, X., and Huang, L. (1996). Potentiation of cationic liposome-mediated gene delivery by polycations. *Biochemistry* 35(3), 1027–1036.

Gardner, M. J., Doolan, D. L., Hedstrom, R. C., Wang, R., Sedegah, M., Gramzinski, R. A., Aguiar, J. C., Wang, H., Margalith, M., Hobart, P., and Hoffman, S. L. (1996). DNA vaccines against malaria: Immunogenicity and protection in a rodent model. *J. Pharm. Sci.* 85(12), 1294–1300.

Gluck, R., Mischler, R., Brantschen, S., Just, M., Althaus, B., and Cryz, S. J., Jr. (1992). Immunopotentiating reconstituted influenza virus virosome vaccine delivery system for immunization against hepatitis A. *J. Clin. Invest.* 90(6), 2491–2495.

Gordon, D. M. (1993). Use of novel adjuvants and delivery systems to improve the humoral and cellular immune response to malaria vaccine candidate antigens. *Vaccine* 11(5), 591–593.

Gregoriadis, G. (1990). Immunological adjuvants: A role for liposomes. *Immunol. Today* 11(3), 89–97.

Gregoriadis, G. (1995). Engineering liposomes for drug delivery: Progress and problems. *Trends Biotechnol.* 13(12), 527–537.

Gregoriadis, G. (1998). Genetic vaccines: Strategies for optimization. *Pharm. Res.* 15(5), 661–670.

Gregoriadis, G., and Allison, A. C. (1974). Entrapment of proteins in liposomes prevents allergic reactions in pre-immunised mice. *FEBS Lett.* 45(1), 71–74.

Gregoriadis, G., Bacon, A., Caparros-Wanderley, W., and McCormack, B. (2002). A role for liposomes in genetic vaccination. *Vaccine* 20(Suppl. 5), B1–B9.

Gregoriadis, G., Saffie, R., and de Souza, J. B. (1997). Liposome-mediated DNA vaccination. *FEBS Lett.* 402(2–3), 107–110.

Gregoriadis, G., Saffie, R., and Hart, S. L. (1996). High yield incorporation of plasmid DNA within liposomes: Effect on DNA integrity and transfection efficiency. *J. Drug Target* 3(6), 469–475.

Gregoriadis, G., Weereratne, H., Blair, H., and Bull, G. M. (1982). Liposomes in Gaucher type I disease: Use in enzyme therapy and the creation of an animal model. *Prog. Clin. Biol. Res.* 95, 681–701.

Harris, D. T., Matyas, G. R., Gomella, L. G., Talor, E., Winship, M. D., Spitler, L. E., and Mastrangelo, M. J. (1999). Immunologic approaches to the treatment of prostate cancer. *Semin. Oncol.* 26(4), 439–447.

Houghton, A. N., Gold, J. S., and Blachere, N. E. (2001). Immunity against cancer: Lessons learned from melanoma. *Curr. Opin. Immunol.* 13(2), 134–140.

Ignatius, R., Mahnke, K., Rivera, M., Hong, K., Isdell, F., Steinman, R. M., Pope, M., and Stamatatos, L. (2000). Presentation of proteins encapsulated in sterically stabilized liposomes by dendritic cells initiates CD8(+) T-cell responses *in vivo*. *Blood* 96(10), 3505–3513.

Karrer, U., Althage, A., Odermatt, B., Roberts, C. W., Korsmeyer, S. J., Miyawaki, S., Hengartner, H., and Zinkernagel, R. M. (1997). On the key role of secondary lymphoid organs in antiviral immune responses studied in alymphoplastic (aly/aly) and spleenless (Hox11(-)/-) mutant mice. *J. Exp. Med.* 185(12), 2157–2170.

Lewis, P. J., and Babiuk, L. A. (1999). DNA vaccines: A review. *Adv. Virus Res.* 54, 129–188.

Li, S., and Huang, L. (1997). *In vivo* gene transfer via intravenous administration of cationic lipid-protamine-DNA (LPD) complexes. *Gene Ther.* 4(9), 891–900.

Li, S., Rizzo, M. A., Bhattacharya, S., and Huang, L. (1998). Characterization of cationic lipid-protamine-DNA (LPD) complexes for intravenous gene delivery. *Gene Ther.* 5(7), 930–937.

Liu, D. W., Tsao, Y. P., Kung, J. T., Ding, Y. A., Sytwu, H. K., Xiao, X., and Chen, S. L. (2000). Recombinant adeno-associated virus expressing human papillomavirus type 16 E7 peptide DNA fused with heat shock protein DNA as a potential vaccine for cervical cancer. *J. Virol.* **74**(6), 2888–2894.

Lodmell, D. L., Ray, N. B., Parnell, M. J., Ewalt, L. C., Hanlon, C. A., Shaddock, J. H., Sanderlin, D. S., and Rupprecht, C. E. (1998). DNA immunization protects nonhuman primates against rabies virus. *Nature Med.* **4**(8), 949–952.

Mackay, C. R., Marston, W. L., and Dudler, L. (1990). Naive and memory T cells show distinct pathways of lymphocyte recirculation. *J. Exp. Med.* **171**(3), 801–817.

Maloy, K. J., Erdmann, I., Basch, V., Sierro, S., Kramps, T. A., Zinkernagel, R. M., Oehen, S., and Kundig, T. M. (2001). Intralymphatic immunization enhances DNA vaccination. *Proc. Natl. Acad. Sci. USA* **98**(6), 3299–3303.

Manesis, E. K., Cameron, C. H., and Gregoriadis, G. (1978). Incorporation of hepatitis-B surface antigen (HBsAg) into liposomes. *Biochem. Soc. Trans.* **6**(5), 925–928.

Manickan, E., Karem, K. L., and Rouse, B. T. (1997). DNA vaccines: A modern gimmick or a boon to vaccinology? *Crit. Rev. Immunol.* **17**(2), 139–154.

Manickan, E., Yu, Z., Rouse, R. J., Wire, W. S., and Rouse, B. T. (1995). Induction of protective immunity against herpes simplex virus with DNA encoding the immediate early protein ICP 27. *Viral Immunol.* **8**(2), 53–61.

Martins, L. P., Lau, L. L., Asano, M. S., and Ahmed, R. (1995). DNA vaccination against persistent viral infection. *J. Virol.* **69**(4), 2574–2582.

McElrath, M. J. (1995). Selection of potent immunological adjuvants for vaccine construction. *Semin. Cancer Biol.* **6**(6), 375–385.

Nabel, G. J. (2004). Genetic, cellular and immune approaches to disease therapy: Past and future. *Nature Med.* **10**(2), 135–141.

Ochsenbein, A. F., Klenerman, P., Karrer, U., Ludewig, B., Pericin, M., Hengartner, H., and Zinkernagel, R. M. (1999). Immune surveillance against a solid tumor fails because of immunological ignorance. *Proc. Natl. Acad. Sci. USA* **96**(5), 2233–2238.

Owens, R. R., and Alving, C. R. (2000). A vaccine for induction of immunity to malaria. U.S. Patent No. 6,093,406.

Perrie, Y., Frederik, P. M., and Gregoriadis, G. (2001). Liposome-mediated DNA vaccination: The effect of vesicle composition. *Vaccine* **19**(23–24), 3301–3310.

Rao, M., and Alving, C. R. (2000). Delivery of lipids and liposomal proteins to the cytoplasm and Golgi of antigen-presenting cells. *Adv. Drug Deliv. Rev.* **41**(2), 171–188.

Schuler, G., and Steinman, R. M. (1997). Dendritic cells as adjuvants for immune-mediated resistance to tumors. *J. Exp. Med.* **186**(8), 1183–1187.

Sedegah, M., Hedstrom, R., Hobart, P., and Hoffman, S. L. (1994). Protection against malaria by immunization with plasmid DNA encoding circumsporozoite protein. *Proc. Natl. Acad. Sci. USA* **91**(21), 9866–9870.

Spier, R. E. (1996). International meeting on the nucleic acid vaccines for the prevention of infectious disease and regulating nucleic acid (DNA) vaccines. *Vaccine* **14**(13), 1285–1288.

Tascon, R. E., Colston, M. J., Ragno, S., Stavropoulos, E., Gregory, D., and Lowrie, D. B. (1996). Vaccination against tuberculosis by DNA injection. *Nature Med.* **2**(8), 888–892.

Toes, R. E., Blom, R. J., Offringa, R., Kast, W. M., and Melief, C. J. (1996). Enhanced tumor outgrowth after peptide vaccination: Functional deletion of tumor-specific CTL induced by peptide vaccination can lead to the inability to reject tumors. *J. Immunol.* **156**(10), 3911–3918.

Turell, M. J., Bunning, M., Ludwig, G. V., Ortman, B., Chang, J., Speaker, T., Spielman, A., McLean, R., Komar, N., Gates, R., McNamara, T., Creekmore, T., Farley, L., and Mitchell, C. J. (2003). DNA vaccine for West Nile virus infection in fish crows (*Corvus ossifragus*). *Emerg. Infect. Dis.* **9**(9), 1077–1081.

Ulmer, J. B., Donnelly, J. J., Parker, S. E., Rhodes, G. H., Felgner, P. L., Dwarki, V. J., Gromkowski, S. H., Deck, R. R., DeWitt, C. M., and Friedman, A. (1993). Heterologous protection against influenza by injection of DNA encoding a viral protein. *Science* **259**(5102), 1745–1749.

van Driel, W. J., Ressing, M. E., Kenter, G. G., Brandt, R. M., Krul, E. J., van Rossum, A. B., Schuuring, E., Offringa, R., Bauknecht, T., Tamm-Hermelink, A., van Dam, P. A., Fleuren, G. J., Kast, W. M., Melief, C. J., and Trimbos, J. B. (1999). Vaccination with HPV16 peptides of patients with advanced cervical carcinoma: Clinical evaluation of a phase I-II trial. *Eur. J. Cancer* **35**(6), 946–952.

Wang, B., Ugen, K. E., Srikantan, V., Agadjanyan, M. G., Dang, K., Refaeli, Y., Sato, A. I., Boyer, J., Williams, W. V., and Weiner, D. B. (1993). Gene inoculation generates immune responses against human immunodeficiency virus type 1. *Proc. Natl. Acad. Sci. USA* **90**(9), 4156–4160.

Wang, R. F., and Wang, H. Y. (2002). Enhancement of antitumor immunity by prolonging antigen presentation on dendritic cells. *Nature Biotechnol.* **20**(2), 149–154.

Whitmore, M., Li, S., and Huang, L. (1999). LPD lipopolyplex initiates a potent cytokine response and inhibits tumor growth. *Gene Ther.* **6**(11), 1867–1875.

Whitmore, M. M., Li, S., Falo, L., Jr., and Huang, L. (2001). Systemic administration of LPD prepared with CpG oligonucleotides inhibits the growth of established pulmonary metastases by stimulating innate and acquired antitumor immune responses. *Cancer Immunol. Immunother.* **50** (10), 503–514.

Xu, L., Sanchez, A., Yang, Z., Zaki, S. R., Nabel, E. G., Nichol, S. T., and Nabel, G. J. (1998). Immunization for Ebola virus infection. *Nature Med.* **4**(1), 37–42.

Yokoyama, M., Zhang, J., and Whitton, J. L. (1995). DNA immunization confers protection against lethal lymphocytic choriomeningitis virus infection. *J. Virol.* **69**(4), 2684–2688.

Zinkernagel, R. M., Ehl, S., Aichele, P., Oehen, S., Kundig, T., and Hengartner, H. (1997). Antigen localisation regulates immune responses in a dose- and time-dependent fashion: A geographical view of immune reactivity. *Immunol. Rev.* **156**, 199–209.

14

Non-Viral Vectors for Gene Therapy: Clinical Trials in Cardiovascular Disease

Pinak B. Shah and Douglas W. Losordo
Divisions of Cardiology, Vascular Medicine, and Cardiovascular Research
Caritas St. Elizabeth's Medical Center
Boston, Massachusetts 02135

0065-2660/05 $35.00
DOI: 10.1016/S0065-2660(05)54014-8

ABSTRACT

The population of patients with end-stage symptomatic coronary and peripheral vascular disease is ever-expanding. Many of these patients no longer have options for mechanical revascularization, and despite maximal medical therapy, they remain physically limited due to angina or critical limb ischemia. The fundamental problem in these patients is insufficient blood supply to muscle due to severely diseased conduit vessels to the target tissue. Therefore, it seems logical that increasing the blood supply to ischemic tissue will relieve symptoms. One potential means to achieving this goal is via therapeutic angiogenesis. The molecular mechanisms behind vascular development are being elucidated, and animal models have shown that mediators of vascular development can be harnessed to produce new capillaries in ischemic tissue. These mediators include cytokines such as vascular endothelial growth factor (VEGF) and fibroblast growth factor (FGF). Angiogenic cytokines can be delivered in several forms including recombinant protein or via gene delivery as a naked plasmid or via viral vector. This chapter will describe the clinical trial experience to date with delivery of non-viral gene therapy for therapeutic angiogenesis in humans with disabling myocardial ischemia and peripheral vascular disease. © 2005, Elsevier Inc.

I. INTRODUCTION

Significant strides have been made over recent decades in the understanding and treatment of atherosclerotic vascular disease. The importance of vascular injury, inflammation, and endothelial dysfunction is now recognized, and as a result, the mechanisms leading to the conversion from stable disease to acute vascular syndromes are being elucidated. With this knowledge, novel therapies have been developed in efforts to attenuate the atherosclerotic process and to reduce morbidity and mortality. Recent advancements in medical therapy for cardiovascular disease that have shown clear benefits in clinical trials include the thienopyridenes, angiotensin-converting enzyme inhibitors, and cholesterol-lowering medications.

Numerous advances have also been made in mechanical revascularization strategies for patients with symptomatic obstructive vascular disease. Surgical procedures such as lower extremity bypass grafting, carotid endarterctomy, and coronary artery bypass grafting (CABG) have been mainstays of therapy for patients with coronary artery disease, lower extremity vascular disease, and carotid artery disease, respectively. With improvements in technology, minimally invasive endovascular procedures have made treatment of obstructive cardiovascular disease simpler and safer under a widening array of circumstances. The first balloon angioplasty for symptomatic coronary artery disease was performed in 1977. In just over a quarter of a century, coronary stenting procedures have emerged as the predominant form of revascularization, performed in an ever-expanding patient population. Additionally, iliac and superficial femoral artery (SFA) angioplasty procedures are becoming the standard form of revascularization for patients with intermittent claudication or rest pain. Carotid artery stenting is also starting to gain acceptance as a treatment modality for patients with symptomatic or critical carotid artery disease.

As medical therapies and mechanical revascularization techniques continue to be refined and the population continues to age, there is an increasingly growing population of patients who, despite optimal medical therapy and risk factor modification, are no longer candidates for conventional mechanical revascularization. They may suffer from severe diffuse atherosclerotic disease not amenable to surgery or angioplasty or they may have had prior revascularization procedures, making future repeat procedures technically prohibitive. These patients continue to live with symptomatic obstructive vascular disease resulting in lifestyle-limiting claudication and limb ischemia, intractable angina, and congestive heart failure. Improving the quality of life, morbidity, and mortality of this growing patient population is a major challenge facing the future of cardiovascular medicine.

II. THERAPEUTIC ANGIOGENESIS

Patients who are no longer candidates for mechanical revascularization yet remain disabled from obstructive vascular disease suffer from an insufficient blood supply to working muscle. Mechanisms exist that allow for adaptation to insufficient blood supply though the compensatory development of new collateral circulation. Ischemic muscle may depend on collateral circulation and flow from healthier vessels to maintain perfusion; however, collateral insufficiency may lead to continued symptoms. Improving the function and supply of collateral vessels to ischemic tissue in order to relieve symptoms is termed *therapeutic angiogenesis.*

Embryonic and adult neovascularization are complex processes involving a variety of cells, cytokines, and growth factors specific to vascular development. Studies have revealed that bone marrow-derived circulating endothelial progenitor cells (EPCs) play a central role in postnatal neovascularization in adult life (Asahara et al., 1997). With a greater understanding of the mediators and processes of angiogenesis, numerous efforts have been made and are underway to harness this process (via therapeutic delivery of growth factors and EPCs) to treat a variety of vascular ischemic syndromes. Extensive preclinical work has been performed evaluating growth factors such as the vascular endothelial growth factor (VEGF) family, fibroblast growth factor (FGF) family, hepatocyte growth factor (HGF) family, and angiopoetin. The VEGF and FGF families, however, have been the most studied growth factors in clinical trials of gene therapy for therapeutic angiogenesis.

A. Vascular endothelial growth factor (VEGF)

The human VEGF proteins that have been identified to date are VEGF-1 or VEGF-A, VEGF-2 or VEGF-C, VEGF-3 or VEGF-B, VEGF-D, VEGF-E, and placental growth factor (PlGF). While each of these factors is encoded by genes on different chromosomes, there is significant genetic homology among these factors. VEGF-1 has four separate isoforms as a result of differential splicing. Each isoform is identified by the number of amino acids in the isoform: $VEGF_{121}$, $VEGF_{165}$, $VEGF_{189}$, and $VEGF_{206}$. These isoforms of VEGF show similar angiogenic potential in animal models (Takeshita et al., 1996) but differ in their solubility and heparin-binding capacity, accounting for differences in target cell binding. The principal cellular target of VEGF is the endothelial cell (EC). There are three known endothelial-specific fms-like tyrosine kinases: VEGFR-1 (Flt-1), VEGFR-2 (Flk-1/KDR), and VEGFR-3. Hypoxia induces the formation of VEGF by the ECs and leads to upregulation of VEGF receptors (Brogi et al., 1996). VEGFR-1 generates signals that organize the assembly of ECs into tubes and functional vessels (Fong et al., 1995). VEGFR-2 is responsible for EC proliferation and migration (Carmeliet et al., 1997; Shalaby et al., 1995). VEGFR-3 (Flt-4) principally mediates lymphangiogenesis (Jeltsch et al., 1997).

Several features of VEGF favor its use as a growth factor for gene transfer. VEGF contains a hydrophobic leader sequence, which is a secretory signal sequence that permits the protein to be secreted naturally from intact cells and enables a sequence of additional paracrine effects to be activated (Leung et al., 1989). The high-affinity binding sites of VEGF are highly expressed on ECs, and therefore, the mitogenic effects of VEGF are most prominent in ECs. This differs from the FGF family, which is known to be mitogenic for smooth muscle cells and fibroblasts as well as ECs (Ferrara et al., 1989). VEGF possesses an autocrine loop that is shared by most angiogenic cytokines and facilitates the modulation of EC behavior (Namiki et al., 1995). When activated under hypoxic conditions, the autocrine loop serves to amplify and thereby protract the response in ECs stimulated by exogenously administered VEGF. Furthermore, myocyte hypoxia stimulates an increase in the expression of VEGF receptors on ECs (Brogi et al., 1996). Such localized receptor expression may explain the finding that angiogenesis does not occur indiscriminately, but is relatively limited to sites of tissue ischemia. An important additional role for VEGF has been described in the augmentation of circulating EPC numbers documented in mice and humans following VEGF gene transfer (Asahara et al., 1999; Kalka et al., 2000a,b). These EPC have been shown to home into areas of myocardial ischemia.

B. Fibroblast growth factor (FGF)

The FGF family consists of several factors, including acidic FGF (FGF-1), basic FGF (bFGF or FGF-2), and FGF 3–9. Acidic FGF and basic FGF are the most extensively characterized members of the FGF family. FGFs are nonsecreted and lack a signal peptide sequence. Cell death or damage results in the extracellular release of FGF. It binds to tyrosine kinase receptors via cell-surface heparin sulfate proteoglycans, and therefore is removed rapidly from the circulation and is localized to cells and the extracellular matrix. While FGF is a potent EC mitogen, it lacks specificity for the EC and is a mitogen for other cell types, including fibroblasts and smooth muscle cells. At least four high-affinity FGF receptors have been identified and their cDNAs cloned. FGFs, like VEGF, also stimulate EC synthesis of proteases, including the plasminogen activator and metalloproteinases, important for extracellular matrix digestion in the process of angiogenesis (Carmeliet, 2000). Unlike VEGF, however, the common forms of FGF (FGF-1 and 2) lack a secretory signal sequence, and therefore, clinical trials of FGF gene transfer have required either modification of the FGF gene or use of another of the FGF gene family with a signal sequence (Giordano *et al.*, 1996; McKirnan *et al.*, 2000; Tabata *et al.*, 1997).

III. DELIVERY OF ANGIOGENIC CYTOKINES

A major challenge facing the therapeutic application of angiogenesis is determining the optimal method of delivery of angiogenic cytokines to ischemic tissue. Ischemic muscle has certain characteristics that are favorable for angiogenic therapy, such as hypoxia-induced expression of angiogenic growth factor receptors (Brogi *et al.*, 1996; Goldberg and Schneider, 1994; Levy *et al.*, 1996; Waltenberger *et al.*, 1996). There are numerous potential methods of angiogenic cytokine delivery. One method involves delivering the cytokine in a recombinant protein form. Recombinant protein therapy is felt to be the most conventional and practical means for the delivery of angiogenic cytokines; however, there are several important limitations to recombinant protein therapy. First, the dose of protein that needs to be administered in order to engender a desirable biological effect may also result in an increase in adverse events. For example, the VEGF protein has been shown to induce nitric oxide-mediated hypotension (Hariawala *et al.*, 1996) and therefore limited dosing in clinical studies (Henry *et al.*, 2003). Second, recombinant proteins have a short half-life in the circulation because of circulating proteases, and therefore, the half-life of activity is limited. Third, recombinant proteins can be difficult and costly to produce.

Angiogenic cytokines can be delivered in the form of a gene that allows for transcription and translation of angiogenic protein at the targeted tissue. Genes can be transferred using non-viral and viral methods. Skeletal and cardiac muscle have been shown to take up and express genes encoded in plasmids (Wolff *et al.*, 1990) as well as transgenes incorporated into viral vectors (Giordano *et al.*, 1996). Further, the efficiency of gene transfection is enhanced significantly in ischemic tissue (Takeshita *et al.*, 1996; Tsurumi *et al.*, 1996).

Viral transfer strategies utilize virus vectors (generally adenoviruses or retroviruses) in order to deliver the gene to the targeted tissue. Viral transfer methods are felt to result in increased transfection efficiency compared to non-viral methods. However, viral vectors induce an inflammatory response that results in undesired side effects and may limit the ability for repeat administration due to the host immune response with repeat administration. (Yang *et al.*, 1994).

Non-viral methods generally involve delivery of a naked plasmid encoding the gene of interest or delivery of DNA material in a liposomal vehicle. Because of circulating nucleases, delivery of DNA is limited by a shorter half-life in targeted tissue; therefore, the time for gene expression may be limited. However, it is felt that delivery of non-virally encoded DNA may allow for meaningful biological effects, as protein is secreted locally because the genes do not remain intracellular. *In vitro* and *in vivo* studies have also shown that despite lower gene transfer efficiency with non-viral methods, site-specific administration of the gene still results in physiological effects within the local tissue (Losordo *et al.*, 1994; Takeshita *et al.*, 1994). An additional advantage of non-viral gene transfer is that it should allow for repeat administration of the therapeutic agent.

In addition to the preparation in which the cytokine is delivered (protein, naked gene, or viral-encoded gene), the route in which the cytokine is delivered also remains controversial. In the case of naked DNA gene therapy, the gene can be delivered intravascularly (intravenously or intraarterially). The intravascular route, however, is limited by short dwell times in ischemic tissue, significant distribution of the gene to nonischemic tissue, and rapid degradation of the gene due to circulating nucleases. Therefore, very large doses of gene are necessary in order to deliver a therapeutic effect.

Gene therapy can also be delivered by direct intramuscular injection. In patients with peripheral arterial disease (PAD), intramuscular gene delivery is simply delivered by percutaneous intramuscular injection into affected muscle (i.e., gastrocnemius for calf claudication). In patients with coronary artery disease and myocardial ischemia, however, the delivery of gene therapy is more complicated. Early studies of myocardial gene therapy required that the patient have a thoracotomy for comcomitant CABG or that thoracotomy was performed solely for myocardial delivery of the vector. The obvious disadvantage of minithoracotomy is the need for general anesthesia and surgery, both of which are associated with risk and morbidity. A second limitation of the surgical approach is that in clinical trials of myocardial gene therapy, a true control group is not possible because of ethical limitations in subjecting patients to surgery and administration of placebo. Therefore, a placebo effect in patients who improve after gene therapy delivery by surgical approaches severely limits the interpretation of efficacy.

A percutaneous approach to myocardial gene therapy utilizing the NOGA system (Biosense-Webster, Warren, NJ) has been utilized to identify ischemic myocardium (Ben-Haim *et al.*, 1996; Gepstein *et al.*, 1998) and to deliver gene therapy to ischemic zones. The NOGA system uses a mapping catheter that is advanced via the femoral artery across the aortic valve into the left ventricle. This catheter is then manipulated to contact the myocardium at numerous points to create an electromechanical map of the left ventricle. Ischemic myocardium is identified as myocardium with normal electrical activity but abnormal mechanical activity, i.e., wall motion. Once the ischemic zone is mapped, a second mapping catheter with a retractable 27-gauge needle is advanced into the left ventricle to the ischemic zone. Once in contact with the ischemic myocardium, the needle is advanced and therapy is delivered. In addition to being less invasive than surgical delivery of the gene, this percutaneous method of gene delivery has the added advantage of safely allowing a placebo group to be included in clinical trials so that the efficacy of myocardial gene therapy can be assessed.

IV. CLINICAL TRIALS OF NON-VIRAL GENE THERAPY FOR THERAPEUTIC ANGIOGENESIS IN CARDIOVASCULAR DISEASE

Clinical trials of angiogenesis have evaluated all forms of angiogeneic growth factors for delivery to ischemic tissue, including recombinant protein, naked gene (generally encoded in a plasmid), and virally encoded gene. Additionally, ongoing studies of angiogenesis are evaluating the delivery of EPCs alone and in combination with angiogenic growth factors. This section focuses on clinical trials performed to date evaluating non-viral vectors of gene therapy for therapeutic angiogenesis (Table 14.1).

A. Studies of VEGF gene therapy for the treatment of peripheral arterial disease

The first attempt in humans to deliver VEGF gene therapy for chronic limb ischemia (CLI) utilized the plasmid encoding $VEGF_{165}$ ($phVEGF_{165}$) driven by a cytomegalovirus promoter applied to the polymer surface of a hydrogel-coated angioplasty balloon. A total of 22 patients with CLI were treated by $phVEGF_{165}$-coated balloon angioplasty in the native vasculature proximal to

Table 14.1. Clinical Trials of Gene Therapy (Non-viral Vectors) for Ischemic Cardiovascular Disease[a]

Reference	Therapy	Patient characteristics	Delivery route	Study design	Active/placebo n/n
Isner et al. (1995)	VEGF$_{165}$	PAD-CLI	Coated PTA balloon	Phase I	22/0
Baumgartner et al. (1998)	VEGF$_{165}$	PAD-CLI, ulcers	Direct intramuscular	Phase I	9/0
Isner et al. (1998)	VEGF$_{165}$	PAD-CLI in Buerger's	Direct intramuscular	Phase I	11/0
Vale et al. (1998)	VEGF$_{165}$	PAD-SFA PTA	Intraarterial	Phase I	30/0
Losordo et al. (1998)	VEGF$_{165}$	CAD	Direct thoracotomy	Phase I	30/0
Vale et al. (2001)	VEGF-2	CAD	Direct myocardial-NOGA	Phase I	3/3
Losordo et al. (2002)	VEGF-2	CAD	Direct myocardial-NOGA	Phase I/II	13/6
Comerota et al. (2002)	FGF1	PAD-CLI, ulcers	Direct intramuscular	Phase I	51/0
Hedman et al. (2003)	VEGF$_{165}$	CAD-PCI	Intracoronary during PCI	Phase II	37/28/38[b]
Kastrup et al. (2005)	VEGF$_{165}$	CAD	Direct myocardial-NOGA	Phase II	40/40
GENASES[c]	VEGF-2	CAD	Direct myocardial	Phase IIb	300/100

[a]VEGF, vascular endothelial growth factor; PAD, peripheral arterial disease; CLI, critical limb ischemia; SFA, superficial femoral artery; PTA, percutaneous transluminal angioplasty; CAD, coronary artery disease; FGF, fibroblast growth factor; PCI, percutaneous coronary intervention.

[b]Three arms (in order): adenovirus encoding VEGF$_{165}$, plasmid encoding VEGF$_{165}$, and placebo.

[c]Trial presently enrolling.

the infrapopliteal vessels in a dose-escalating fashion (Isner *et al.*, 1995). Magnetic resonance angiography (MRA) and contrast angiography both showed increases in collateral vasculature in treated patients (Isner, 1998).

Subsequent preclinical studies demonstrated the efficacy of VEGF gene delivery using direct intramuscular injection. Using the rabbit hindlimb ischemia model, Tsurumi and colleagues (1996) provided evidence that administration of naked phVEGF$_{165}$ could induce therapeutic angiogenesis similar to previous experience with recombinant VEGF. Rabbits treated with direct intramuscular injection of phVEGF$_{165}$ experienced an increase in the Doppler pressure ratio of ischemic limb to nonischemic limb compared to control rabbits (0.70 $\pm$ 0.08 vs 0.50 $\pm$ 0.18, $P < 0.05$). Angiography and necropsy studies confirmed an increase in capillary density in animals treated with phVEGF$_{165}$. A finding of interest in this study was that angiogenesis was not indiscriminate but more localized to areas of ischemia. Similar findings were seen with VEGF-2 (VEGF-C) in a rabbit hindlimb model of ischemia (Witzenbichler *et al.*, 1998). Additionally, VEGF-2 was also shown to increase nitric oxide (NO) release from endothelial cells.

In a phase I clinical trial, direct injection of phVEGF$_{165}$ was performed in 10 limbs of 9 patients with rest pain or nonhealing ulcers due to PAD (Baumgartner *et al.*, 1998). Patients were treated with a direct injection of 2000 μg of phVEGF$_{165}$ into ischemic muscle followed by a repeat treatment 4 weeks later. Using ELISA, VEGF protein levels were measured in the bloodstream and were shown to peak at 1 to 3 weeks after injection. There was an overall increase in the ankle brachial index (ABI) (0.33 $\pm$ 0.05 to 0.48 $\pm$ 0.03, $P = 0.02$) as well as increased collateral flow on angiography at follow-up. Ischemic ulcers healed in four of seven limbs, including three patients who were originally referred for below knee amputations. Tissue specimens obtained from an amputee 10 weeks after gene therapy showed foci of proliferating endothelial cells by immunohistochemistry. This small study showed that gene therapy with phVEGH$_{165}$ for PAD is safe and feasible and provided preliminary evidence that transfection with phVEGF$_{165}$ could result in sufficient production of VEGF for therapeutic angiogenesis.

In a follow-up study, 55 patients with ischemic rest pain or ischemic ulcers were treated with intramuscular injections of phVEGF$_{165}$. Clinical improvement was evident in 72% of patients with rest pain and 63% of patients with ischemic ulcers over a follow-up period of 4 to 36 months. Patients with rest pain and patient under the age of 50 were more likely to experience an improvement of clinical symptoms (Rauh *et al.*, 1999).

Intramuscular injection of phVEGF$_{165}$ has also been used to treat 11 patients with Buerger's disease and CLI. Nine patients experienced clinical improvement with resolution of rest pain. Improvements in objective measures were also seen, including improvements of ischemic ulcers, increases in ABI, and enhancement of collateral vasculature on MRA and contrast angiography (Isner *et al.*, 1998).

These studies confirmed the safety and feasibility of intramuscular injection of phVEGF$_{165}$ in patients with CLI. Adverse events were limited to lower extremity edema, which is seen in one-third of patients treated with VEGF gene therapy (Baumgartner *et al.*, 2000). The edema was treated readily with diuretic therapy. Edema in these patients may be due to increased vascular permeability (van der Zee *et al.*, 1997) and nitric oxide-induced vasodilatation (Roberts and Palade 1995), both mediated by VEGF, although the assessment of this effect, as well as the positive effects noted, is limited by the lack of a placebo group.

B. Studies of VEGF gene therapy for the prevention of restenosis after balloon angioplasty

Patients with intermittent claudication (IC) and CLI often have severe disease of the superficial femoral artery. The SFA has proven to be a difficult area to treat with percutaneous methods due to restenosis, with rates as high as 60% in patients treated with balloon angioplasty and stenting. A number of postulates have been put forth as to why the SFA responds poorly to percutaneous procedures, but the exact mechanisms are not known. No medical therapy has been proven to attenuate restenosis after SFA intervention.

Delay of reendothelialization at the site of angioplasty is believed to play a vital role in the development of restenosis in any vessel. VEGF is a specific EC mitogen without migratory effects on smooth muscle cells (Ferrara *et al.*, 1989; Leung *et al.*, 1989). Animal models have shown that restoration of endothelial integrity with an EC mitogen such as VEGF may attenuate restenosis after vascular interventional procedures.

In a preclinical study, rabbits were subjected to balloon injury and gene transfer with phVEGF$_{165}$ or LacZ in one femoral artery (Asahara *et al.*, 1996). The contralateral femoral artery was also subjected to balloon injury but without gene transfection. In vessels treated with phVEGF$_{165}$, reendothelialization was nearly complete in the injured arteries, whereas reendothelialization was not complete in LacZ-treated vessels. In phVEGF$_{165}$-treated rabbits, injured vessels showed a reduced intima to media ratio, a decreased frequency of thrombotic occlusion, anda normal vasomotor response to EC-dependent vasodilators.

The favorable response in preclinical studies set the framework for clinical trials evaluating the use of VEGF gene therapy in patients with PAD treated with PTA. Our center performed a phase 1, dose-escalating, open-label, unblinded gene therapy trial of local VEGF gene transfer to accelerate reendothelialization at the site of PTA-induced endothelial disruption as a novel means to inhibit restenosis following PTA (Vale *et al.*, 1998). The primary objective of this study was to document the safety of percutaneous catheter-based delivery of the gene encoding VEGF in patients with claudication due to SFA obstruction.

Arterial VEGF gene transfer has been performed in 30 patients: 21 males and 9 females with a mean age of 68 years. All patients had two or more cardiovascular risk factors. Gene expression was documented by a rise in plasma levels of VEGF. Peak plasma levels were recorded at a mean of 12 days following gene transfer. Mean claudication time increased from 2 min at baseline to 5 min up to 18 months postgene transfer. Prior to gene transfer, all patients were classified as Rutherford class 3. At 12–18 months following gene transfer, 15 patients were asymptomatic and 8 patients were class 1. After an initial improvement in two Rutherford classes following revascularization, 6 patients returned to class 3. One patient developed critical limb ischemia and required salvage therapy with an intramuscular gene transfer of naked plasmid DNA encoding VEGF.

There was a significant and sustained improvement in the ankle brachial index, postgene transfer compared to baseline. Prior to gene transfer, the mean ABI was 0.70, which increased to 0.92 at 18 months after gene transfer and was sustained at 0.91 at 48 months after gene transfer. SFA stenosis in 24 patients dropped from a mean of 94% at baseline to 30% at an average of 9 months following gene transfer. These results were supported by IVUS findings at the time of follow-up angiography. Six patients had evidence of restenosis at angiography performed 6–12 months following gene transfer. Target vessel revascularization was required in all 6 patients. Histology from 3 out of 4 patients undergoing directional atherectomy at the time of repeat revascularization for restenosis demonstrated active smooth muscle cell proliferation and high levels of proliferating cell nuclear antigen, indicating extensive proliferative activity.

Thus, in the 30 patients that have been treated with arterial VEGF gene transfer for the prevention of restenosis, VEGF expression has been documented by ELISA assay. At 48 months follow-up, 6 out of 30 patients (20%) required target vessel revascularization for angiographic and ultrasound evidence of restenosis. When compared to historical controls, at 6 months, the restenosis rate was 20% in the gene transfer group, 29% in patients undergoing brachytherapy, and 55% in patients undergoing percutaneous angioplasty alone (Minar et al., 2000). This preliminary study has suggested that gene therapy designed to accelerate reendothelialization at the site of PTA-induced endothelial disruption can be performed safely. Importantly, no evidence of accelerated atherosclerosis or increased restenosis was observed following gene transfer.

Hedman et al. (2003) reported similar findings of safety as well as evidence of improved perfusion after VEGF gene delivery in the coronary circulation. In this study, 103 patients with CCS class II or III angina and coronary artery disease amenable to percutaneous coronary intervention were treated with balloon angioplasty followed by gene delivery or placebo to the target vessel with a perfusion catheter. Patients were randomized to receive $VEGF_{165}$ encoded by an adnenovirus, plasmid-encoded $VEGF_{165}$, or lactated Ringer's solution. The primary end point was minimum lumen diameter and diameter stenosis as measured by quantitative coronary angiography at 6 months follow-up. There were no significant differences in angiographic outcomes at 6 months among the three groups. However, there was a significant improvement in the perfusion defect score with nuclear perfusion testing in patients receiving $VEGF_{165}$ encoded by adenovirus at 6 months compared to baseline. Similar improvements were not seen in the plasmid-encoded $VEGF_{165}$ group or the placebo group.

C. Studies of VEGF gene therapy for the treatment of myocardial ischemia

VEGF gene therapy may have advantages over protein therapy for two important reasons. First, VEGF protein therapy requires doses of protein that may result in systemic hypotension (Henry *et al.*, 2000, 2003). This is a result of VEGF-mediated release of NO (Hariawala *et al.*, 1996, 1996). Second, while the effects of protein therapy are short-lived, gene therapy may provide local expression of VEGF for a period of weeks after transfection. (Gal *et al.*, 1993; Lee *et al.*, 2000; Mack *et al.*, 1998).

A phase I, single-center, dose-escalating trial of direct intramyocardial delivery of $phVEGF_{165}$ in patients with myocardial ischemia was performed (Losordo *et al.*, 1998; Symes *et al.*, 1999) Patients were included in this study if they had symptomatic myocardial ischemia, objective evidence of ischemia on single photon emission computed tomography (SPECT) scanning, and multivessel coronary artery disease with no option for revascularization. Thirty patients were selected for this study and all received $phVEGF_{165}$ administered by direct myocardial injection in four aliquots of 2.0 ml via a minithoracotomy. Ten patients received a total dose of $125\mu g$ of plasmid, 10 received 250 μg, and 10 received 500 μg. An immobile field for plasmid injection in the beating heart was secured using a myocardial-stabilizing device. Monitoring with transesophageal echocardiography was used to monitor the development of wall motion abnormalities associated with injections and to ensure that plasmid was not injected into the left ventricle (Esakof *et al.*, 1999).

This study provided evidence for the safety of intramyocardial injection of plasmid-encoding $VEGF_{165}$ via minithoracotomy. No operative complications occurred and no patient experienced myocardial infarction or decrement in left ventricular function as a result of therapy. There were two deaths (4.5 and 28.5 months), and one patient underwent a cardiac transplant at 13 months (Symes *et al.*, 1999). This study also demonstrated proof of concept that direct gene transfection can result in gene expression. Plasma VEGF protein levels were noted to increase significantly as monitored by ELISA.

This study was not designed to assess the efficacy of this treatment for chronic myocardial ischemia as there was no control group. A control group would not be feasible because of the need for patients to undergo minithoracotomy and sham treatment. Nevertheless, all patients experienced symptomatic improvement and/or objective evidence of improved myocardial perfusion. At 1 year, half of the patients were free of angina and had a significant decrease in nitrate use. Exercise time, a more objective measure of efficacy, had also increased from baseline. Evidence of reduced ischemia on SPECT-Sestamibi myocardial perfusion scanning was documented in 22/29 patients with a significant reduction in both stress and rest mean perfusion/ischemia score at day 60. Improvements in rest SPECT scans after gene therapy support the concept that SPECT defects at rest contain foci of hibernating myocardium that have improved mechanical activity as a result of therapeutic angiogenesis (Dilsizian and Bonow, 1993; Shen and Vatner, 1995; Wijns et al., 1998).

NOGA was used in 13 of the patients in this study to further demonstrate efficacy of $VEGF_{165}$ gene transfer (Vale et al., 2000). Prior to gene transfer, these patients underwent electromechanical mapping to identify areas of ischemic, but viable myocardium. Myocardial viability did not change after gene transfer; however, there was significant improvement of linear shortening. The area of ischemic myocardium was reduced from $6.45 \pm 1.37 \text{ cm}^2$ before gene transfer to $0.95 \pm 0.41 \text{ cm}^2$ after gene transfer ($P = 0.001$).

An endovascular approach to direct myocardial gene transfer using NOGA electromechanical mapping was shown to be safe and feasible in a preclinical study using pigs (Vale et al., 1999). In two of the pigs, methylene blue was injected into the myocardium. At necropsy, the location of the methylene blue staining corresponded to the location of injection on endocardial mapping. Six pigs were treated with plasmid using a cytomegalovirus promoter/enhancer encoding the nuclear-specific LacZ gene (pCMV-nlsLacZ). Peak β-galactosidase activity was noted after 5 days from the myocardial injection site. In order to determine transfection capability in ischemic myocardium, two pigs underwent ameroid constrictor placement on the left circumflex artery prior to gene transfer. They then underwent NOGA-directed gene transfer using pCMV-nlsLacZ. Again, peak β-galactosidase activity was noted at the injection site 5 days after gene transfer. There were no complications during endocardial mapping and gene transfer.

Subsequently, a pilot study of NOGA-guided direct myocardial gene transfer in humans using plasmid encoding VEGF-2 (phVEGF-2) was performed (Losordo et al., 2002; Vale et al., 2001). Six patients with chronic symptomatic myocardial ischemia and no option for conventional mechanical revascularization were randomized to receive 200 μg of phVEGF-2 or placebo. The treatment was administered in six injections into ischemic myocardium. Patients treated with phVEGF-2 experienced fewer anginal episodes after gene therapy

$(36.2 \pm 2.3$ episodes/week before treatment versus 3.5 ± 1.2 episodes/week after treatment) and reduced nitroglycerin consumption $(33.8 \pm 2.3$ tablets/week before treatment versus 4.1 ± 1.5 tablets/week after treatment) for 1 year after treatment. Similar improvements were also seen in patients treated with placebo; however, these improvements persisted only 30 days after treatment, at which point patients returned to baseline in terms of angina and nitrate use. Repeat electromechanical mapping was performed in all patients 90 days after gene therapy. The area of ischemic territory was improved in patients treated with phVEGF-2 compared to baseline (mean area of ischemia, 10.2 ± 3.5 cm^2 versus 2.8 ± 1.6 cm^2, $P = 0.04$). No adverse events were noted in any patient treated in the study. This study confirmed the safety and feasibility of endovascular delivery of gene therapy in humans.

Safety and efficacy of this strategy were further evaluated in a phase I/II randomized, double-blind, placebo-controlled trial of VEGF-2 therapy (Losordo et al., 2002). A total of 27 patients were planned for randomization into one of three treatment groups: 200, 800, or 2000 μg of phVEGF-2. In each dose cohort, patients were randomized to receive the specified dose of phVEGF-2 or placebo in a 2:1 ratio. The study was stopped early due to a moratorium on all gene therapy protocols put in place by the Food and Drug Administration. Only one patient was treated in the third dose cohort and received active treatment. At 12 weeks after treatment, there was a statistically significant improvement in the Canadian Cardiovascular Society (CCS) angina class in phVEGF-2 treated versus placebo-treated patients (-1.3 versus -0.1, $P = 0.04$). There was a trend toward improvement in change in exercise duration (91.8 s versus 3.9 s), functional improvement by $>$ or $=2$ CCS classes (9 of 12 versus 1 of 6), and Seattle Angina Questionnaire data in phVEGF-2 patients compared to placebo-treated patients. Importantly, no adverse events were noted as a result of either NOGA-guided gene therapy delivery or the gene therapy itself.

The recently reported Euroinject One study was a trial of 80 patients with CCS class III or IV angina and no option for conventional revascularization randomized to receive direct myocardial injection of phVEGF$_{165}$ or placebo using NOGA guidance (Kastrup et al., 2005). The primary end point of this study was change in perfusion defect scores from baseline at 3 months. In the phVEGF$_{165}$ group, there was a significant improvement in stress-induced perfusion defects at 3 months compared to baseline while a similar improvement was not seen in the control group. The difference between the two groups, however, was not significant. There were also significant improvements in wall motion as assessed by NOGA and contrast ventriculography in the phVEGF$_{165}$ group compared to the control group. Overall, there were no adverse events due to gene therapy in this study, although there were five adverse events related to NOGA mapping.

A phase IIb pivotal trial of endovascular delivery of phVEGF-2 in patients with chronic myocardial ischemia is presently underway. This trial will enroll 400 patients with CCS class III or IV angina to receive 20, 200, or 800 μg or placebo injected directly into the myocardium. NOGA will not be used in this trial. Instead, therapy will be delivered via the Stiletto (Boston Scientific, Natick, MA) endocardial injection catheter. While electromechanical mapping will not be used in this study for the identification of ischemic myocardium, animal studies have suggested that adequate neovascularization can be achieved with blind delivery of gene therapy. The primary end point of this study will be a change in exercise duration from baseline 3 months after therapy. Changes in CCS class, functional status, and SPECT scans will also be assessed in this study.

D. Studies FGF gene therapy for cardiovascular disease

The majority of clinical trials of angiogenesis for ischemic vascular disease utilizing FGF have used FGF in recombinant protein form incorporated into a viral vector. Phase I trials have been performed using FGF protein injected intraarterially for the treatment of intermittent claudication (Lederman et al., 2002), direct intramyocardial delivery during CABG surgery or via minithoracotomy (Schumacher et al., 1998; Stegmann et al., 2000), delivered periadventitially during CABG surgery (Laham et al., 1999; Stegmann, et al., 2000), and by intracoronary injection during cardiac catheterization (Udelson et al., 2000; Unger et al., 2000). The FIRST trial was a phase I/II evaluation of FGF-2 recombinant protein delivered via intracoronary injection for the treatment of ischemic heart disease (Simons et al., 2002). The recently published and reported AGENT trials evaluated FGF-4 incorporated into an adenoviral vector for the treatment of ischemic heart disease (Grines et al., 2002; presented at Late-Breaking Clinical Trials, Transcatheter Therapeutics 2004).

The only published study to date of naked FGF gene therapy evaluated naked plasmid encoding FGF1 (NV1FGF) delivered by intramuscular injection in patients with lower extremity rest pain or tissue necrosis and no option for conventional revascularization (Comerota et al., 2002). Fifty-one patients were treated with an escalating single dose of NV1FGF (500, 1000, 2000, 4000, 8000, and 16000 μg) or escalating repeated doses of NV1FGF (2×500, 2×1000, 2×2000, 2×4000, and 2×8000 μg). All patients underwent angiography before treatment and 12 weeks after treatment. Overall, treatment with NV1FGF was well tolerated with only minor adverse events reported at the injection site. There were no increases in urine levels of plasmid and only a transient increase in the plasma level of plasmid. There was no increase in plasma FGF-1 protein levels. In 15 patients in which clinical outcomes were evaluated, there were significant improvements compared to baseline in pain and ulcer size and significant improvements in transcutaneous oxygen pressure and ABI.

V. ISSUES IN CLINICAL TRIAL DESIGN IN GENE THERAPY FOR CARDIOVASCULAR DISEASE

There are several major challenges facing clinical investigation of gene therapy for cardiovascular disease. The first is determining the optimal method of gene delivery. While there are advantages of both approaches, whether delivery of a gene using a viral vector versus delivery of a gene in the form of a naked plasmid remains unclear. There also still remains considerable debate as to whether angiogenic cytokines are better delivered as recombinant proteins or in the form of gene therapy. Preclinical data suggest that angiogenesis with gene therapy may be augmented with concurrent treatment with cytokines to mobilize EPCs. This strategy has yet to be tested in humans. Finally, the optimal route of delivery (direct intramuscular vs intravenous vs intraarterial vs via coronary sinus or pericardium) remains to be elucidated.

Second, choosing appropriate end points for clinical trials for cardiovascular gene therapy remains problematic. Ultimately, the goal of cardiovascular gene therapy is to alleviate ischemic symptoms (rest pain, claudication, and angina), improve exercise capacity, and improve the quality of life. However, numerous studies of angiogenesis have shown that patients in control arms of studies also experience a significant benefit in these subjective measures, indicating the presence of a significant placebo effect. These measures may prove more useful in larger scale phase II and III studies, but are difficult to assess in smaller phase I trials.

The search for the ideal objective measure of angiogenesis continues. Follow-up arteriography increases the cost and risk to the patient. Further, increased vascularity is difficult to quantify on arteriography, as the diameter of newly formed vessels is under the level of resolution of conventional contrast angiography. Rest and stress SPECT perfusion imaging with technetium-99m sestamibi or thallium-201 also provides an objective measure of perfusion in myocardial angiogenesis, but similarly may suffer from a lack of sensitivity. Follow-up NOGA mapping is instrumental in identifying areas of neovascularization after angiogenesis therapy, but suffers from the lack of widespread expertise in the technique as well as the requirement of a repeat invasive procedure. Positron emission tomography can quantify coronary blood flow at rest and peak hyperemia, while advancements in magnetic resonance imaging technology may aid in the detection of foci of neovascularization (Pearlman et al., 1995, 2000). However, there are limited centers with expertise with these modalities.

VI. SAFETY CONCERNS REGARDING GENE THERAPY FOR ISCHEMIC VASCULAR DISEASE

Several safety concerns regarding gene therapy for ischemic vascular disease have been raised. To date, however, no clear evidence has emerged in clinical trials of adverse consequences of vascular gene therapy. Concerns exist regarding the induction of angiogenesis in humans, as well as concerns of adverse effects of the cytokines encoded by the gene therapy, particularly VEGF (Isner et al., 2001).

Concern for an increased risk of neoplastic disease and malignancy with therapeutic angiogenesis is the natural result of the fact that angiogenic cytokines were discovered as factors responsible for promoting tumor vascularization (Folkman et al., 1971) At the present time, however, there are no in vitro or in vivo data to suggest that gene therapy with angiogenic cytokines increases the risk of malignancy. In nearly 80 patients treated with vascular gene therapy for angiogenesis at a single center, the 7-year incidence of cancer has been limited to 2 patients who developed bladder cancer and 1 patient with liver and brain metastases from an unknown primary source (Isner et al., 2001).

There additionally has been concern that induction of angiogenesis can lead to the development of retinopathy. High levels of VEGF have been demonstrated in ocular fluid of patients with proliferative retinopathy, resulting in a loss of vision (Aiello et al., 1994). This concern has not been borne out in clinical trials of angiogenesis. In nearly 100 patients treated with VEGF gene therapy at a single institution, no patient developed retinopathy over a 4-year follow-up period as assessed by independent retina specialists, despite the fact that nearly one-third of the patients were diabetic (Vale et al., 1998).

Data have shown that treatment with inhibitors of angiogenesis in apolipoprotein E-deficient mice inhibits plaque growth and intimal thickening (Moulton et al., 1999) and have led to concerns that inducing angiogenesis would lead to accelerated plaque growth and the potential of worsening vascular ischemic syndromes. However, preclinical and clinical studies have failed to support the notion that accelerated atherosclerosis occurs after the induction of vascular angiogenesis (Asahara et al., 1996; Van Belle et al., 1997a,b). In fact, angiogenic therapy may accelerate reendothelialization and lead to reductions in intimal thickening. In patients enrolled in clinical trials of angiogenesis, there has been no significant increase in anginal symptoms or acute vascular syndromes.

The administration of VEGF protein has been shown to result in significant hypotension due to the upregulation of nitric oxide synthase (Hariawala et al., 1996; Horowitz et al., 1997). This complication, however, has not been demonstrated with VEGF gene therapy in animals or humans. Experiments in transgenic mice engineered to overexpress VEGF ± angiopoetin have demonstrated lethal permeability-enhancing effects of VEGF (Thurston et al.,

1999). However, even though VEGF has been reported to cause local edema, which manifests as pedal edema in patients treated with VEGF for critical limb ischemia, it responds well to treatment with diuretics (Baumgartner *et al.*, 2000).

It is important to remember that patients with end-stage vascular disease are generally elderly and are at risk for malignancy and retinopathy because of their underlying risk factors and comorbidities (age, smoking, diabetes). Therefore, candidates for trials of vascular gene therapy are likely to develop such complications as part of the natural history of their disease states and not necessarily because of therapy with angiogenic agents. Nevertheless, painstaking efforts must be taken continually to ensure that candidates for clinical trials of angiogenesis are screened thoroughly for malignant and premalignant conditions, as well as retinopathy, until these risks are completely understood.

VII. CONCLUSION

As the population ages, the burden of ischemic vascular disease will continue to expand. Therefore, strategies will be necessary to treat a growing population of end-stage vascular disease patients who have no further options for conventional therapy. To date, no form of therapy for this patient population has provided indisputable evidence of efficacy. Therapeutic angiogenesis may prove to be a solution.

Results of phase I and II trials of gene therapy for therapeutic angiogenesis in patients with ischemic vascular disease are encouraging. In trials performed to date, gene therapy appears to be safe and feasible. Furthermore, endovascular techniques for gene delivery continue to be refined and, as a result, gene therapy is delivered in a minimally invasive fashion.

The field of therapeutic angiogenesis with gene therapy now hinges on the performance of larger scale, randomized, placebo-controlled trials to assess the true efficacy of this approach. As these strategies evolve, so will our understanding of the best means of measuring the therapeutic efficacy of vascular gene therapy. Other important questions that will be answered include the specific utility of individual cytokine(s), the appropriate routes of delivery, the form of the angiogenic cytokine, dose, and whether combination therapy (e.g., with EPCs) may provide advantages compared to monotherapy.

References

Aiello, L. P., Avery, R. L., Arrigg, P. G., Keyt, B. A., Jampel, H. D., Shah, S. T., Pasquale, L. R., Theme, H., Iwamoto, M. A., Parke, J. E., Nguyen, M. D., Aiello, L. M., Ferrara, N., and King, G. L. (1994). Vascular endothelial growth factor in ocular fluids of patients with diabetic retinopathy and other retinal disorders. *N. Engl. J. Med.* **331**, 1480–1487.

Asahara, T., Chen, D., Tsurumi, Y., Kearney, M., Rossow, S., Passeri, J., Symes, J., and Isner, J. (1996). Accelerated restitution of endothelial integrity and endothelium-dependent function following phVEGF$_{165}$ gene transfer. *Circulation* **94,** 3291–3302.

Asahara, T., Murohara, T., Sullivan, A., Silver, M., van der Zee, R., Li, T., Witzenbichler, B., Schatteman, G., and Isner, J. M. (1997). Isolation of putative progenitor endothelial cells for angiogenesis. *Science* **275,** 964–967.

Asahara, T., Takahashi, T., Masuda, H., Kalka, C., Chen, D., Iwaguro, H., Inai, Y., Silver, M., and Isner, J. M. (1999). VEGF contributes to postnatal neovascularization by mobilizing bone marrow-derived endothelial progenitor cells. *EMBO J.* **18,** 3964–3972.

Baumgartner, I., Pieczek, A., Manor, O., Blair, R., Kearney, M., Walsh, K., and Isner, J. M. (1998). Constitutive expression of phVEGF165 after intramuscular gene transfer promotes collateral vessel development in patients with critical limb ischemia. *Circulation* **97,** 1114–1123.

Baumgartner, I., Rauh, G., Pieczek, A., Wuensch, D., Magner, M., Kearney, M., Schainfeld, R., and Isner, J. M. (2000). Lower-extremity edema associated with gene transfer of naked DNA vascular endothelial growth factor. *Ann. Intern. Med.* **132,** 880–884.

Ben-Haim, S. A., Osadchy, D., Schuster, I., Gepstein, L., Hayam, G., and Joeephson, M. E. (1996). Nonfluoroscopic, *in vivo* navigation and mapping technology. *Nature Med.* **2,** 1393–1395.

Brogi, E., Schatteman, G., Wu, T., Kim, E. A., Varticovski, L., Keyt, B., and Isner, J. M. (1996). Hypoxia-induced paracrine regulation of vascular endothelial growth factor receptor expression. *J. Clin. Invest.* **97,** 469–476.

Carmeliet, P. (2000). Mechanisms of angiogenesis and arteriogenesis. *Nature Med.* **6,** 389–395.

Carmeliet, P., and Collen, D. (1997). Molecular analysis of blood vessel formation and disease. *Am. J. Physiol.* **273,** H2091–H2104.

Comerota, A. J., Throm, R. C., Miller, K. A., Henry, T., Chronos, N., Laird, J., Sequeira, R., Kent, C. K., Bacchetta, M., Goldman, C., Salenius, J. P., Schmieder, F. A., and Pilsudski, R. (2002). Naked plasmid DNA encoding fibroblast growth factor type 1 for the treatment of end-stage unreconstructible lower extremity ischemia: Preliminary results of a phase I trial. *J. Vasc. Surg.* **35,** 930–936.

Dilsizian, V., and Bonow, R. O. (1993). Current diagnostic techniques of assessing myocardial viability in patients with hibernating and stunned myocardium. *Circulation* **87,** 1–20.

Esakof, D. D., Maysky, M., Losordo, D. W., Vale, P. R., Lathi, K., Pastore, J. O., Symes, J. F., and Isner, J. M. (1999). Intraoperative multiplane transesophageal echocardiography for guiding direct myocardial gene transfer of vascular endothelial growth factor in patients with refractory angina pectoris. *Hum. Gene Ther.* **10,** 2315–2323.

Ferrara, N., and Henzel, W. J. (1989). Pituitary follicular cells secrete a novel heparin-binding growth factor specific for vascular endothelial cells. *Biochem. Biophys. Res. Commun.* **161,** 851–855.

Folkman, J., Merler, E., Abernathy, C., and Williams, G. (1971). Isolation of a tumor factor responsible for angiogenesis. *J. Exp. Med.* **133,** 275–288.

Fong, G. H., Rossant, J., Gertsenstein, M., and Breitman, M. L. (1995). Role of flt-1 receptor tyrosine kinase in regulating the assembly of vascular endothelium. *Nature* **376,** 66–70.

Gal, D., Weir, L., Leclerc, G., Pickering, J. G., Hogan, J., and Isner, J. M. (1993). Direct myocardial transfection in two animal models: Evaluation of parameters affecting gene expression and percutaneous gene delivery. *Lab. Invest.* **68,** 18–25.

Gepstein, L., Goldin, A., Lessick, J., Hayam, G., Shpun, S., Schwartz, Y., Hakim, G., Shofty, R., Turgeman, A., Kirshenbaum, D., and Ben-Haim, S. A. (1998). Electromechanical characterization of chronic myocardial infarction in the canine coronary occlusion model. *Circulation* **98,** 2055–2064.

Giordano, F. J., Ping, P., McKirnan, M. D., Nozaki, S., DeMaria, A. N., Dillmann, W. H., Mathieu-Costello, O., and Hammond, H. K. (1996). Intracoronary gene transfer of fibroblast growth factor-5 increases blood flow and contractile function in an ischemic region of the heart. *Nature Med.* **2,** 534–539.

Goldberg, M. A., and Schneider, T. J. (1994). Similarities between the oxygen-sensing mechanisms regulating the expression of vascular endothelial growth factor and erythropoietin. *J. Biol. Chem.* **269,** 4355–4359.

Grines, C. L., Watkins, M. W., Helmer, G., Penny, W., Brinker, J., Marmur, J. D., West, A., Rade, J. J., Marrott, P., Hammond, H. K., and Engler, R. L. (2002). Angiogenic gene therapy (AGENT) trial in patients with stable angina pectoris. *Circulation* **105,** 1291–1297.

Hariawala, M. D., Horowitz, J. R., Esakof, D., Sheriff, D. D., Walter, D. H., Keyt, B., Isner, J. M., and Symes, J. F. (1996). VEGF improves myocardial blood flow but produces EDRF-mediated hypotension in porcine hearts. *J. Surg. Res.* **63,** 77–82.

Hedman, M., Hartikainen, J., Syvanne, M., *et al.* (2003). Safety and feasibility of catheter-based local intracoronary VEGF transfer in the prevention of postangioplasty and in-stent restenosis and in the treatment of chronic myocardial ischemia: Phase II results of the KAT-trial. *Circulation* **107,** 2677–2683.

Henry, T. D., Annex, B. H., McKendall, G. R., Azrin, M. A., Lopex, J. J., and Giordano, F. J. (2003). The VIVA trial: Vascular endothelial growth factor in ischemia or vascular angiogenesis. *Circulation* **107,** 1359–1365.

Henry, T. D., McKendall, G. R., Azrin, M. A., Lopez, J. J., Benza, R., Willerson, J. T., Giacomini, J., Olson, R., Bart, B. A., and Roel, J. P. (2000). VIVA trial: One year follow up. *Circulation* **102,** II–309.

Horowitz, J. R., Rivard, A., van der Zee, R., Hariawala, M. D., Sheriff, D. D., Esakof, D. D., Chaudhry, M., Symes, J. F., and Isner, J. M. (1997). Vascular endothelial growth factor/vascular permeability factor produces nitric oxide-dependent hypotension. *Arterioscler. Thromb. Vasc. Biol.* **17,** 2793–2800.

Isner, J. M. (1998). Arterial gene transfer for naked DNA for therapeutic angiogenesis: Early clinical results. *Adv. Drug Deliv.* **30,** 185–197.

Isner, J. M., Baumgartner, I., Rauh, G., Schainfeld, R., Blair, R., Manor, O., Razvi, S., and Symes, J. F. (1998). Treatment of thromboangiitis obliterans (Buerger's disease) by intramuscular gene transfer of vascular endothelial growth factor: Preliminary clinical results. *J. Vasc. Surg.* **28,** 964–973; discussion 973–965.

Isner, J. M., Vale, P. R., Symes, J. F., and Losordo, D. W. (2001). Assessment of risks associated with cardiovascular gene therapy in human subjects. *Circ. Res.* **89,** 389–400.

Isner, J. M., Walsh, K., Symes, J. F., Pieczek, A., Takeshita, S., Lowry, J., Rosenfield, K., Weir, L., Brogi, E., and Jurayj, D. (1995). Arterial gene therapy for therapeutic angiogenesis in patients with peripheral artery disease. *Circulation* **91,** 2687–2692.

Jeltsch, M., Kaipainen, A., Joukov, V., Meng, X., Lakso, M., Rauvala, H., Swartz, M., Fukumura, D., Jain, R. K., and Alitalo, K. (1997). Hyperplasia of lymphatic vessels in VEGF-C transgenic mice. *Science* **276,** 1423–1425.

Kalka, C., Masuda, H., Takahashi, T., Gordon, R., Tepper, O., Gravereaux, E., Pieczek, A., Iwaguro, H., Hayashi, S. I., Isner, J. M., and Asahara, T. (2000a). Vascular endothelial growth factor(165) gene transfer augments circulating endothelial progenitor cells in human subjects. *Circ. Res.* **86,** 1198–1202.

Kalka, C., Tehrani, H., Laudenberg, B., Vale, P. R., Isner, J. M., Asahara, T., and Symes, J. F. (2000b). Mobilization of endothelial progenitor cells following gene therapy with VEGF$_{165}$ in patients with inoperable coronary disease. *Ann. Thorac. Surg.* **70,** 829–834.

Kastrup, J., Jorgensen, E., Ruck, A, Tagil, K., Glogar, D., Ruzyllo, W., Botker, H. E., Dudek, D., Dryota, V., Hesse, B., Thuesen, L., Blomberg, P., Gyonggyosi, M., and Sylven, C. (2005). Direct intramyocardial plasmid vascular endothelial growth factor-A$_{165}$ gene therapy in patients with stable severe angina pectoris: The Euroinject One Trial. *J. Am. Coll. Cardiol.* **45,** 982–988.

Laham, R. J., Sellke, F. W., Edelman, E. R., Pearlman, J. D., Ware, J. A., Brown, D. L., Gold, J. P., and Simons, M. (1999). Local perivascular delivery of basic fibroblast growth factor in patients undergoing coronary bypass surgery: Results of a phase 1 randomized, double-blind, placebo-controlled trial. *Circulation* **100,** 1865–1871.

Lederman, R. J., Mendelsohn, F. O., Anderson, R. D., Saucedo, J. F., Tenaglia, A. N., Hermiller, J. B., Hillegass, W. B., Rocha-Singh, K., Moon, T. E., Whitehouse, M. J., and Annex, B. H. (2002). Therapeutic angiogenesis with recombinant fibroblast growth factor-2 for intermittent claudication (the TRAFFIC study): A randomised trial. *Lancet* **359,** 2053–2058.

Lee, L. Y., Patel, S. R., Hackett, N. R., Mack, C. A., Polce, D. R., El-Sawy, T., Hachamovitch, R., Zanzonico, P., Sanborn, T. A., Parikh, M., Isom, O. W., Crystal, R. G., and Rosengart, T. K. (2000). Focal angiogen therapy using intramyocardial delivery of an adenovirus vector coding for vascular endothelial growth factor 121. *Ann. Thorac. Surg.* **69,** 14–23; discussion 23–14.

Leung, D. W., Cachianes, G., Kuang, W. J., Goeddel, D. V., and Ferrara, N. (1989). Vascular endothelial growth factor is a secreted angiogenic mitogen. *Science* **246,** 1306–1309.

Levy, A. P., Levy, N. S., and Goldberg, M. A. (1996). Post-transcriptional regulation of vascular endothelial growth factor by hypoxia. *J. Biol. Chem.* **371,** 2746–2753.

Losordo, D. W., Pickering, J. G., Takeshita, S., Leclerc, G., Gal, D., Weir, L., Kearney, M., Jekanowski, J., and Isner, J. M. (1994). Use of the rabbit ear artery to serially assess foreign protein secretion after site specific arterial gene transfer *in vivo*: Evidence that anatomic identification of successful gene transfer may underestimate the potential magnitude of transgene expression. *Circulation* **89,** 785–792.

Losordo, D. W., Vale, P. R., Hendel, R. C., Milliken, C. E., Fortuin, F. D., Cummings, N., Schatz, R. A., Asahara, T., Isner, J. M., and Kuntz, R. E. (2002). Phase 1/2 placebo-controlled, double-blind, dose-escalating trial of myocardial vascular endothelial growth factor 2 gene transfer by catheter delivery in patients with chronic myocardial ischemia. *Circulation* **105,** 2012–2018.

Losordo, D. W., Vale, P. R., Symes, J. F., Dunnington, C. H., Esakof, D. D., Maysky, M., Ashare, A. B., Lathi, K., and Isner, J. M. (1998). Gene therapy for myocardial angiogenesis: Initial clinical results with direct myocardial injection of phVEGF165 as sole therapy for myocardial ischemia. *Circulation* **98,** 2800–2804.

Mack, C. A., Patel, S. R., Schwarz, E. A., Zanzonico, P., Hahn, R. T., Ilercil, A., Devereux, R. B., Goldsmith, S. J., Christian, T. F., Sanborn, T. A., Kovesdi, I., Hackett, N., Isom, O. W., Crystal, R. G., and Rosengart, T. K. (1998). Biologic bypass with the use of adenovirus-mediated gene transfer of the complementary deoxyribonucleic acid for vascular endothelial growth factor 121 improves myocardial perfusion and function in the ischemic porcine heart. *J. Thorac. Cardiovasc. Surg.* **115,** 168–176; discussion 176–167.

McKirnan, M. D., Guo, X., Waldman, L. K., Dalton, N., Lai, N. C., Gao, M. H., Roth, D. A., and Hammond, H. K. (2000). Intracoronary gene transfer of fibroblast growth factor-4 increases regional contractile function and responsiveness to adrenergic stimulation in heart failure. *Card. Vasc. Regen.* **1,** 11–21.

Minar, E., Pokrajac, B., Maca, T., Ahmadi, R., Fellner, C., Mittlbock, M., Seitz, W., Wolfram, R., and Potter, R. (2000). Endovascular brachytherapy for prophylaxis of restenosis after femoropopliteal angioplasty: Results of a prospective randomized study. *Circulation* **102,** 2694–2699.

Moulton, K. S., Heller, E., Konerding, M. A., Flynn, E., Palinski, W., and Folkman, J. (1999). Angiogenesis inhibitors endostatin or TNP-470 reduce intimal neovascularization and plaque growth in apolipoprotein E-deficient mice. *Circulation* **99,** 1726–1732.

Namiki, A., Brogi, E., Kearney, M., Kim, E. A., Wu, T., Couffinhal, T., Varticovski, L., and Isner, J. M. (1995). Hypoxia induces vascular endothelial growth factor in cultured human endothelial cells. *J. Biol. Chem.* **270,** 31189–31195.

Pearlman, J. D., Hibberd, M. G., Chuang, M. L., Harada, K., Lopez, J. J., Gladston, S. R., Friedman, M., Sellke, F. W., and Simons, M. (1995). Magnetic resonance mapping demonstrates benefits of VEGF-induced myocardial angiogenesis. *Nature Med.* **1**, 1085–1089.

Pearlman, J. D., Laham, R. J., and Simons, M. (2000). Coronary angiogenesis: Detection *in vivo* with MR imaging sensitive to collateral neocirculation. *Radiology* **214**, 801–807.

Rauh, G., Gravereaux, E. C., Pieczek, A. M., Radley, S., Schainfeld, R. M., and Isner, J. M. (1999). Age <50 years and rest pain predict positive clinical outcome after intramuscular gene transfer of phVEGF$_{165}$ in patients with critical limb ischemia. *Circulation* **100**, I–319.

Roberts, W. G., and Palade, G. E. (1995). Increased microvascular permeability and endothelial fenestration induced by vascular endothelial growth factor. *J. Cell Sci.* **108**, 2369–2379.

Schumacher, B., Pecher, P., vonSpecht, B. U., and Stegmann, T. (1998). Induction of neoangiogenesis in ischemic myocardium by human growth factors: First clinical results of a new treatment of coronary heart disease. *Circulation* **97**, 645–650.

Shalaby, F., Rossant, J., Yamaguchi, T. P., Gertsenstein, M., Wu, X.-F., Breitman, M. L., and Schuh, A. C. (1995). Failure of blood-island formation and vasculogenesis in Flk-1 deficient mice. *Nature* **376**, 62–66.

Shen, Y.-T., and Vatner, S. F. (1995). Mechanism of impaired myocardial function during progressive coronary artery stenosis in conscious pigs: Hibernation versus stunning? *Circ. Res.* **76**, 479–488.

Simons, M., Annex, B. H., Laham, R. J., Kleiman, N., Henry, T., Dauerman, H., Udelson, J. E., Gervino, E. V., Pike, M., Whitehouse, M. J., Moon, T., and Chronos, N. A. (2002). Pharmacological treatment of coronary artery disease with recombinant fibroblast growth factor-2: Double-blind, randomized, controlled clinical trial. *Circulation* **105**, 788–793.

Stegmann, T. J., Hoppert, T., Schlurmann, W., and Gemeinhardt, S. (2000). First angiogenic treatment of coronary heart disease by FGF-1: Long-term results after 3 years. *Card. Vasc. Regen.* **1**, 5–10.

Symes, J. F., Losordo, D. W., Vale, P. R., Lathi, K., Esakof, D. D., Maysky, M., and Isner, J. M. (1999). Gene therapy with vascular endothelial growth factor for inoperable coronary artery disease: Preliminary clinical results. *Ann. Thorac. Surg.* **68**, 830–837.

Tabata, H., Silver, M., and Isner, J. M. (1997). Arterial gene transfer of acidic fibroblast growth factor for therapeutic angiogenesis *in vivo*: Critical role of secretion signal in use of naked DNA. *Cardiovasc. Res.* **35**, 470–479.

Takeshita, S., Isshiki, T., and Sato, T. (1996). Increased expression of direct gene transfer into skeletal muscles observed after acute ischemic injury in rats. *Lab. Invest.* **74**, 1061–1065.

Takeshita, S., Losordo, D. W., Kearney, M., and Isner, J. M. (1994). Time course of recombinant protein secretion following liposome-mediated gene transfer in a rabbit arterial organ culture model. *Lab. Invest.* **71**, 387–391.

Takeshita, S., Weir, L., Chen, D., Zheng, L. P., Riessen, R., Bauters, C., Symes, J. F., Ferrara, N., and Isner, J. M. (1996). Therapeutic angiogenesis following arterial gene transfer of vascular endothelial growth factor in a rabbit model of hindlimb ischemia. *Biochem. Biophys. Res. Commun.* **227**, 628–635.

Thurston, G., Suri, C., Smith, K., McClain, J., Sato, T. N., Yancopoulos, G. D., and McDonald, D. M. (1999). Leakage-resistant blood vessels in mice transgenically overexpressing angiopoietin-1. *Science* **286**, 2511–2514.

Tsurumi, Y., Takeshita, S., Chen, D., Kearney, M., Rossow, S. T., Passeri, J., Horowitz, J. R., and Symes, J. F. (1996). Direct intramuscular gene transfer of naked DNA encoding vascular endothelial growth factor augments collateral development and tissue perfusion. *Circulation* **94**, 3281–3290.

Udelson, J. E., Dilsizian, V., Laham, R. J., Chronos, N., Vansant, J., Blais, M., Galt, J. R., Pike, M., Yoshizawa, C., and Simons, M. (2000). Therapeutic angiogenesis with recombinant fibroblast growth factor-2 improves stress and rest myocardial perfusion abnormalities in patients with severe symptomatic chronic coronary artery disease. *Circulation* **102,** 1605–1610.

Unger, E. F., Goncalves, L., Epstein, S. E., Chew, E. Y., YTrapnell, C. B., and Cannon, R. O., III (2000). Effects of a single intracoronary injection of basic fibroblast growth factor in stable angina pectoris. *AJC* **85,** 1414–1419.

Vale, P. R., Rauh, G., Wuensch, D. I., Pieczek, A., and Schainfeld, R. M. (1998). Influence of vascular endothelial growth factor on diabetic retinopathy. *Circulation* **17,** I-353.

Vale, P. R., Wuensch, D. I., Rauh, G. F., Rosenfield, K., Schainfeld, R. M., and Isner, J. M. (1998). Arterial gene therapy for inhibiting restenosis in patients with claudication undergoing superficial femoral artery angioplasty. *Circulation* **98,** I-66.

Vale, P. R., Losordo, D. W., Tkebuchava, T., Chen, D., Milliken, C. E., and Isner, J. M. (1999). Catheter-based myocardial gene transfer utilizing nonfluoroscopic electromechanical left ventricular mapping. *J. Am. Coll. Cardiol.* **34,** 246–254.

Vale, P. R., Losordo, D. W., Milliken, C. E., Maysky, M., Esakof, D. D., Symes, J. F., and Isner, J. M. (2000). Left ventricular electromechanical mapping to assess efficacy of phVEGF(165) gene transfer for therapeutic angiogenesis in chronic myocardial ischemia. *Circulation* **102,** 965–974.

Vale, P. R., Losordo, D. W., Milliken, C. E., McDonald, M. C., Gravelin, L. M., Curry, C. M., Esakof, D. D., Maysky, M., Symes, J. F., and Isner, J. M. (2001). Randomized, single-blind, placebo-controlled pilot study of catheter-based myocardial gene transfer for therapeutic angiogenesis using left ventricular electromechanical mapping in patients with chronic myocardial ischemia. *Circulation* **103,** 2138–2143.

Van Belle, E., Tio, F. O., Chen, D., Maillard, L., Kearney, M., and Isner, J. M. (1997a). Passivation of metallic stents following arterial gene transfer of phVEGF$_{165}$ inhibits thrombus formation and intimal thickening. *J. Am. Coll. Cardiol.* **29,** 1371–1379.

Van Belle, E., Tio, F. O., Couffinhal, T., Maillard, L., Passeri, J., and Isner, J. M. (1997b). Stent endothelialization: Time course, impact of local catheter delivery, feasibility of recombinant protein administration, and response to cytokine expedition. *Circulation* **95,** 438–448.

van der Zee, R., Murohara, T., Luo, Z., Zollmann, F., Passeri, J., Lekutat, C., and Isner, J. M. (1997). Vascular endothelial growth factor (VEGF)/vascular permeability factor (VPF) augments nitric oxide release from quiescent rabbit and human vascular endothelium. *Circulation* **95,** 1030–1037.

Waltenberger, J., Mayr, U., Pentz, S., and Hombach, V. (1996). Functional upregulation of the vascular endothelial growth factor receptor KDR by hypoxia. *Circulation* **94,** 1647–1654.

Wijns, W., Vatner, S. F., and Camici, P. G. (1998). Hibernating myocardium. *N. Engl. J. Med.* **3,** 173–181.

Witzenbichler, B., Asahara, T., Murohara, T., Silver, M., Spyridopoulos, I., Magner, M., Principe, N., Kearney, M., Hu, J. S., and Isner, J. M. (1998). Vascular endothelial growth factor-C (VEGF-C/VEGF-2) promotes angiogenesis in the setting of tissue ischemia. *Am. J. Pathol.* **153,** 381–394.

Wolff, J. A., Malone, R. W., Williams, P., Chong, W., Acsadi, G., Jani, A., and Felgner, P. L. (1990). Direct gene transfer into mouse muscle. *in vivo. Science* **247,** 1465–1468.

Yang, Y., Nunes, F. A., Berencsi, K., Furth, E. E., Gonczol, E., and Wilson, J. M. (1994). Cellular immunity to viral antigens limits E1-deleted adenoviruses for gene therapy. *Proc. Natl. Acad. Sci. USA* **91,** 4407–4411.

Index

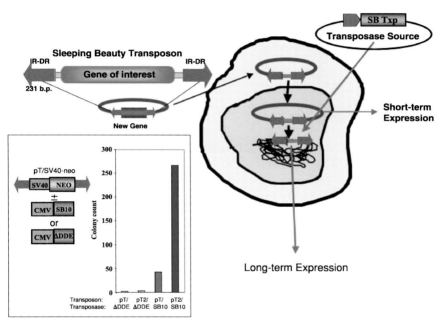

Chapter 9, Figure 9.2. SB transposons for gene therapy. Delivery of a transposon with a transposase-encoding sequence, shown here on two plasmid vectors, can provide long-term expression of the transposed gene compared with shorter durations of expression when the gene remains as an episome in transformed cells *in vivo*. (Inset) The increase in levels of gene expression as a result of transposition in the presence of SB transposase compared to levels observed when a defective form (ΔDDE) of the enzyme is supplied (Ivics *et al.*, 1997). T is the original transposon and T2 is an improved version (Cui *et al.*, 2002) as measured by the frequency of G418-resistant HeLa cell colony formation following transposition of an SV40-Neo construct in either of the two transposons and with or without active SB transposase. The numbers of colonies obtained in the ΔDDE experiments are about equivalent to the levels found following delivery without any transposase and represent random, illegitimate recombination into chromosomes.

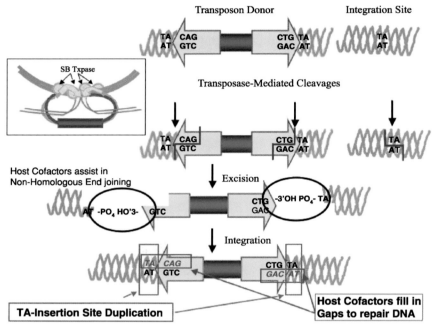

Chapter 9, Figure 9.3. SB-mediated transposition from a donor site (green lines) to an integration site (purple lines). Two SB transposase molecules, shown as yellow circles in the boxed insert on the left, bind on each of the inverted terminal repeat (arrows) to introduce three cleaves: two flanking the transposon (pink structure with inverted arrows representing the inverted terminal repeats) and one in the target integration site (second line). The insert emphasizes that SB transposase molecules act in concert in a complex of transposon donor and target integration site. The excision step is shown on the third line with integration occurring by the invasion of the 3′ ends of the transposon joining the exposed TA nucleotides at the integration site (shown in the ellipsoids in the third line). Following ligation of the single strands on each side, DNA repair enzymes fill in the remaining five nucleotide gaps (shown in red in the fourth line). TA target site duplication is indicated in the last line by the boxed TA dinucleotide base pairs. The transposon-donor sequence is resealed and the single base mismatch is repaired by cellular enzymes (lower right corner).

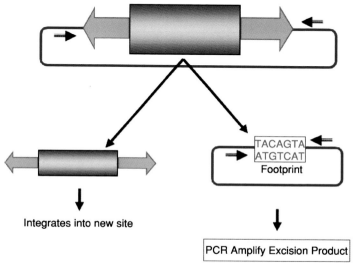

Chapter 9, Figure 9.4. The excision assay for quantifying one step in the transposition process. A pair of primers (black arrowheads) is designed to flank a transposon in a donor molecule, shown here in a plasmid vector (blue line). Following excision of the plasmid, a resealed vector is produced that will be considerably smaller than the donor DNA and often will have a precise, canonical footprint, shown here in the box. The middle A-T base pair shown could be T-A as well. Following delivery of transposons to a multicellular tissue, each of the integration sites will be unique but the remaining plasmid will be the same so that a single set of primers can record all excision (and correlated integration) events.

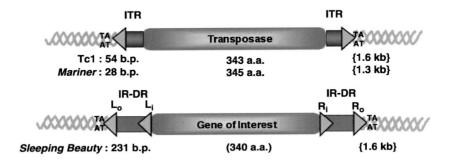

```
Lo:  TA/CAGTTGAAGTCGGAAGTTTACATACACTTAAG
Li:      TCCAGTGGGTCAGAAGTTTACATACACTAAGT
Ri:      CCCAGTGGGTCAGAAGTTTACATACACTCAAT
Ro:  TA/CAGTTGAAGTCGGAAGTTTACATACACCTTAG
         ===========================     SB footprint
Consensus:  TA/CAGTTGAAGTCGGAAGTTTACATACACTTAAG  DRo's
            CAGTGG--GTCAGAAGTTTACATACACTMART  DRi's
```

Chapter 9, Figure 9.5. Comparative structures of *Tc1/mariner*-like transposons and SB transposons. ITR, inverted terminal repeat sequence; IR-DR, inverted repeat containing direct repeated sequences. The lengths of the repeat sequences are noted for each transposon as well as the transposase size (or reconstructed size in the case of SB—the transposase is never supplied from a gene flanked by two IR-DRs). Consensus sizes of the transposons are shown in the brackets on the right. The specific sequences of the four DRs in the original SB transposon, *T*, are shown at the bottom along with the consensus sequences that were used to build an improved transposon called T2. The SB footprint refers to the portion of the DR sequence that is protected from DNase hydrolysis when bound by SB transposase (Ivics *et al.*, 1997).

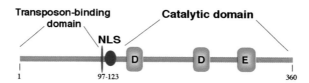

Transposon-Binding Domain
- 1 Transposase molecule binds to each DR sequence
- Coiled-coil structure
- Dimerization activity between two SB transposase molecules

Nuclear Localization Sequence (NLS)
- Bipartite structure

Catalytic Domain
- Mediates cleavage at the ends of the Inverted Terminal Repeats
- Recognizes integration sites at TA dinucleotide basepairs
- Mediates non-homologous end-joining - the "paste" step

Chapter 9, Figure 9.6. Diagram of SB transposase. The three functional domains are identified at the top; numbers below the structure are the approximate amino acid residue boundaries of the domains. The transposon-binding domain, often called the DNA-binding domain, binds to DRs and can protect the nested sequence from degradation by DNase I as identified in Fig. 9.4. The transposon-binding domain is also responsible for dimerization of transposase molecules to form the complex shown in Fig. 9.3. The NLS sequence comprises amino acids 79–123 and has two clusters of basic amino acids separated by a 10 amino acid spacer. The catalytic domain characterized by the DDE motif is commonly found in all cut-and-paste recombination enzymes.

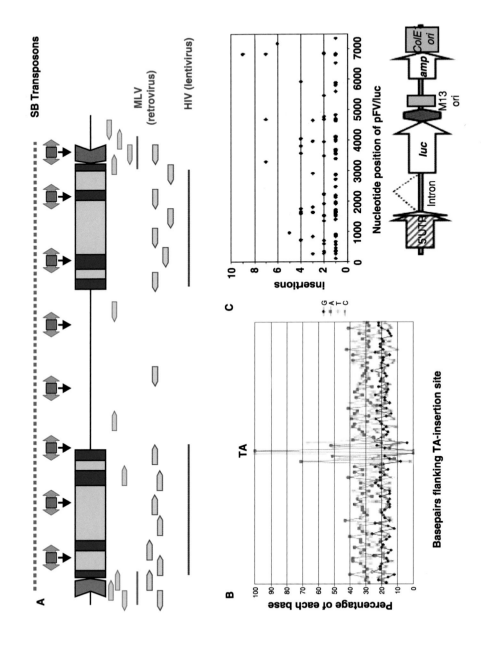

A

SB Transposons

MLV
(retrovirus)

HIV (lentivirus)

B

Percentage of each base

TA

Basepairs flanking TA-insertion site

G
A
T
C

C

Insertions

Nucleotide position of pFV/luc

5'UTR   Intron   *luc*   M13 ori   *amp*   *ColE1 ori*

Chapter 9, Figure 9.7. Integration site preferences of SB transposase. (A) Integration preferences are shown for SB transposase (red double-sided arrows and dotted lines), retroviruses (green arrowheads and lines), and lentiviruses (blue arrowheads and lines). The preferences of the three classes of vectors are shown by the horizontal lines (dotted for SB transposons) with clustering indicated by the symbols. Separating transposons and viruses is a schematic of chromosomal DNA with transcriptional motifs shown as green chevrons and transcriptional units shown by the dark blue (exons) and light blue (introns) boxes. (B) Percentages of bases over 50 bp on each side of TA insertion sites in mouse genomes; the TA insertion sites in the center of the chart are invariant (G. Liu and Y. Horie, unpublished result). (C) Preferential TA insertion sites in the pFV/luc plasmid. Integration sites with two or more hits vary from an expected Poisson distribution; 29% of the total hits were between base pairs 6815 and 6854, which comprises less than 0.5% of the plasmid (Liu et al., 2004b).

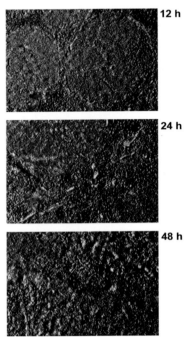

Chapter 13, Figure 13.4. Distribution of LPD/E7 particles in the spleen following iv administration. LPD/E7 particles containing Cy3-labeled ODN were visualized at the indicated times after administration.